Laser Synthesis of Nanomaterials

Laser Synthesis of Nanomaterials

Editors

Mohamed Boutinguiza
Antonio Riveiro
Jesús del Val

MDPI • Basel • Beijing • Wuhan • Barcelona • Belgrade • Manchester • Tokyo • Cluj • Tianjin

Editors

Mohamed Boutinguiza
Applied Physics
University of Vigo
Vigo
Spain

Antonio Riveiro
Dept. Materials Engineering,
Applied Mechanics and
Construction
University of Vigo
Vigo
Spain

Jesús del Val
Defense University Center at
the Spanish Naval Academy
University of Vigo
Vigo
Spain

Editorial Office
MDPI
St. Alban-Anlage 66
4052 Basel, Switzerland

This is a reprint of articles from the Special Issue published online in the open access journal *Nanomaterials* (ISSN 2079-4991) (available at: www.mdpi.com/journal/nanomaterials/special_issues/laser).

For citation purposes, cite each article independently as indicated on the article page online and as indicated below:

LastName, A.A.; LastName, B.B.; LastName, C.C. Article Title. *Journal Name* **Year**, *Volume Number*, Page Range.

ISBN 978-3-0365-6929-1 (Hbk)
ISBN 978-3-0365-6928-4 (PDF)

© 2023 by the authors. Articles in this book are Open Access and distributed under the Creative Commons Attribution (CC BY) license, which allows users to download, copy and build upon published articles, as long as the author and publisher are properly credited, which ensures maximum dissemination and a wider impact of our publications.

The book as a whole is distributed by MDPI under the terms and conditions of the Creative Commons license CC BY-NC-ND.

Contents

About the Editors vii

Mohamed Boutinguiza Larosi, Jesús del Val García and Antonio Riveiro Rodríguez
Laser Synthesis of Nanomaterials
Reprinted from: *Nanomaterials* **2022**, *12*, 2903, doi:10.3390/nano12172903 1

Alexey Rybaltovsky, Evgeniy Epifanov, Dmitriy Khmelenin, Andrey Shubny, Yuriy Zavorotny and Vladimir Yusupov et al.
Two Approaches to the Laser-Induced Formation of Au/Ag Bimetallic Nanoparticles in Supercritical Carbon Dioxide
Reprinted from: *Nanomaterials* **2021**, *11*, 1553, doi:10.3390/nano11061553 5

Roman Romanov, Vyacheslav Fominski, Maxim Demin, Dmitry Fominski, Oxana Rubinkovskaya and Sergey Novikov et al.
Application of Pulsed Laser Deposition in the Preparation of a Promising $MoS_x/WSe_2/C(B)$ Photocathode for Photo-Assisted Electrochemical Hydrogen Evolution
Reprinted from: *Nanomaterials* **2021**, *11*, 1461, doi:10.3390/nano11061461 17

Ashish Nag, Laysa Mariela Frias Batista and Katharine Moore Tibbetts
Synthesis of Air-Stable Cu Nanoparticles Using Laser Reduction in Liquid
Reprinted from: *Nanomaterials* **2021**, *11*, 814, doi:10.3390/nano11030814 35

Izumi Takayama, Akito Katayama and Mitsuhiro Terakawa
Fabrication of Hollow Channels Surrounded by Gold Nanoparticles in Hydrogel by Femtosecond Laser Irradiation
Reprinted from: *Nanomaterials* **2020**, *10*, 2529, doi:10.3390/nano10122529 47

Vyacheslav Fominski, Dmitry Fominski, Roman Romanov, Mariya Gritskevich, Maxim Demin and Petr Shvets et al.
Specific Features of Reactive Pulsed Laser Deposition of Solid Lubricating Nanocomposite Mo–S–C–H Thin-Film Coatings
Reprinted from: *Nanomaterials* **2020**, *10*, 2456, doi:10.3390/nano10122456 57

Sergey Syubaev, Stanislav Gurbatov, Evgeny Modin, Denver P. Linklater, Saulius Juodkazis and Evgeny L. Gurevich et al.
Laser Printing of Plasmonic Nanosponges
Reprinted from: *Nanomaterials* **2020**, *10*, 2427, doi:10.3390/nano10122427 81

María J. Rivera-Chaverra, Elisabeth Restrepo-Parra, Carlos D. Acosta-Medina, Alexandre. Mello and Rogelio. Ospina
Synthesis of Oxide Iron Nanoparticles Using Laser Ablation for Possible Hyperthermia Applications
Reprinted from: *Nanomaterials* **2020**, *10*, 2099, doi:10.3390/nano10112099 95

Li-Hsiou Chen, Huan-Ting Shen, Wen-Hsin Chang, Ibrahim Khalil, Su-Yu Liao and Wageeh A. Yehye et al.
Photocatalytic Properties of Graphene/Gold and Graphene Oxide/Gold Nanocomposites Synthesized by Pulsed Laser Induced Photolysis
Reprinted from: *Nanomaterials* **2020**, *10*, 1985, doi:10.3390/nano10101985 107

Kristin Charipar, Heungsoo Kim, Alberto Piqué and Nicholas Charipar
ZnO Nanoparticle/Graphene Hybrid Photodetectors via Laser Fragmentation in Liquid
Reprinted from: *Nanomaterials* **2020**, *10*, 1648, doi:10.3390/nano10091648 119

Muhammad Abdullah Butt, Daria Mamonova, Yuri Petrov, Alexandra Proklova, Ilya Kritchenkov and Alina Manshina et al.
Hybrid Orthorhombic Carbon Flakes Intercalated with Bimetallic Au-Ag Nanoclusters: Influence of Synthesis Parameters on Optical Properties
Reprinted from: *Nanomaterials* **2020**, *10*, 1376, doi:10.3390/nano10071376 131

Yanwei Huang, Yu Qiao, Yangyang Li, Jiayang He and Heping Zeng
Zn-Doped Calcium Copper Titanate Synthesized via Rapid Laser Sintering of Sol-Gel Derived Precursors
Reprinted from: *Nanomaterials* **2020**, *10*, 1163, doi:10.3390/nano10061163 141

Mónica Fernández-Arias, Mohamed Boutinguiza, Jesús del Val, Antonio Riveiro, Daniel Rodríguez and Felipe Arias-González et al.
Fabrication and Deposition of Copper and Copper Oxide Nanoparticles by Laser Ablation in Open Air
Reprinted from: *Nanomaterials* **2020**, *10*, 300, doi:10.3390/nano10020300 153

V. Fominski, M. Demin, D. Fominski, R. Romanov, A. Goikhman and K. Maksimova
Comparative Study of the Structure, Composition, and Electrocatalytic Performance of Hydrogen Evolution in $MoS_{x\text{-}2+\delta}/Mo$ and $MoS_{x\text{-}3+\delta}$ Films Obtained by Pulsed Laser Deposition
Reprinted from: *Nanomaterials* **2020**, *10*, 201, doi:10.3390/nano10020201 169

Maurizio Muniz-Miranda, Francesco Muniz-Miranda and Emilia Giorgetti
Spectroscopic and Microscopic Analyses of Fe_3O_4/Au Nanoparticles Obtained by Laser Ablation in Water
Reprinted from: *Nanomaterials* **2020**, *10*, 132, doi:10.3390/nano10010132 189

About the Editors

Mohamed Boutinguiza

Mohamed Boutinguiza Larosi received his Ph.D in Applied Physics from the University of Vigo, Spain (2001), where he began his academic career as an Assistant Professor and his research in the field of laser processing materials. He is currently an associate professor in the Department of Applied Physics at School of Engineering (University of Vigo). His research is focused on the laser synthesis and processing of nanomaterials, the production and the deposition of nanoparticles through laser ablation in different media, photo-catalysis, plasmonic nanostructures and biomaterials. He has over 17 years of experience in synthesizing nanostructured materials using laser ablation and their characterization as well as their applications.

He is listed in the ranking from Stanford University featuring the World Top 2% Scientist classification released by PlosBiology in 2019, 2020 and 2021. He has published more than 100 research papers in peer-reviewed journals and co-authored several patents and book chapters.

Antonio Riveiro

Dr. Antonio Riveiro is assistant prof. in the Department of Materials Engineering, Applied Mechanics and Construction (School of Engineering) at University of Vigo (Spain), where he teaches courses on solid mechanics, and laser processing. He received his PhD in Applied Physics (2009) from the University of Vigo, and did postdoctoral stays at University of Manchester (School of Materials), Imperial College London (Department of Materials) and University of Minho (3B's Research Group in Biomaterials, Biodegradables and Biomimetics). He specializes in laser materials processing, focusing on the application of laser processes for biomedical applications. His research interests also include, laser surface engineering for modification of wetting properties of advanced materials, laser synthesis of micro- and nanomaterials, synthesis and laser processing of bioactive glasses.

Jesús del Val

Dr. Jesús del Val is an assistant prof. at the Defense University Center at the Spanish Naval Academy, at University of Vigo (Spain), where he teaches courses on solid mechanics, physics, and laser processing. He received his PhD in Applied Physics (2015) from the University of Vigo, and did postdoctoral stays at Imperial College London (Department of Materials) and Columbia University in New York City (Advanced Manufacturing Laboratory, Department of Mechanical Engineering). He specializes in processing laser materials, focusing on the application of laser processes for biomedical applications. His research interests also include: laser micro-cladding, laser surface engineering for the modification of wetting properties of advanced materials, laser synthesis of micro- and nanomaterials, synthesis and laser processing of bioactive glasses.

Editorial

Laser Synthesis of Nanomaterials

Mohamed Boutinguiza Larosi [1,2,*], Jesús del Val García [2,3] and Antonio Riveiro Rodríguez [2,4]

1. Department of Applied Physics, University of Vigo, EEI, Lagoas-Marcosende, 36310 Vigo, Spain
2. LaserON Laser Applications Research Group, Research Center in Technologies, University of Vigo, Energy and Industrial Processes, Rúa Maxwel, 36310 Vigo, Spain
3. Defense University Center at the Spanish Naval Academy, University of Vigo, Plaza de España 2, Marín, 36920 Pontevedra, Spain
4. Materials Engineering, Applied Mechanics and Construction Department, University of Vigo, EEI, Lagoas-Marcosende, 36310 Vigo, Spain
* Correspondence: mohamed@uvigo.es

Citation: Larosi, M.B.; García, J.d.V.; Rodríguez, A.R. Laser Synthesis of Nanomaterials. *Nanomaterials* **2022**, *12*, 2903. https://doi.org/10.3390/nano12172903

Received: 4 August 2022
Accepted: 21 August 2022
Published: 24 August 2022

Publisher's Note: MDPI stays neutral with regard to jurisdictional claims in published maps and institutional affiliations.

Copyright: © 2022 by the authors. Licensee MDPI, Basel, Switzerland. This article is an open access article distributed under the terms and conditions of the Creative Commons Attribution (CC BY) license (https://creativecommons.org/licenses/by/4.0/).

Nanomaterials, defined as materials with typical dimensions of less than 100 nm in at least one dimension, exhibit very special physicochemical properties that are highly dependent on their size and shape. These new properties, different from those of the corresponding bulk material, are behind the revolution in both nanomaterial fabrication techniques and the application of nanomaterials in many areas of science and technology. The possibility of tailoring the properties of materials by manipulating them at the nanoscale to obtain materials with highly specific surface area and improved mechanical, optical, magnetic, etc., properties has opened up and promoted new and promising lines of research. Assembling nanostructures by different methods, called nanofabrication, has led to enormous growth in the majority of industrial sectors.

There are many techniques and methods for producing nanostructures. These can be broadly classified into: "top down" methods, in which the starting material is reduced to the desired sizes following physical or chemical methods (such as electron or ion beam, milling, laser ablation, and reactive ion etching), and "bottom-up" methods, in which nanofabrication is carried out by assembling individual atoms or molecules to obtain the final nanostructure; this strategy includes chemical and physical vapor deposition (CVD and PVD), epitaxial growth, self-assembly, etc. [1,2].

Among the wide range of nanofabrication techniques, laser ablation is a nanomaterial synthesis technique with outstanding advantages, such as high production efficiency, low cost, good stability, and reliable processing quality, in addition to being environmentally friendly. This technique has been used to fabricate a wide range of nanomaterials and nanostructures with improved chemical, optical, magnetic, and electronic properties. Laser ablation technique has been used in a variety of experimental setups and configurations to synthesize different nanomaterials in a wide range of atmospheres and conditions. Laser-based techniques, such as laser ablation, laser vaporization, pulsed laser deposition (PLD), laser–chemical vapor deposition (LCVD), etc., are being used to fabricate nanoscale materials with a controlled size, shape, and specific properties [3]. Another laser-based nanomaterial processing technique with major impact in nanotechnology is laser synthesis and processing of colloids (LSPC), a scalable and versatile technique for the synthesis of ligand-free nanomaterials in controlled liquid environments. With this method, nanoparticles with high surface purity can be synthesized and alloys or series of doped nanomaterials can also be obtained. This technique can be classified into three approaches: laser ablation in liquids (LAL), where the laser beam is kept scanning a bulk target in liquid to produce colloidal nanoparticles.; laser fragmentation in liquids (LFL), in which microparticle suspensions or colloidal nanoparticles are irradiated with lasers to be fractured; and laser melting in liquids (LML), in which the laser beam is used to melt primary nanoparticles into bigger ones [4]. In this context, the present Special Issue has been organized to include works focused on obtaining nanostructured materials by laser-based synthesis methods. Combined properties of different starting materials can be also obtained by laser synthesis, as is reported by Alexey Rybaltovsky et al. [5]. Two approaches are used to synthesize bimetallic Au/Ag nanoparticles, using a laser to ablate targets of gold and silver in a medium of supercritical carbon dioxide. In the first configuration, Ag and Au targets are placed side-by-side, vertically, on the side wall of a high-pressure reactor, and the ablation of the target plates occurs alternately with a stationary "wide" horizontal beam with a laser pulse repetition rate of 50 Hz. With this configuration, Ag/Au alloy nanoparticles are obtained. Meanwhile, "core–shell" bimetallic Au/Ag nanoparticles with a gold core and a silver shell are synthesized by placing the targets horizontally at the bottom of a reactor, and the ablation of their parts is carried out by scanning from above with a vertical "narrow" laser beam with a pulse repetition rate of 60 kHz.

Roman Romanov et al. [6] conducted an investigation to produce MoSx~4/WSe2/C(B)/Al$_2$O$_3$ photocathodes using PLD, which can be employed for effective solar water splitting to produce hydrogen. The fabricated photocathode presents the following characteristics as regards hydrogen evolution reaction in 0.5 M H$_2$SO$_4$ acid solution during light irradiation with an intensity of 100 mW/cm^2: the current density at 0 V (RHE) is ~3 mA/cm^2; the onset potential reaches 400 mV (RHE).

The synthesis of air-stable Cu nanoparticles by irradiating solutions of copper acetylacetone in a mixture of methanol and isopropyl alcohol by femtosecond laser is reported by Ashish Nag et al. [7] following the method of bottom-up laser reduction in liquid. The obtained Cu NPs exhibit remarkable stability over 7 days on the basis of the lack of significant changes observed in the UV-vis absorbance and XRD features, and their photocatalytic performance was maintained.

Izumi Takayama et al. [8] reported the fabrication of hollow channels surrounded by gold nanoparticles in poly(ethylene glycol) diacrylate (PEGDA). The hollow channels and gold nanoparticles were formed in a single step of irradiating a femtosecond laser pulse in PEGDA hydrogels which contained gold ions. Taking into account that hydrogels and gold nanoparticles present high biocompatibility, this research could open the door to many applications in tissue engineering, microfluidics, and drug delivery.

The structure and chemical states of thin-film coatings obtained by pulsed laser co-deposition of Mo and C in a reactive gas (H$_2$S) are investigated [9]. The performance of these coatings was analyzed for their prospective use as solid lubricating coatings for friction units operating in extreme conditions, showing that the nanophase composition in Mo–S–C–H_5.5 coatings has good antifriction properties and increased wear resistance, even at -100 °C.

An approach for producing porous plasmonic nanostructures using direct ns-laser ablation followed by Ar-ion beam etching is reported in Ref. [10]. The nanopores were found to form through the explosive evaporation/boiling of the nitrogen-rich metal film areas irradiated by an ns laser pulse. This scalable method for producing 3D plasmonic nanosponges represents a promising technique to control and improve their performances for various nonlinear optical and sensing applications.

Iron oxide nanoparticles are synthesized by laser ablation in water by María J. Rivera-Chaverra et al. [11]. Their characteristics, as a function of the laser energy and for the possible application in magnetic hyperthermia, were evaluated. It was shown that the temperature rise in iron oxide nanoparticles was not greatly influenced by the energy change in magnetic hyperthermia measurements. Experiments show that, for hyperthermia applications, low values of laser energy give better results, as these produce higher specific absorption rate (SAR) values.

Li-Hsiou Chen et al. [12] reported the synthesis of Graphene (Gr)/gold (Au) and graphene-oxide (GO)/Au nanocomposites (NCPs) by pulsed-laser-induced photolysis (PLIP) on hydrogen peroxide and chloroauric acid (HAuCl4) that coexisted with Gr or GO in an aqueous solution. Both kinds of nanoparticles exhibited photocatalytic degradation of methylene blue under solar light illumination with removal efficiencies over 92% and showed good stability and a large potential in the practical treatment of dye-contaminated wastewater through an ecofriendly fabrication process.

Kristin Charipar et al. [13] used pulsed laser fragmentation in liquid as a ligand-free alternative to traditional nanoparticle synthesis techniques for the fabrication of ZnO nanoparticles. They also demonstrated the possibility of producing hybrid ZnO nanoparticle/graphene phototransistors, exhibiting a responsivity of up to 4×10^4 AW^{-1} with a maximum gain of 1.3×10^5 and superior spectral selectivity below 400 nm, which make them ideal for solar-blind UV photodetectors.

The carbonaceous flake-like structures self-assemble during a laser-induced growth process is studied in Ref. [14], with particular emphasis to the dependence of the optical and geometrical properties of these hybrid carbon–metal flakes on the fabrication parameters. This study shows that the geometrical parameters of orthorhombic metal–carbon hybrid flakes can be tailored during the fabrication process by controlling various fabrication parameters, such as irradiation time and the application of an external field along the substrate.

Yanwei Huang et al. [15] reported the synthesis of Zn-doped calcium copper titanate (CCTO) by rapid laser sintering of sol–gel-derived precursors without the conventional long-time heat treatment. The used technique overcomes the shortcomings of long-time thermal energy supply by a furnace and presents references to synthesize the ceramic materials through a combination of soft-chemical methods.

Mónica Fernández-Arias et al. [16] reported the synthesis of Cu and Cu oxide nanoparticles by laser ablation in open air and in argon atmosphere using 532 and 1064 nm radiation generated by nanosecond and picosecond Nd:YVO$_4$ lasers, respectively, to be directly deposited onto Ti substrates. The coatings were tested as an antibacterial agent, showing strong antibacterial activity of the obtained copper nanoparticles against *S. aureus*. The best inhibitory effects are provided by Cu nanoparticles obtained by laser ablation in argon with a laser radiation of 1064 nm in wavelength. These results confirm the influence of size, crystallographic structure, and oxidation state in the bactericidal effects of copper nanoparticles.

Two modes of PLD from a MoS2 target were reported by V. Fominski et al. [17] to obtain amorphous MoSx-based catalysts with different compositions and morphologies. The mode of off-axis PLD differs from the mode of on-axis PLD in terms of a higher deposition rate for catalytic MoSx~3 + δ films and a larger S concentration in the amorphous MoSx~3 + δ (δ ~0.8–1.1) phase.

Maurizio Muniz-Miranda et al. [18] reported the synthesis of magneto-plasmonic nanoparticles constituted of gold and iron oxide in an aqueous environment by laser ablation of iron and gold targets in two successive steps without the presence of surfactants, stabilizers, or any contaminants. The plasmonic properties of the obtained colloids, as well as their adsorption capability, were tested by surface-enhanced Raman scattering (SERS) spectroscopy using 2,2'-bipyridine as a probe molecule.

Laser-based synthesis techniques emerged as powerful and versatile technology for nanomaterials synthesis, proving their feasibility for the synthesis of nanoparticles and nanostructures from different starting materials (metals, oxides, semiconductors, etc.) in a wide range of liquid and gas environments. This Special Issue presents different works using different laser-based techniques and approaches to tailor the final nanomaterial in order to obtain specific properties to improve certain applications, such as SERS, photocatalysis, photodetectors, hyperthermia, or antibacterial agents.

However, despite the relatively simple experimental setup of laser-based synthesis of nanomaterials, there are many limitations and challenges to be overcome, such as the limited productivity, adequate nanomaterials adhesion onto the appropriate substrate, better understanding laser beam–target interactions and removal mechanisms to better control the final product, higher stability in the case of colloidal solutions, etc.

Given the increasing number of researches and the rapid evolution of the laser synthesis of nanomaterials field, it is expected that most of the mentioned disadvantages will be reduced or overcome in the near future by improving the aforementioned techniques and/or integrating them with other synthesis routes.

Funding: This research received no external funding.

Conflicts of Interest: The authors declare no conflict of interest.

References

1. Abid, N.; Khan, A.M.; Shujait, S.; Chaudhary, K.; Ikram, M.; Imran, M.; Haider, J.; Khan, M.; Khan, Q.; Maqbool, M. Synthesis of nanomaterials using various top-down and bottom-up approaches, influencing factors, advantages, and disadvantages: A review. *Adv. Colloid Interface Sci.* **2022**, *300*, 102597. [CrossRef] [PubMed]
2. Baig, N.; Kammakakam, I.; Falath, W. Nanomaterials: A review of synthesis methods, properties, recent progress, and challenges. *Mater. Adv.* **2021**, *2*, 1821–1871. [CrossRef]
3. Yang, L.; Wei, J.; Ma, Z.; Song, P.; Ma, J.; Zhao, Y.; Huang, Z.; Zhang, M.; Yang, F.; Wang, X. The Fabrication of Micro/Nano Structures by Laser Machining. *Nanomaterials* **2019**, *9*, 1789. [CrossRef] [PubMed]
4. Zhang, Z.; Gökce, B.; Barcikowski, S. Laser Synthesis and Processing of Colloids: Fundamentals and Applications. *Chem. Rev.* **2017**, *117*, 3990–4103. [CrossRef] [PubMed]
5. Rybaltovsky, A.; Epifanov, E.; Khmelenin, D.; Shubny, A.; Zavorotny, Y.; Yusupov, V.; Minaev, N. Two Approaches to the Laser-Induced Formation of Au/Ag Bimetallic Nanoparticles in Supercritical Carbon Dioxide. *Nanomaterials* **2021**, *11*, 1553. [CrossRef]
6. Romanov, R.; Fominski, V.; Demin, M.; Fominski, D.; Rubinkovskaya, O.; Novikov, S.; Volkov, V.; Doroshina, N. Application of Pulsed Laser Deposition in the Preparation of a Promising MoSx/WSe2/C(B) Photocathode for Photo-Assisted Electrochemical Hydrogen Evolution. *Nanomaterials* **2021**, *11*, 1461. [CrossRef] [PubMed]
7. Nag, A.; Frias Batista, L.M.; Tibbetts, K.M. Synthesis of Air-Stable Cu Nanoparticles Using Laser Reduction in Liquid. *Nanomaterials* **2021**, *11*, 814. [CrossRef] [PubMed]
8. Takayama, I.; Katayama, A.; Terakawa, M. Fabrication of Hollow Channels Surrounded by Gold Nanoparticles in Hydrogel by Femtosecond Laser Irradiation. *Nanomaterials* **2020**, *10*, 2529. [CrossRef] [PubMed]
9. Fominski, V.; Fominski, D.; Romanov, R.; Gritskevich, M.; Demin, M.; Shvets, P.; Maksimova, K.; Goikhman, A. Specific Features of Reactive Pulsed Laser Deposition of Solid Lubricating Nanocomposite Mo–S–C–H Thin-Film Coatings. *Nanomaterials* **2020**, *10*, 2456. [CrossRef] [PubMed]
10. Syubaev, S.; Gurbatov, S.; Modin, E.; Linklater, D.P.; Juodkazis, S.; Gurevich, E.L.; Kuchmizhak, A. Laser Printing of Plasmonic Nanosponges. *Nanomaterials* **2020**, *10*, 2427. [CrossRef] [PubMed]
11. Rivera-Chaverra, M.J.; Restrepo-Parra, E.; Acosta-Medina, C.D.; Mello, A.; Ospina, R. Synthesis of Oxide Iron Nanoparticles Using Laser Ablation for Possible Hyperthermia Applications. *Nanomaterials* **2020**, *10*, 2099. [CrossRef] [PubMed]
12. Chen, L.-H.; Shen, H.-T.; Chang, W.-H.; Khalil, I.; Liao, S.-Y.; Yehye, W.A.; Liu, S.-C.; Chu, C.-C.; Hsiao, V.K.S. Photocatalytic Properties of Graphene/Gold and Graphene Oxide/Gold Nanocomposites Synthesized by Pulsed Laser Induced Photolysis. *Nanomaterials* **2020**, *10*, 1985. [CrossRef] [PubMed]
13. Charipar, K.; Kim, H.; Piqué, A.; Charipar, N. ZnO Nanoparticle/Graphene Hybrid Photodetectors via Laser Fragmentation in Liquid. *Nanomaterials* **2020**, *10*, 1648. [CrossRef] [PubMed]
14. Butt, M.A.; Mamonova, D.; Petrov, Y.; Proklova, A.; Kritchenkov, I.; Manshina, A.; Banzer, P.; Leuchs, G. Hybrid Orthorhombic Carbon Flakes Intercalated with Bimetallic Au-Ag Nanoclusters: Influence of Synthesis Parameters on Optical Properties. *Nanomaterials* **2020**, *10*, 1376. [CrossRef]
15. Huang, Y.; Qiao, Y.; Li, Y.; He, J.; Zeng, H. Zn-Doped Calcium Copper Titanate Synthesized via Rapid Laser Sintering of Sol-Gel Derived Precursors. *Nanomaterials* **2020**, *10*, 1163. [CrossRef]
16. Fernández-Arias, M.; Boutinguiza, M.; del Val, J.; Riveiro, A.; Rodríguez, D.; Arias-González, F.; Gil, J.; Pou, J. Fabrication and Deposition of Copper and Copper Oxide Nanoparticles by Laser Ablation in Open Air. *Nanomaterials* **2020**, *10*, 300. [CrossRef] [PubMed]
17. Fominski, V.; Demin, M.; Fominski, D.; Romanov, R.; Goikhman, A.; Maksimova, K. Comparative Study of the Structure, Composition, and Electrocatalytic Performance of Hydrogen Evolution in $MoS_{x\sim2+\delta}/Mo$ and $MoS_{x\sim3+\delta}$ Films Obtained by Pulsed Laser Deposition. *Nanomaterials* **2020**, *10*, 201. [CrossRef] [PubMed]
18. Muniz-Miranda, M.; Muniz-Miranda, F.; Giorgetti, E. Spectroscopic and Microscopic Analyses of Fe_3O_4/Au Nanoparticles Obtained by Laser Ablation in Water. *Nanomaterials* **2020**, *10*, 132. [CrossRef] [PubMed]

Article

Two Approaches to the Laser-Induced Formation of Au/Ag Bimetallic Nanoparticles in Supercritical Carbon Dioxide

Alexey Rybaltovsky [1,2], Evgeniy Epifanov [2], Dmitriy Khmelenin [3], Andrey Shubny [2], Yuriy Zavorotny [1], Vladimir Yusupov [2] and Nikita Minaev [2,*]

[1] Skobeltsyn Science Research Institute of Nuclear Physics, Lomonosov Moscow State University, 119991 Moscow, Russia; alex19422008@rambler.ru (A.R.); ummagumma44@yahoo.com (Y.Z.)
[2] Institute of Photon Technologies, Federal Scientific Research Centre "Crystallography and Photonics" RAS, 108840 Moscow, Russia; rammic0192@gmail.com (E.E.); deerhunter9136@gmail.com (A.S.); iouss@yandex.ru (V.Y.)
[3] FSRC "Crystallography and Photonics" RAS, 119333 Moscow, Russia; xorrunn@gmail.com
* Correspondence: minaevn@gmail.com; Tel.: +7-915-053-2103

Abstract: Two approaches are proposed for the synthesis of bimetallic Au/Ag nanoparticles, using the pulsed laser ablation of a target consisting of gold and silver plates in a medium of supercritical carbon dioxide. The differences between the two approaches related to the field of "green chemistry" are in the use of different geometric configurations and different laser sources when carrying out the experiments. In the first configuration, the Ag and Au targets are placed side-by-side vertically on the side wall of a high-pressure reactor and the ablation of the target plates occurs alternately with a stationary "wide" horizontal beam with a laser pulse repetition rate of 50 Hz. In the second configuration, the targets are placed horizontally at the bottom of a reactor and the ablation of their parts is carried out by scanning from above with a vertical "narrow" laser beam with a pulse repetition rate of 60 kHz. The possibility of obtaining Ag/Au alloy nanoparticles is demonstrated using the first configuration, while the possibility of obtaining "core–shell" bimetallic Au/Ag nanoparticles with a gold core and a silver shell is demonstrated using the second configuration. A simple model is proposed to explain the obtained results.

Keywords: laser ablation; supercritical fluid; supercritical carbon dioxide; plasmonic nanoparticles

1. Introduction

The recent increase in interest in the synthesis of bimetallic nanoparticles of silver and gold (Ag/Au BMNPs) is associated with their unique plasmonic and catalytic properties [1–3]. It is possible to purposefully change their characteristics and give them new functional properties by varying the elemental composition and morphology of such particles [4–7]. Due to this behavior, Ag/Au BMNPs can serve as the main elements in sensitive spectroscopic systems based on SERS, which can be used for detecting small amounts of bioorganic molecules or can be used in optical systems as detectors for recording changes in certain characteristics of liquid media, such as the refractive index [4,6].

One of the first methods for the synthesis of Ag/Au BMNPs was the chemical method, which is implemented through the preparation of a mixture of solutions of inorganic compounds of gold and silver, such as $AgNO_3$ and $HAuCl_4$ [8]. Recently, synthesis technologies based on the methods of "green chemistry" have been developed. These include methods using plant extracts as the biogenic agents, allowing Au NPs, Ag NPs, and Ag/Au BMNPs of various types to be obtained [3,9,10]; methods of biosynthesis of functionalized nanoparticles using biomolecules from microorganisms as the reactants, and capping agents of noble metals of microorganisms [11,12].

Simultaneously with the above approaches, methods based on laser ablation in various media have been developed. Their advantages include higher productivity and flexibility compared to other methods for synthesizing metal nanoparticles [5,13–17]. Currently, many methods for the synthesis of Ag/Au BMNPs using laser ablation are already known. For example, in some cases, the effect is achieved due to the action of laser radiation on thin films of silver and gold [14]. In other experiments, the required concentrations of gold and silver nanoparticles are first produced in separate volumes by laser ablation of the corresponding massive targets, and then the mixture of colloids is subjected to laser action [6]. Some articles [3,6,8,14] have noted the possibility of varying the molar composition of the obtained Ag/Au BMNPs by changing the content of the initial components for synthesis. In this way, it is possible to obtain not only nanoparticles with different plasmonic characteristics, but also with different morphologies—either the "alloy" type or the "core–shell" type [3].

The efficiency of the ablation process, the particle size distribution, their morphology, and other characteristics are influenced by the environment, the geometry of the experiment, and the parameters of the laser action (wavelength; energy, duration, and frequency of pulses; focusing parameters). In most works devoted to the formation of Au NPs, Ag NPs, and Ag/Au NPs, ablation was carried out in liquid, mainly in water [5,8,13–15]. Vacuum [18], gas [19,20], and supercritical carbon dioxide ($scCO_2$) [7,21] have also been used as ablation media.

The use of a $scCO_2$ medium for laser ablation is of interest because this medium has a number of unique properties that affect the formation, modification, and deposition of nanoparticles. The fundamental difference between a supercritical fluid medium and liquid media is the absence of an overheating regime and, as a consequence, the virtual absence of explosive boiling and the formation of vapor gas bubbles during pulsed laser heating. An important advantage of supercritical fluids is the ability to adjust the density and a number of other characteristics of the medium within a wide range by changing the temperature and pressure. This allows the ablation process to be controlled by selecting the optimal parameters of the medium, including during laser exposure [22]. The supercritical state of CO_2 is characterized by extremely low viscosity at the level of gases, and high mobility of molecules, which allows ablation products to penetrate into hard-to-reach cavities and pores of various materials. At the same time, $scCO_2$ is a strong nonpolar solvent and a cheap and environmentally friendly material, so it is widely used in the development of new technologies in the field of "green" chemistry [23–25]. Unlike organic media, $scCO_2$ creates practically no byproducts arising from the photodegradation of organic molecules as a result of laser action.

This work is devoted to the development of approaches to the synthesis of Ag/Au BMNPs using the method of laser ablation of massive targets in supercritical fluid media. As in our first experiments on this topic [7,22], we used targets consisting of silver and gold plates. The $scCO_2$ medium was chosen as the working medium in which the production of nanoparticles occurred.

We aimed to find specific experiment configurations to synthesize Ag/Au BMNPs by pulsed laser ablation of a target, which would make it possible to implement different morphologies of Ag/Au BMNPs. Two ablation modes were implemented with different sources of laser radiation and optics schemes, different geometrical configurations of the target, and different laser beam locations. In the first regime, at a peak power density of laser pulses of ~33 GW/cm^2 and a large spot size on the target surface (Ø: ~0.3 mm), scanning by radiation over the target surface was not performed. In this case, the pulse repetition rate was low (50 Hz), and the laser radiation was supplied to the target horizontally. In the second regime using the same laser peak power density but a small spot size (Ø: ~45 μm), continuous scanning was carried out over the target surface. In this case, the pulse repetition rate was high (60 kHz), and the laser radiation was supplied to the target vertically.

2. Materials and Methods

Experiments on the synthesis of Ag/Au BMNPs using the pulsed laser ablation of a double target were carried out in two configurations of the experimental setup (Figure 1), corresponding to two different ablation modes. The facilities were based on a high-pressure reactor (1) with a set of optical ports. The reactor had a modular design that made it possible to implement various configurations of laser action on the target with simultaneous diagnostics of the ongoing physical and chemical processes, using visual observation and spectral measurement methods. The reactor was equipped with six transparent 10-mm-thick quartz windows arranged in a hexagonal pattern in the horizontal plane, as well as an additional window in the upper cover of the reactor. Control over the process of laser ablation of targets in all experiments was carried out visually through the windows (15) using a digital camera. In the first configuration, the target was fixed on the side wall of the reactor, while in the second configuration, the target was mounted on the bottom of the reactor.

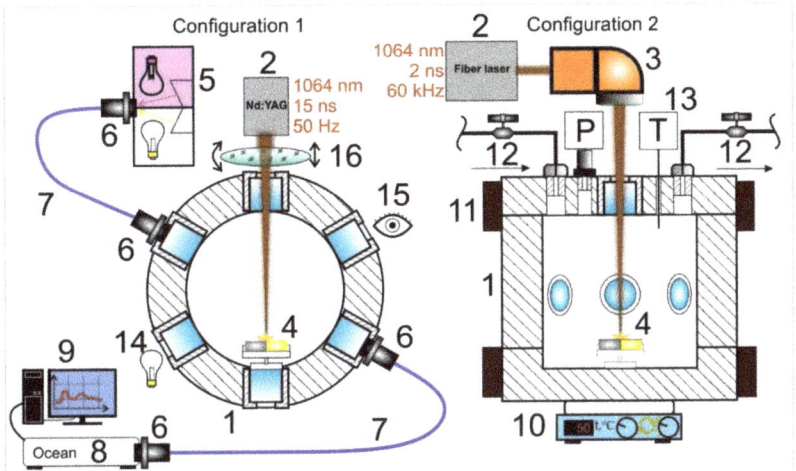

Figure 1. Schemes of installations for the synthesis of Ag/Au BMNPs show two different models of ablation. The schematic sections of the reactor, showing top (for configuration 1) and side views (for configuration 2): 1—high-pressure reactor; 2—laser sources; 3—galvo scanner with F-theta lens; 4—gold and silver targets; 5—UV and visible light sources; 6—collimator lens; 7—optical fibers; 8—spectrometer; 9—PC; 10—heating plate; 11—ring heaters; 12—needle valves for CO_2 inlet and outlet; 13—pressure and temperature sensors; 14—backlight; 15—observation window; 16—quartz lens.

To achieve the first ablation regime (configuration 1), a Lotis LS-2138TF Nd:YAG laser (Minsk, Belarus) with an average power of P_L = 11 W operating in Q-switching mode was used as the source of laser radiation. Laser radiation (λ = 1064 nm, τ = 15 ns, E_p = 220 mJ, f = 50 Hz) was introduced into the internal volume of the high-pressure reactor through a side window and focused on the target using quartz lenses with a focal length of 8 cm (16). We used non-sharp focusing of the laser radiation with a spot size on the target surface of Ø 0.30 ± 0.02 mm. The target was gradually moving towards the focal plane of the laser beam along the path of the beam. In this case, with this spot size at the time of the onset of ablation, the optical breakdown of the medium occurred directly on the target surface. The peak power density of laser pulses on the target was 33 ± 5 GW/cm^2.

To implement the second ablation regime (configuration 2), we used the radiation of a YLPP-1-150V-30 fiber laser (IPG Laser GmbH, Burbach, Germany) with an average power of P_L = 30 W. Laser radiation (λ = 1064 nm, τ = 2 ns, E_p = 0.5 mJ, f = 60 kHz) was introduced into the reactor through the upper window. Focusing and movement of the laser beam were carried out using an LscanH-10 galvanic scanning system (Ateko-TM, Moscow, Russia) with an SL-1064-50-63 F-theta lens (Ronar-Smith, Singapore). The optical system made it possible to form a laser spot with Ø 45 ± 5 µm on the Au–Ag target. In the experiments, we used targets made of gold (99.95%) and silver (99.8%) with dimensions of $10 \times 5 \times 1$ mm^3, located side-by-side inside the reactor on a PTFE holder. The peak power density of the laser pulses in the focusing region, as in the first regime, was 34 ± 8 GW/cm^2.

In the second regime (configuration 2), the laser radiation spot was moved using a galvanoscanner over the target surface at a speed of 100 mm/s, filling a rectangle of the selected size with a fill density of 100 lines per mm. The dimensions of these rectangles in the case of ablation of individual parts of the target made of gold or silver were 1×1 mm^2, while in the case of simultaneous ablation of two parts of the target were 1×5 mm^2.

To record the extinction spectra in both configurations, a fiber-optic spectroscopic system was assembled (5–9). The spectra were recorded using MAYA2000 PRO (spectral range: 200–1100 nm) and QE65000 (350–1100 nm) (Ocean Optics, Orlando, FL, USA) fiber spectrophotometers. Radiation from a halogen lamp and a deuterium gas discharge lamp (5) was introduced into the volume of the high-pressure reactor using 74UV collimators (Ocean Optics) (6) and an optical fiber (7). The radiation that passed through the reactor volume was collected by a collimator and entered the spectrometer (8). The spectra were recorded during the entire experiment with intervals ranging from 1 to 5 s.

The formation of Ag/Au BMNPs in both configurations consisted of successive laser action cycles on parts of the Au and Ag targets in scCO$_2$ (P = 200 bar, T = 50 °C). In configuration 1 (Figure 1), a sequence of three ablation cycles of the golden part of the target was carried out as Au → Au → Au, then a sequence of three ablation cycles of the silver part of the target was carried as Ag → Ag → Ag. The duration of each cycle was 1 min, with 1 min intervals between cycles. In configuration 2 (Figure 1), one ablation cycle for the golden part of the target, one cycle for the silver part of the target, then one cycle simultaneously ablating both parts (three cycles in total) were carried out. The duration of each cycle was 10 min, with 20 min intervals between cycles.

The structure of the obtained nanoparticles was studied by transmission electron microscopy (TEM), scanning transmission electron microscopy with a wide-angle dark field detector (HAADF STEM), and energy-dispersive analysis (EDX) using FEI Osiris (FEI, Lincoln, NE, USA), FEI Scios (FEI), and Phenom PROX (Thermo Fisher, Eindhoven, Netherlands) electron microscopes. When using the TEM method, samples in the form of suspensions in organic solvents (after pretreatment in an ultrasonic bath for 15 min) were applied onto carbon-coated copper grids using the drop method. The collection of nanoparticles in the form of a suspension was carried out by washing them off the walls of the reactor with isopropyl alcohol. Measurements with an electron microscope were carried out on either the day of synthesis or the next day.

3. Results and Discussion

The first results were obtained using a Lotis LS-2138TF laser source in configuration 1 (Figure 1), with a horizontal introduction of the beam onto the target. As a result of the laser action, the amplitude and shape of the extinction spectra gradually changed. The spectra were recorded continuously. Figure 2a shows the characteristic spectra obtained during the ablation of gold (times: 70 s and 240 s) and during the subsequent transition to the ablation of silver (times: 10 s, 80 s, and 195 s).

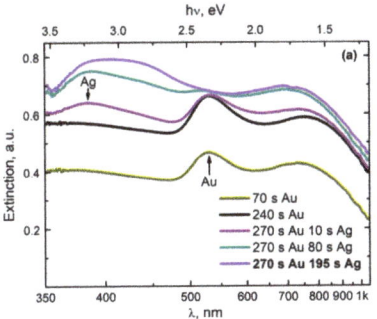 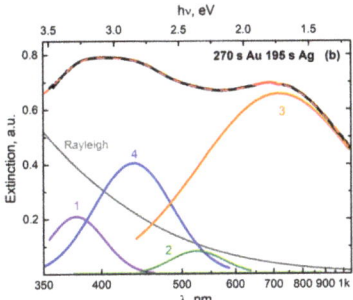

Figure 2. (a) Transformation of the extinction spectra of a colloidal solution of scCO$_2$ during successive cycles of ablation of Au and Ag plates in configuration 1. The numbers corresponding to the curves show the times since the beginning of the ablation of the indicated (Au or Ag) target. (b) Variant of decomposition of the final spectrum (red line) into the sum (black dashed line) of individual Gaussian components (1–4) and the Rayleigh scattering component (Rayleigh).

As a result of three successive gold target ablation cycles, an Au NP colloid was formed in the reactor. This led to the appearance of a plasmon resonance (PR) band with a maximum at the wavelength of 520 nm (Figure 2a) [22,26–28]. During the laser action for ~1 min ("70 s Au" curve, Figure 2a), the plasmon resonance band reached intensity saturation. At the same time, during the 1 min pause, the intensity hardly decreased, indicating low rates of aggregation and sedimentation processes of the Au NP colloid in scCO$_2$. In the next stage, three consecutive cycles of laser ablation of a silver target were carried out, involving 1 min of ablation and a 1 min pause. This led to the efficient formation of a colloidal solution of plasmonic Ag NP. As a result, an increase in the absorption of the medium inside the reactor was observed and a PR band appeared on the extinction spectrum with a maximum at the wavelength of 380 nm ("240 s Au" curve, Figure 2a), corresponding to Ag NPs with a size of 4–10 nm [7,22,26,29]. However, it should be noted that during the pause, the intensity of this band decreased much faster than in the case of the colloid with Au NPs. We believe that this was due to the more efficient aggregation of small silver nanoparticles into large ones as compared to gold nanoparticles, followed by their gravitational sedimentation to the bottom of the reactor [22]. The faster aggregation of Ag NPs can be explained by the higher energy activation ability of silver atoms to form chemical compounds and complexes in comparison with gold atoms [28]. At the same time, it has been argued [3] that the gold atoms in liquids are able to participate more efficiently in comparison with silver atoms in the processes of their self-assembly into nanoparticles. It is likely that at the initial stage of the synthesis of such metal nanoparticles, this process is primarily influenced by the interactions of metal atoms with molecules from the environment.

Upon ablation of a silver target, simultaneously with an increase in the PR band with the maximum at 380 nm, a rapid degradation of the PR band with a maximum at 520 nm occurred, which was even previously retained during pauses. At the same time, a new band with a maximum at 420 nm appeared on the long-wavelength wing of the PR band, with a maximum at 380 nm. This effect was especially pronounced during pauses, when the intensity of the component from the PR of silver dropped and another component remained, which we believe was from Ag/Au BMNPs. In Figure 2b, the decomposition of the final spectrum into separate components can be seen. The experimental spectrum is shown with a black dashed line, while the sum of the five proposed components is shown with a solid red line. It was observed that in addition to the PR bands corresponding to Ag NPs (curve 1, maximum at 380 nm), there were Au NPs (curve 2, maximum at 550 nm) and possibly AuNP aggregates (curve 3, maximum at 750 nm); curve 4 (maximum at 420 nm) can be attributed to Ag/Au BMNPs. Regarding differences in the positions of the maximum and the widths of the PR band for single Au NPs obtained in this expansion in comparison with the results in Figure 2a, these can most likely be associated with the manifestation of the PR band from some residual Au NP fraction (larger), which had not yet entered the "hot zone" near the silver target.

A similar transformation of the extinction spectra was observed (Figure 3) when using a YLM-1-150V-30 laser source in configuration 2 (Figure 1), with a vertical insertion of the beam onto a double target of gold and silver located at the bottom of the high-pressure reactor.

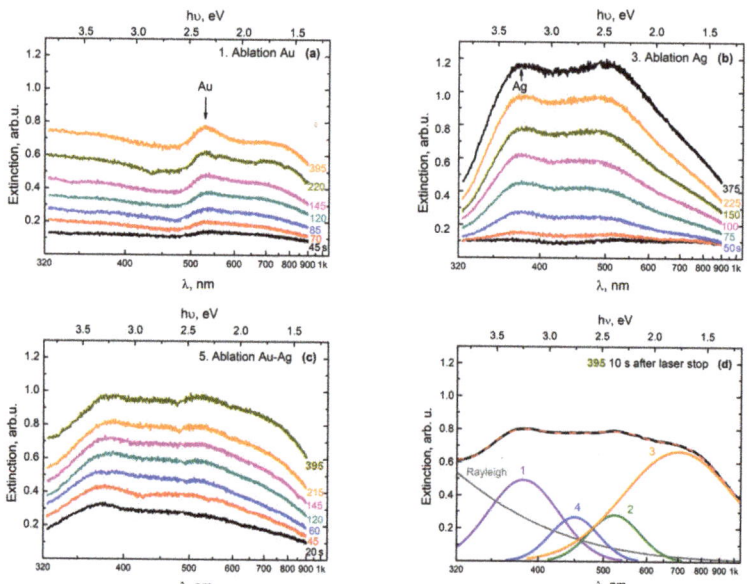

Figure 3. Transformation of the extinction spectra of a colloidal solution scCO$_2$ during successive cycles of ablation of the (**a**) gold and (**b**) the silver parts of the target. (**c**) Simultaneous ablation of both parts in configuration 2. The numbers on the curves show the times from the beginning of the ablation in seconds. The numbers to the right of the curves show the times in seconds since the start of the ablation cycle. (**d**) Variation of the decomposition of the spectrum obtained 10 s after the end of the last stage of ablation (red line) into the sum (black dashed line) of individual Gaussian components (1–4) and the Rayleigh scattering component (Rayleigh).

In the first stage, an Au NP colloid was formed during the ablation of a gold target, as in the previous case. This led to the appearance in the extinction spectrum of the colloid of the PR band, with the maximum at 520 nm (Figure 3a). In the next stage of ablation of the silver target, a broad band appeared with a maximum at 380 nm (Figure 3b), which corresponded to the PR of Ag NPs with a size of 4–10 nm [22]. Note that simultaneously to the growth of this maximum during the second stage, the amplitude of the maximum corresponding to the PR from gold nanoparticles also increased.

We believe that this effect was due to the fact that at the location of the target at the bottom of the reactor, the large gold nanoparticles that formed in the colloid during the first stage of ablation were also effectively deposited onto the surface of the silver target. Therefore, in the second stage (during the ablation of the silver target), the silver target was actually ablated together with the surface layer consisting of Au NPs, which were deposited onto the surface as a result of the ablation of the gold target. The presence of this deposited Au NP film on the target plates was confirmed by visual observations when the reactor was opened after the first cycle of the experiment. Figure 3c presents the extinction spectra, which were obtained in the process of ablation of gold and silver targets simultaneously, when scanning a laser beam over the surfaces of the two parts (Au and Ag) of the target. With a gradual increase in extinction in the entire observed wavelength range, the shape of the curves underwent transformation, in which an absorption band corresponding to Ag/Au BMNPs appeared against the background of PR bands from Ag NPs and Au NPs, as well as a long-wavelength absorption band.

Figure 3d shows a variation of the decomposition of the extinction spectrum of a colloid, obtained as a result of the simultaneous ablation of two parts of the target (Au and Ag) in configuration 2. It was observed that in addition to the PR bands of silver nanoparticles (curve 1, the peak with the maximum at 380 nm) and gold (curve 2, the peak with the maximum at 520 nm), there were also long-wave absorption bands (curve 3, the peak with the maximum at 700 nm), which can be classified as large nanoparticles and their aggregates. Additionally, in Figure 3d, band 4 (the peak with the maximum at 450 nm) stands out; in our opinion, it comprises Ag/Au BMNPs, but most likely is of a slightly different type, meaning it differs from the particles obtained in configuration 1.

The large width of the obtained spectra is associated with a rather strong scattering of NPs in terms of the composition (Ag/Au ratio), shape, and size. However, the obtained results can be considered preliminary, which makes it possible to reveal the specifics of the experimental parameters and the regularities for the preferential synthesis of bimetallic NPs of a certain type.

The TEM analysis of the obtained nanoparticles using the EDX method turned out to be informative and confirmed the stated provisions on the formation of bimetallic nanoparticles of various types. A similar method using the EDX system for the analysis of the compositions of composite nanoparticles was successfully used by the authors of another study [30] in the synthesis of carbon-coated Au NPs in the process of pulsed laser ablation of a gold target in a $scCO_2$ medium. Figure 4 shows a TEM image of nanoparticles, which was obtained using the EDX method during laser ablation of a double target in configuration 1.

In the upper region of this figure, nanoparticles measuring less than 10–20 nm can be observed, the PR bands of which are in the visible and near-UV wavelength ranges. Indeed, the maxima of the PR bands shown in Figure 2 for NP are in the range of 380–520 nm. The band with a maximum at 380 nm corresponds to small silver nanoparticles in $scCO_2$. Usually, for such particles dispersed in organic solvents, solid matrices, or water, the maximum of the PR band is observed in the range of 410–420 nm [26,28,29,31], and the short-wavelength shift for such a PR arises from other dielectric properties of the $scCO_2$ medium. It is likely that a certain shift in the PR band for small Au NPs will also occur when they are placed in a supercritical medium. In our situation, the maximum of the PR band for such particles based on Figure 2 is in the region of 520 nm, while for similar particles in organic or aqueous media, the maximum is at 540 nm [27,31]. According to [3], with a homogeneous distribution of Au and Au atoms during the formation of bimetallics of the "alloy" type, the maximum of the PR band smoothly shifts to the short-wavelength region from the position at 540 nm (at 100 mol % Au), with an increase in the molar content of the silver atoms contained within. The appearance in the PR spectrum of a component with a maximum at 420 nm (see Figure 2b) may well correspond to the synthesis of bimetallic nanoparticles of the "alloy" type, whereby the content of silver atoms, taking into account the above considerations, should be in the range of 60–80 mol %. The observed color gamut, which corresponds to the mixed elemental composition of the Ag/Au BMNPs, one part of which is indicated by an arrow, also does not contradict the reasoning given here.

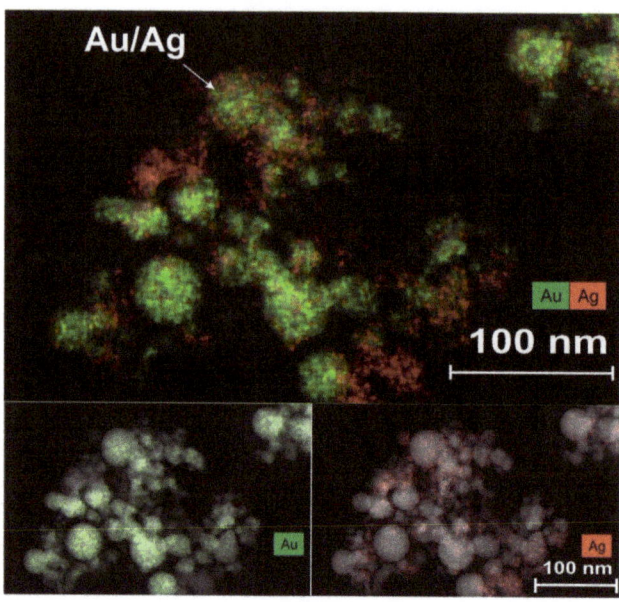

Figure 4. TEM images of nanoparticles obtained using the EDX method. The arrow indicates one of the largest Ag/Au-BMNP-type "alloys", measuring up to 40 nm in size. Nanoparticles were synthesized in the setup of configuration 1, using a low-frequency laser source with a high pulse energy.

Figure 5 shows TEM images of nanoparticles that appeared during the ablation of a double target in configuration 2. In this figure, using the EDX technique, two particles can be distinguished, which can be attributed to the bimetallics. Figure 5b shows the profiles of the intensity distribution of pixels corresponding to Ag and Au elements from two nanoparticles from Figure 5a. Comparison with the corresponding distribution for a model nanoparticle with a core–shell structure (Figure 5b, top) shows that the two isolated nanoparticles also have a core–shell structure. It can be observed that these Ag/Au BMNPs are composed of a gold core and a silver shell. The core and cladding diameters are 17 ± 1 and 19 ± 1 nm for the larger Au/Ag BMNPs and 13 ± 1 and 15 ± 1 nm for the smaller ones, respectively.

In these experimental studies, we used two configurations of installations (Figure 1), allowing two different ablation regimes at the same peak power densities (~33 GW/cm^2).

In the first configuration (Figure 1), laser action with a low pulse repetition rate (f = 50 Hz) was carried out horizontally without moving the laser spot over the target surface. As a result of the ablation, a crater with a diameter of 300 µm formed on the target. When implementing this regime, it was planned that the synthesis of Ag/Au BMNPs would mainly occur during the action of a laser pulse in a sufficiently wide laser beam (Figure 6). In the case of large laser spots, the ejection of the silver target material occurred mainly perpendicular to its surface towards the beam. As shown in [32], with increases in the size of the plasma source over 20 µm, its density on the optical axis gradually increased. The formation of a flux of atoms, ions, and electrons directed perpendicular to the target surface was also facilitated by the formation of a crater on the surface of the silver target. Such a directed flux of particles from the target led to numerous optical breakdowns in the region of the laser beam (Figure 6). A visual confirmation of this effect was the observed bright laser track in the scCO$_2$ medium in the first configuration of the experiment. Such a breakdown was accompanied by a sharp increase in temperature throughout the entire region and the formation of an ion atom plasma with Ag and Au. Further self-assembly of these particles led to the predominant formation of Au/Ag BMNPs of the "alloy" type, which were recorded using TEM (see Figure 4).

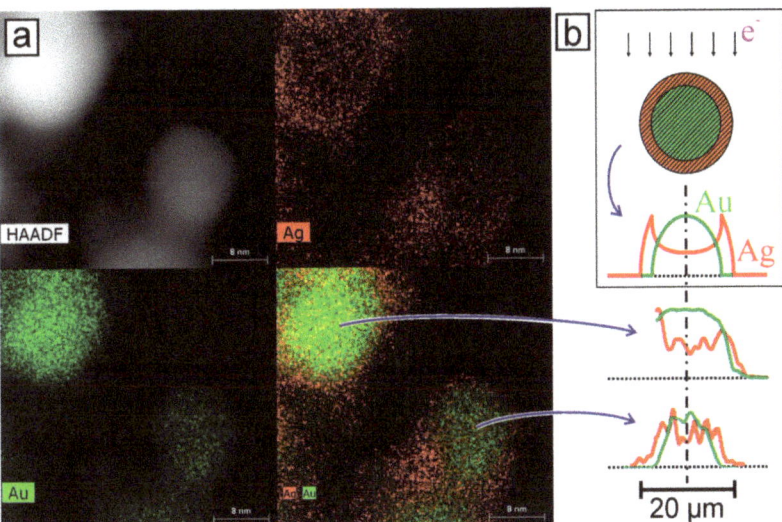

Figure 5. (**a**) TEM images of Au/Ag "core–shell" BMNPs, demonstrated using the EDX technique. Nanoparticles were obtained by laser synthesis in configuration N2. (**b**) Top: A cross-sectional image of a nanoparticle model with a core–shell structure and the pixel intensity profile of the elements from the core and shell for the corresponding EDX TEM image. Arrows show the direction of the flow of electrons (e-) during image acquisition. Bottom: Pixel intensity profiles for Au and Ag for TEM images of two nanoparticles.

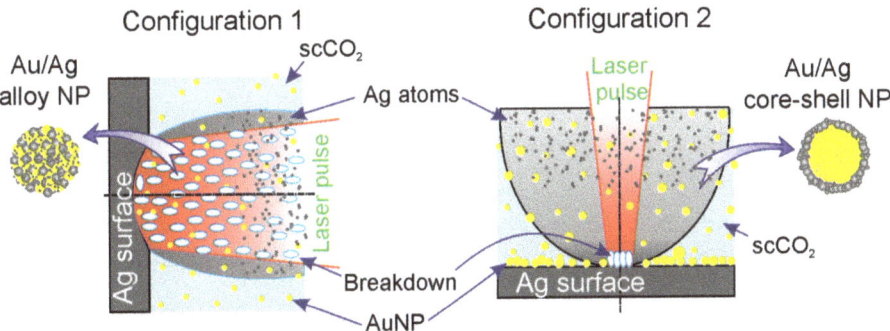

Figure 6. Model representation of the processes occurring during the ablation of a silver target in two different configuration setups (Figure 1) and leading to the synthesis of Au/Ag BMNP "alloy" (configuration 1) and "core–shell" (configuration 2) types.

In the second configuration (Figures 1 and 6), laser action with a pulse frequency of 60 kHz was carried out vertically. In this case, the spot size on the target surface was relatively small (Ø: 20 µm) and constantly moved over the target surface. With this configuration, during the ablation of the gold target (at the first stage), a layer of Au NPs was deposited on the surface of the silver part of the docked target due to the effect of gravitational sedimentation. In this case, with further ablation of the silver target, the removed target material interacted not only with the colloid but also with the gold nanoparticles on its surface. It should also be taken into account that due to the small size of the laser spot on the target and the absence of a crater comparable in area to the spot size, the substance in this case was carried into the colloid at a wide solid angle [32]; thus, when this configuration was implemented, the synthesis of Ag/Au BMNPs mainly occurred not in a relatively narrow laser beam but in a sufficiently large volume of the colloid into which the target material was ejected (Figure 6). In this volume, the deposition of Ag ions and atoms on the surfaces of Au NPs contained in $scCO_2$ took place with the formation of Au/Ag BMNPs of the "core–shell" type (see Figure 5). It is noted that in this configuration, due to the wide scattering angle of the silver target material, breakdown events occurred only in the immediate vicinity of the target surface at a relatively small volume, as was observed visually during the experiment. Therefore, the probability of the formation of Au/Ag BMNPs of the "alloy" type in configuration 2 was several orders of magnitude lower than in configuration 1.

The mechanisms of the formation of small bimetallic nanoparticles in the $scCO_2$ colloid in the two presented configurations were different. These different mechanisms led to the predominant formation of Au/Ag BMNPs of the "alloy" type (Figure 4) in configuration 1 and Au/Ag BMNPs of the "core–shell" type in configuration 2.

In configuration 1 (Figure 6), Au/Ag BMNPs of the "alloy" type (Figure 4) were synthesized in $scCO_2$ from a cloud of Ag and Au ions and atoms in a wide laser beam. The formation of this cloud occurred as a result of numerous breakdown events caused by the interaction of a laser pulse with a strongly directed flux of particles from an Ag plate flying towards it. In configuration 2 (Figure 6), Au/Ag BMNPs of the "core–shell" type (Figure 5) were synthesized, mostly from a cloud of Ag ions and atoms and Au NPs in a wide region formed by a weakly directed flow of particles from an Ag plate [33–35].

In general, the formation of bimetallic nanoparticles in scCO2 occurred as a result of (1) the self-assembly of atoms, (2) the aggregation (or coalescence) of nanoparticles and atoms, or (3) autocatalytic growth [36]. Some of these structures in the process of enlargement can form Ag or Au nanoparticles. The others can be transformed into bimetallic nanoparticles of the "core–shell" type [3,4,6,35] (Figure 5) or form nanoparticles of the "alloy" type mixed from Au and Ag atoms (Figure 4). It should be noted that the proposed simple model shows possible ways of obtaining NPs with different Ag/Au ratios. For example, if it is necessary to obtain gold nanoparticles of the core–shell type with a thinner silver layer on the surface of the gold core, configuration 2 will reduce the laser intensity or the silver ablation time.

Due to the vortex and convection flows arising from pulsed laser heating, the formed nanoparticles are scattered throughout the reactor volume and enter the observation zone, causing the appearance of characteristic plasmon resonance absorption bands. As noted in our first experiment [7], as well as in other papers on this topic [4,5,33], the PR bands for Au/Ag BMNPs are located in the interval between the known PR bands of pure silver and gold. Our explanation for the mechanism of the predominant formation of bimetallic nanoparticles of different types in two configurations does not contradict the considerations expressed in a previous [3], where the authors argue that the formation of Ag/Au BMNPs of one type or another largely depends on the ratio of silver and gold atoms in the reaction zone; with a significant excess of gold atoms in a liquid medium, these primarily combine with each other due to higher rates of movement in this environment in comparison with silver atoms. Indeed, in our situation, in the presence of an Au NP layer formed on the surface of a silver target in a $scCO_2$ medium, this phenomenon is possible.

It is important to note that the proposed specific approaches to the implementation of Au/Ag BMNP synthesis using a supercritical fluid medium belong to the field of "green chemistry". First of all, scCO2 is usually used as a substitute for organic solvents, which allows one to get rid of large volumes of liquid waste. In addition, $scCO_2$ can be converted into a gaseous state during the synthesis process, which makes it possible to implement a closed production cycle without the emission of pollutants.

4. Conclusions

Two different modes of laser action on a target consisting of two plates of Au and Ag in supercritical carbon dioxide in two different configurations were considered. The configuration for the experiment and the mode of laser action—with the same peak power density for the laser pulses (~33 GW/cm^2)—played fundamentally important roles in the production of bimetallic Au/Ag nanoparticles of both the "alloy" and "core–shell" types (with a gold core and silver cladding).

In configuration 1, when the double target was located vertically on the side wall of the chamber and its components were ablated alternately by a stationary wide beam of a low-frequency laser with a high pulse energy, Au/Ag BMNPs of the "alloy" type were predominantly formed. In the case of configuration 2, in which the double target was located horizontally at the bottom of the chamber and the ablation of its components occurred alternately by scanning a narrow beam of a high-frequency laser, Au/Ag BMNPs of the "core–shell" type were predominantly formed. A simple model was proposed that explains the predominant formation of the two types of nanoparticles (core–shell type or alloy type) with different experiment configurations and laser action parameters.

Author Contributions: Conceptualization, A.R., V.Y., and N.M.; data curation, E.E., D.K., A.S., and N.M.; formal analysis, E.E. and A.S.; funding acquisition, V.Y.; methodology, A.R., E.E., and Y.Z.; project administration, V.Y. and N.M.; resources, A.S. and N.M.; supervision, E.E.; validation, Y.Z.; visualization, D.K.; writing—original draft, A.R., A.S., V.Y., and N.M.; writing—review and editing, A.R., V.Y., and N.M. All authors have read and agreed to the published version of the manuscript.

Funding: This work was partially supported by the Russian Foundation for Basic Research (project No. 18-29-06056mk), as well as by the Ministry of Science and Higher Education in the framework of the work of the State Task of the Federal Research Center "Crystallography and Photonics" of the Russian Academy of Sciences in the analysis of nanoparticles. The TEM was performed using the equipment from the Shared Research Center of the Federal Scientific Research Center "Crystallography and Photonics" of the Russian Academy of Sciences.

Data Availability Statement: The data presented in this study are available on request from the corresponding author.

Conflicts of Interest: The authors declare no conflict of interest.

References

1. Jellinek, J. Nanoalloys: Tuning properties and characteristics through size and composition. *Faraday Discuss.* **2008**, *138*, 11–35. [CrossRef] [PubMed]
2. Eng, N.B. Synergetic Antibacterial Effects of Silver Nanoparticles @ Aloe Vera Prepared via a Green Method. *Nano Biomed. Eng.* **2010**, *2*, 267–274. [CrossRef]
3. Meena Kumari, M.; Jacob, J.; Philip, D. Green synthesis and applications of Au-Ag bimetallic nanoparticles. *Spectrochim. Acta Part A Mol. Biomol. Spectrosc.* **2015**, *137*, 185–192. [CrossRef] [PubMed]
4. Boote, B.W.; Byun, H.; Kim, J.H. Silver-gold bimetallic nanoparticles and their applications as optical materials. *J. Nanosci. Nanotechnol.* **2014**, *14*, 1563–1577. [CrossRef] [PubMed]
5. Intartaglia, R.; Das, G.; Bagga, K.; Gopalakrishnan, A.; Genovese, A.; Povia, M.; di Fabrizio, E.; Cingolani, R.; Diaspro, A.; Brandi, F. Laser synthesis of ligand-free bimetallic nanoparticles for plasmonic applications. *Phys. Chem. Chem. Phys.* **2013**, *15*, 3075–3082. [CrossRef]
6. Sharma, G.; Kumar, A.; Sharma, S.; Naushad, M.; Prakash Dwivedi, R.; ALOthman, Z.A.; Mola, G.T. Novel development of nanoparticles to bimetallic nanoparticles and their composites: A review. *J. King Saud Univ. Sci.* **2019**, *31*, 257–269. [CrossRef]
7. Tsypina, S.I.; Epifanov, E.O.; Shubny, A.G.; Arakcheev, V.G.; Minaev, N.V.; Rybaltovskii, A.O. Single-Stage Formation of Film Polymer Composites in Supercritical Colloid Solutions of Nanoparticles Obtained by Laser Ablation. *Russ. J. Phys. Chem. B* **2019**, *13*, 1235–1244. [CrossRef]
8. Chen, D.H.; Chen, C.J. Formation and characterization of Au-Ag bimetallic nanoparticles in water-in-oil microemulsions. *J. Mater. Chem.* **2002**, *12*, 1557–1562. [CrossRef]
9. Jacob, J.; Mukherjee, T.; Kapoor, S. A simple approach for facile synthesis of Ag, anisotropic Au and bimetallic (Ag/Au) nanoparticles using cruciferous vegetable extracts. *Mater. Sci. Eng. C* **2012**, *32*, 1827–1834. [CrossRef]
10. Shankar, S.S.; Rai, A.; Ahmad, A.; Sastry, M. Rapid synthesis of Au, Ag, and bimetallic Au core-Ag shell nanoparticles using Neem (Azadirachta indica) leaf broth. *J. Colloid Interface Sci.* **2004**, *275*, 496–502. [CrossRef]

11. Weng, Y.; Li, J.; Ding, X.; Wang, B.; Dai, S.; Zhou, Y.; Pang, R.; Zhao, Y.; Xu, H.; Tian, B.; et al. Functionalized gold and Silver bimetallic nanoparticles using Deinococcus radiodurans protein extract mediate degradation of toxic dye malachite green. *Int. J. Nanomed.* **2020**, *15*, 1823–1835. [CrossRef] [PubMed]
12. Li, J.; Tian, B.; Li, T.; Dai, S.; Weng, Y.; Lu, J.; Xu, X.; Jin, Y.; Pang, R.; Hua, Y. Biosynthesis of Au, Ag and Au–Ag bimetallic nanoparticles using protein extracts of Deinococcus radiodurans and evaluation of their cytotoxicity. *Int. J. Nanomed.* **2018**, *13*, 1411–1424. [CrossRef]
13. Han, H.; Fang, Y.; Li, Z.; Xu, H. Tunable surface plasma resonance frequency in Ag core/Au shell nanoparticles system prepared by laser ablation. *Appl. Phys. Lett.* **2008**, *92*, 023116. [CrossRef]
14. Nikov, R.G.; Nedyalkov, N.N.; Nikov, R.G.; Karashanova, D.B. Nanosecond laser ablation of Ag–Au films in water for fabrication of nanostructures with tunable optical properties. *Appl. Phys. A Mater. Sci. Process.* **2018**, *124*, 847. [CrossRef]
15. Navas, M.P.; Soni, R.K. Laser-Generated Bimetallic Ag-Au and Ag-Cu Core-Shell Nanoparticles for Refractive Index Sensing. *Plasmonics* **2015**, *10*, 681–690. [CrossRef]
16. Heinz, M.; Srabionyan, V.V.; Avakyan, L.A.; Bugaev, A.L.; Skidanenko, A.V.; Kaptelinin, S.Y.; Ihlemann, J.; Meinertz, J.; Patzig, C.; Dubiel, M.; et al. Formation of bimetallic gold-silver nanoparticles in glass by UV laser irradiation. *J. Alloys Compd.* **2018**, *767*, 1253–1263. [CrossRef]
17. Chau, J.L.H.; Chen, C.Y.; Yang, M.C.; Lin, K.L.; Sato, S.; Nakamura, T.; Yang, C.C.; Cheng, C.W. Femtosecond laser synthesis of bimetallic Pt-Au nanoparticles. *Mater. Lett.* **2011**, *65*, 804–807. [CrossRef]
18. Dikovska, A.O.; Alexandrov, M.T.; Atanasova, G.B.; Tsankov, N.T.; Stefanov, P.K. Silver nanoparticles produced by PLD in vacuum: Role of the laser wavelength used. *Appl. Phys. A Mater. Sci. Process.* **2013**, *113*, 83–88. [CrossRef]
19. Kawakami, Y.; Seto, T.; Yoshida, T.; Ozawa, E. Gold nanoparticles and films produced by a laser ablation/gas deposition (LAGD) method. *Appl. Surf. Sci.* **2002**, *197–198*, 587–593. [CrossRef]
20. Machmudah, S.; Wahyudiono; Takada, N.; Kanda, H.; Sasaki, K.; Goto, M. Fabrication of gold and silver nanoparticles with pulsed laser ablation under pressurized CO_2. *Adv. Nat. Sci. Nanosci. Nanotechnol.* **2013**, *4*. [CrossRef]
21. Machmudah, S.; Sato, T.; Wahyudiono; Sasaki, M.; Goto, M. Silver nanoparticles generated by pulsed laser ablation in supercritical CO_2 medium. *High Press. Res.* **2012**, *32*, 60–66. [CrossRef]
22. Minaev, N.V.; Arakcheev, V.G.; Rybaltovskii, A.O.; Firsov, V.V.; Bagratashvili, V.N. Dynamics of formation and decay of supercritical fluid silver colloid under pulse laser ablation conditions. *Russ. J. Phys. Chem. B* **2015**, *9*, 1074–1081. [CrossRef]
23. Eckert, C.A.; Knutson, B.L.; Debenedetti, P.G. Supercritical fluids as solvents for chemical and materials processing. *Nature* **1996**, *383*, 313–318. [CrossRef]
24. Sihvonen, M.; Järvenpää, E.; Hietaniemi, V.; Huopalahti, R. Advances in supercritical carbon dioxide technologies. *Trends Food Sci. Technol.* **1999**, *10*, 217–222. [CrossRef]
25. Stauss, S.; Urabe, K.; Muneoka, H.; Terashima, K. Pulsed Laser Ablation in High-Pressure Gases, Pressurized Liquids and Supercritical Fluids: Generation, Fundamental Characteristics and Applications. In *Applications of Laser Ablation—Thin Film Deposition, Nanomaterial Synthesis and Surface Modification*; InTech: London, UK, 2016; p. 221.
26. Kelly, K.L.; Coronado, E.; Zhao, L.L.; Schatz, G.C. The Optical Properties of Metal Nanoparticles: The Influence of Size, Shape, and Dielectric Environment. *J. Phys. Chem. B* **2003**, *107*, 668–677. [CrossRef]
27. Klimov, V. *Nanoplasmonics*; CRC Press: Boca Raton, FL, USA, 2014; ISBN 9814267422.
28. Pomogailo, A.D.; Rozenberg, A.S.; Uflyand, I.E. *Metal Nanoparticles in Polymers*; Khimiya: Moscow, Russia, 2000; ISBN 572451107X.
29. Khlebtsov, N.G.; Dykman, L.A. Optical properties and biomedical applications of plasmonic nanoparticles. *J. Quant. Spectrosc. Radiat. Transf.* **2010**, *111*, 1–35. [CrossRef]
30. Mardis, M.; Wahyudiono; Takada, N.; Kanda, H.; Goto, M. Formation of Au–carbon nanoparticles by laser ablation under pressurized CO_2. *Asia Pacific J. Chem. Eng.* **2018**, *13*, e2176. [CrossRef]
31. Bagratashvili, V.N.; Rybaltovsky, A.O.; Minaev, N.V.; Timashev, P.S.; Firsov, V.V.; Yusupov, V.I. Laser-induced atomic assembling of periodic layered nanostructures of silver nanoparticles in fluoro-polymer film matrix. *Laser Phys. Lett.* **2010**, *7*, 401–404. [CrossRef]
32. Blonskyy, I.V.; Danko, A.Y.; Kadan, V.N.; Orieshko, E.V.; Puzikov, V.M. Effect of the transverse size of a laser-induced plasma torch on material processing. *Tech. Phys.* **2005**, *50*, 358–363. [CrossRef]
33. Yang, G. *Laser Ablation in Liquids: Principles and Applications in the Preparation of Nanomaterials*; CRC Press: Boca Raton, FL, USA, 2012; ISBN 9814241520.
34. Becker, M.F.; Brock, J.R.; Cai, H.; Henneke, D.E.; Keto, J.W.; Lee, J.; Nichols, W.T.; Glicksman, H.D. Metal nanoparticles generated by laser ablation. *Nanostruct. Mater.* **1998**, *10*, 853–863. [CrossRef]
35. Barcikowski, S.; Devesa, F.; Moldenhauer, K. Impact and structure of literature on nanoparticle generation by laser ablation in liquids. *J. Nanoparticle Res.* **2009**, *11*, 1883–1893. [CrossRef]
36. Urabe, K.; Kato, T.; Stauss, S.; Himeno, R.; Kato, S.; Muneoka, H.; Baba, M.; Suemoto, T.; Terashima, K. Dynamics of pulsed laser ablation in high-density carbon dioxide including supercritical fluid state. *J. Appl. Phys.* **2013**, *114*. [CrossRef]

Article

Application of Pulsed Laser Deposition in the Preparation of a Promising MoS$_x$/WSe$_2$/C(B) Photocathode for Photo-Assisted Electrochemical Hydrogen Evolution

Roman Romanov [1], Vyacheslav Fominski [1,*], Maxim Demin [2], Dmitry Fominski [1], Oxana Rubinkovskaya [1], Sergey Novikov [3], Valentin Volkov [3] and Natalia Doroshina [3]

[1] National Research Nuclear University MEPhI (Moscow Engineering Physics Institute), Kashirskoe sh., 31, 115409 Moscow, Russia; limpo2003@mail.ru (R.R.); dmitryfominski@gmail.com (D.F.); oxygenofunt@gmail.com (O.R.)
[2] Immanuel Kant Baltic Federal University, A. Nevskogo St 14, 236016 Kaliningrad, Russia; sterlad@mail.ru
[3] Center for Photonics and 2D Materials, Moscow Institute of Physics and Technology (MIPT), 141700 Dolgoprudny, Russia; novikov.s@mipt.ru (S.N.); vsv.mipt@gmail.com (V.V.); doroshina.nv@phystech.edu (N.D.)
* Correspondence: vyfominskij@mephi.ru

Abstract: We studied the possibility of using pulsed laser deposition (PLD) for the formation of a MoS$_x$/WSe$_2$ heterostructure on a dielectric substrate. The heterostructure can be employed for effective solar water splitting to produce hydrogen. The sapphire substrate with the conducting C(B) film (rear contact) helped increase the formation temperature of the WSe$_2$ film to obtain the film consisting of 2H-WSe$_2$ near-perfect nanocrystals. The WSe$_2$ film was obtained by off-axis PLD in Ar gas. The laser plume from a WSe$_2$ target was directed along the substrate surface. The preferential scattering of selenium on Ar molecules contributed to the effective saturation of the WSe$_2$ film with chalcogen. Nano-structural WSe$_2$ film were coated by reactive PLD with a nanofilm of catalytically active amorphous MoS$_{x\sim4}$. It was established that the mutual arrangement of energy bands in the WSe$_2$ and MoS$_{x\sim4}$ films facilitated the separation of electrons and holes at the interface and electrons moved to the catalytically active MoS$_{x\sim4}$. The current density during light-assisted hydrogen evolution was above ~ 3 mA/cm^2 (at zero potential), whilst the onset potential reached 400 mV under irradiation with an intensity of 100 mW/cm^2 in an acidic solution. Factors that may affect the HER performance of MoS$_{x\sim4}$/WSe$_2$/C(B) structure are discussed.

Keywords: hydrogen evolution; pulsed leaser deposition; heterostructure; photoelectrocatalysis; semiconductors

1. Introduction

Transition metals chalcogenides have received considerable attention from scientists involved in the development of photoelectrochemical cells for producing hydrogen by solar water splitting [1–3]. These semiconducting materials have physio-chemical properties that enable their usage as both photo-active materials and hydrogen evolution electrocatalysts [4–6]. The good catalytic properties of metal chalcogenides (particularly amorphous molybdenum sulfides MoS$_x$) allow the compounds to replace expensive platinum. Moreover, they can ensure high efficiency of photo-assisted hydrogen evolution when using silicon-based heterostructures (n+p-Si) [7–9]. The photoactivity of crystalline transition metal dichalcogenides is sufficient for creating photocathodes based on these materials. Efficient hydrogen evolution is usually achieved by using an expensive (Pt/Ru) cocatalyst [10,11].

It is essential to coalesce the useful semiconductive and catalytic properties of transition metal chalcogenides in creating hybrid or heterostructures. These structures consist entirely of thin-film metal chalcogenides that have been selected for their structure and chemical composition. These compounds also act as photo-assistant agents in water splitting for hydrogen production [12].

Tungsten diselenide is a promising photo-active transition metal dichalcogenide [12–15]. Crystalline WSe_2 is a p-type semiconductor with a small band gap (~1.1 eV). If the conductivity is sufficiently high, this compound can be combined with catalytically active n-type metal chalcogenides to form photocathodes. The central requirement for WSe_2 films used in such photocathodes arises from the need to obtain a nearly perfect structure with minimum defects, including edge states. The recombination rate of nonequilibrium carriers (electrons and holes), which form during irradiation, decreases in the process. However, nanostructured WSe_2 films can have a greater area for hydrogen evolution, while edge states can be passivated by a co-catalyst [10,13]. WSe_2 films with a nearly perfect crystal lattice are usually obtained by chemical synthesis (from the vapour phase or in a special solution) or by the selenization of thin-film precursors (for instance, [16,17]). These techniques have both advantages and disadvantages. Thus, finding alternative techniques to obtain thin WSe_2 film with targeted properties remains a challenge.

In its traditional on-axis configuration, pulsed layer deposition (PLD) makes it possible to create WSe_2 with a crystal structure and good catalytic properties [18–21]. However, when using the on-axis PLD to obtain WSe_2 films, problems were revealed with obtaining a stoichiometric composition with a perfect chemical state of atoms in the film. This situation can be attributed to several factors, including the preferential sputtering of selenium on a growing WSe_2 when exposed to laser-plasma and the propensity of selenium to form pure Se nanoparticles at room temperature or to be desorbed at higher substrate temperatures [20,22–24]. Submicron- and nanoscale particles of metal W can be introduced into the film [25]. These particles form upon laser irradiation of a WSe_2 target.

In the case of on-axis PLD geometry, the substrate is placed normally to the axis of the laser plume expansion. Fominski et al. [26] have established that, under some on-axis PLD regimes, using a buffer gas makes it possible to decrease the efficiency of the preferential self-sputtering of chalcogen atoms. However, this technique suffers from considerable limitations: under some regimes of ablation of a dichalcogenide metal target, a laser plume may form and localize in a narrow solid angle. It has been revealed that chalcogen atoms can move to the plume periphery, while a buffer gas cannot preclude the metallization of the centre of the film deposition area [27].

Experiential studies of amorphous molybdenum sulfide (a-MoS_x) catalytic film creation have shown that, during off-axis PLD, the buffer gas has the conditions necessary for the effective saturation of films with chalcogen atoms [28]. These conditions are a result of the difference between the S and Mo atomic masses. During off-axis PLD, the substrate is placed parallel to the plume expansion. The film grows chiefly through the deposition of atoms that have collided with buffer gas molecules and changed their direction. The same effect can be achieved during the off-axis PLD of WSe_2 films due to the substantial difference between the W and Se atomic masses. One of the goals of this study was to test whether the off-axis PLD technique could be applied to a WSe_2 target to obtain WSe_2 nanocrystal films that are nearly perfect in terms of structure and chemical state.

Films based on a-MoS_x-nanomaterial have a high electrocatalytic activity during hydrogen evolution reaction (HER) [7,13,29–31]. The most common technique to obtain such films is a chemical synthesis or chemical deposition in the solutions of special precursors. The applicability of a laser-based technique to create amorphous a-MoS_x films with good electrocatalytic properties has been described in the literature [28,32–35]. Fominski et al. demonstrate that the PLD technique is associated with the highest catalytic activity in $MoS_{x\sim4}$ films because they contain Mo_3S_{12}/Mo_3S13 clusters [35]. It is suggested that reactive PLD (RPLD) in H_2S gas be used to create thin homogenous a-MoS_x films with a high content of catalytically active states of sulfur [34]. Performing on-axis RPLD from a Mo target prevents the formation of a substantial number of particles of various sizes during ablation. It also ensures a relatively conformal coating of a rough surface, which is typical of catalysts. This beneficial effect is possible because deposition is carried out using a flux of Mo atoms, which scatter at different angles once collided with H_2S molecules. Given the difficulty of predicting the conductivity type in a-MoS_x films with increased S content, it was necessary to establish whether the mutual arrangement of energy bands in the $MoS_{x\sim4}$/WSe_2 heterostructure is optimal for the effective separation of electron-hole pairs when light irradiated during the photo-assisted HER.

It is widely accepted that the material of the rear contact to the semiconductor can have a pronounced effect on the current transport in the semiconductor photovoltaic structure and probably also the photoelectrocatalyst [36]. We investigated boron-doped carbon films as rear contacts. A preliminary study showed that the introduction of boron atoms could produce p-type conductivity of C(B) films [37]. Films with good conductivity and mechanical strength were created by PLD from a mixed boron/graphite target. The substrate of the heterostructure was a sapphire plate. However, sapphire can be replaced by a cheaper material—glass, quartz, etc.

Our study aimed to form a multi-layered $MoS_{x\sim4}$/WSe_2/C(B) by PLD. The structure had to contain thin-film nanomaterials with properties sufficient for an effective photo-assisted HER in an acidic solution. When selecting PLD conditions for obtaining these nanomaterials, we used the results of a preliminary investigation of each nanomaterial. After the heterostructure have been assembled (i.e., after layer-by-layer nanomaterial deposition), chosen PLD conditions may prove to be non-optimal for efficient photoelectrocatalysis of hydrogen evolution. Nonetheless, the findings made it possible to produce recommendations on how laser-based processes may be improved and the structure and composition of selected nanomaterials modified.

2. Materials and Methods

2.1. Experimental Methods for On-Axis and Off-Axis PLD of Functional Nanolayers for $MoS_{x\sim4}$/WSe_2/C(B) Heterostructure Formation

Figure 1 shows the mutual arrangement of the target and the substrate when using on-axis and off-axis PLD for the formation of a $MoS_{x\sim4}$/WSe_2/C(B) heterostructure. The on-axis PLD configuration was used to deposit a C(B) film. The target, which consisted of a carbon (soot) and boron powder mixture in the proportion C/B~6, was concurrently ablated. For more detail on target manufacturing and the selection of laser ablation, see [37]. A Solar LQ529 laser (Minsk, Belarus) was used to ablate the target. The pulse duration and energy were 10 ns and 100 mJ respectively. The pulse repetition rate was 20 Hz. The energy density on the surface of the C(B) target was 9 J/cm^2. The substrate was placed to the laser plume axis, 3 cm away from the target, and heated to 500 °C. The deposition was performed in a vacuum at a residual pressure of 5×10^{-4} Pa. The deposition period of C(B) films was 10 min. The film thickness did not exceed 150 nm.

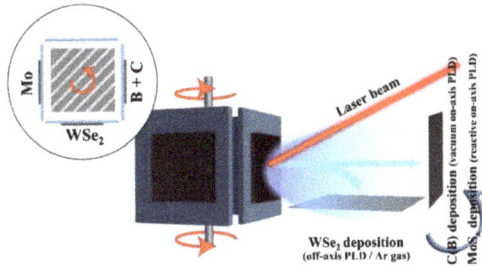

Figure 1. A schematic of the PLD technique employed to form functional layers in a $MoS_{x\sim4}$/WSe_2/C(B) heterostructure on a sapphire substrate. Comments are given in the text.

A WSe_2 film was deposited by off-axis PLD on the surface of the substrate coated with a C(B) film. The substrate was rotated 90° and placed along the laser plume axis 2 cm away from the WSe_2 target, which was manufactured by cold pressing of WSe_2 powder [22].

During the WSe$_2$ target ablation, the laser fluence was reduced to 4 J/cm^2. The substrate temperature was 700 °C. WSe$_2$ film deposition was performed in an Ar + 5% H$_2$ mixture at a pressure of 15 Pa. The gas mixture was introduced into a chamber that had been evacuated to a residual gas pressure of 5 × 10^{-4} Pa or less. The deposition time of a WSe$_2$ film of a thickness of ~200 nm was 20 min.

After the formation of WSe$_2$ films, the sample was allowed to cool down to room temperature. Then it was rotated 90° for subsequent MoS$_x$ film deposition by PLD in the reaction gas. The Ar + 5% H$_2$ gas mixture was pumped out with the help of a turbo-molecular pump, and H$_2$S gas was introduced into the chamber until the pressure reached ~26 Pa. The Mo target ablation was carried out using 100 mJ pulses. The MoS$_x$ film deposition time was set at 6 min. The thickness of a deposited MoS$_x$ film on a smooth substrate did not exceed 20 nm. The choice of the Mo ablation conditions and the H$_2$S pressure was motivated by the results of preliminary investigation of RPLD of MoS$_x$ films [38]. Under the chosen conditions of on-axis RPLD, the expected ratio was = S/Mo~4.

2.2. Structural, Chemical, Electrical, Optical, and Photoelectrochemical Characterization Techniques

In this study, WSe$_2$ films were produced by off-axis PLD for the first time, and thus they require further examination. Yet, C(B) and MoS$_x$ films obtained by on-axis (R)PLD have been studied extensively. We also discussed their structural and chemical properties in several publications. Therefore, in this article, we will focus on the information that will give a comprehensive picture of the components (layers) of the MoS$_{x\sim4}$/WSe$_2$/C(B) heterostructure.

The surface morphologies of the prepared films and heterostructures were examined by scanning electron microscopy (SEM, Tescan LYRA 3, Brno, Czech Republic). Using this microscope, the surface distribution of elements was studied by energy dispersive X-ray spectroscopy (EDS). The structure of the films was investigated by micro-Raman spectroscopy (MRS, Horiba, Kyoto, Japan), using a 632.8-nm (He-Ne) laser. The cross-section of the laser beam was <1 μm. To explore the structural features of WSe$_2$ films obtained by off-axis PLD, the films were separated from the substrate and transferred onto metal grids to study by high-resolution transmission electron microscopy (HRTEM) and selected area electron diffraction (SAED) with the help of a JEM-2100, JEOL microscope (Toyo, Japan).

Band gaps (E_g) in the prepared films were measured optically by processing absorption spectra. To this end, a Tauc plot was constructed to describe the dependence between $(\alpha h\nu)^{1/r}$ and $(h\nu)$, where α is the absorption coefficient, $h\nu$ is the photon energy, and r is a parameter that is taken to be 2 for indirect transitions. The optical absorption and transmission spectra were measured using an Agilent Technologies Cary Series UV-Vis-NIR spectrophotometer. Special samples were manufactured to explore the optical properties of WSe$_2$ and MoS$_x$ films. In these samples, WSe$_2$ and MoS$_x$ films were deposited on transparent sapphire substrates in the selected conditions.

The chemical states of WSe$_2$ and MoS$_x$ films were studied by XPS. XPS spectra were obtained using a Theta Probe Thermo Fisher Scientific spectrometer with a monochromatic Al Kα X-ray source ($h\nu$ = 1486.7 eV) and a 400 μm X-ray spot. The spectrometer energy scale was calibrated using Au4f$_{7/2}$ core level lines located at a binding energy of 84.0 eV. The Advantage Data Spectrum Processing program was used for deconvolution of the experimental XPS spectra. The Shirley background is an approximation method that was used for determining the background under an XPS peak. The peaks were fitted by symmetric convolution of Gaussian and Lorentzian functions. The ratios of atomic concentrations of elements (x = S/Mo) were calculated considering the intensities of Mo 3d and S 2p peaks and the corresponding Scofield's Relative Sensitivity Factor.

The thickness of quite thick MoS$_x$ films (thickness is $\geq$100 nm) was measured by SEM. For this, the Si substrate with the deposited thick MoS$_x$ film was cleaved and the vertical cross section was investigated by SEM. These measurements made it possible to estimate the deposition rate of the MoS$_x$ films during reactive PLD. The deposition rate was used to determine the time for preparation of a very thin MoS$_x$ film (thickness is ~3 nm). The thickness of this thin film was then estimated from the results of XPS studies for MoS$_x$/WSe$_2$ heterostructure. The thickness of thin MoS$_x$ film was estimated as quite adequate if the XPS spectra of both films (MoS$_x$ and WSe$_2$) could be detected at the same time.

The XPS measurements were used to determine the mutual arrangement of valence bands (VB) in the semiconductor heterostructures. The employed technique is widely used to study the band structure in heterojunction the formation of which can cause a change in the energy distribution of electrons [35,39,40]. The leading edge of the valence band spectrum was approximated by a linear function using the least-square fit of the leading edge of the VB spectra. The position of the valence band maximum (VBM) was determined as the intersection of the approximating linear function and the baseline. Determining the shift between the core levels of semiconductors in the heterojunction made it possible to calculate the valence band offset (VBO).

To calculate the VBO in a MoS$_x$/WSe$_2$ heterostructure, a series of measurements was performed. Firstly, the XPS spectra of the Mo3d and W4f core levels were measured along with the spectra of the valence bands of quite thick MoS$_x$ and WSe$_2$ films. Secondly, the spectra of the Mo3d and W4f core levels were measured for a MoS$_{x\sim 4}$/WSe$_2$ heterostructure, in which the thickness of the upper layer (MoS) did not exceed 3 nm. Thirdly, the VBO value for heterojunctions was calculated based on the formula:

$$VBO = (E_{Mo3d5/2} - E_{W4f7/2})_{interface} + (E_{W4f7/2} - VBM_W)_{bulk} - (E_{Mo3d5/2} - VBM_{Mo})_{bulk},$$

where VBM_W and VBM_{Mo} are the energies of the upper edge of the valence band for WSe$_2$ and MoS, respectively. 'Interface' stands for spectra for heterojunctions, and 'bulk' for the spectra of thicker films on C(B)/Al$_2$O$_3$ substrates.

The work function (φ) needed to withdraw an electron from a WSe$_2$ film was calculated using the formula $\varphi = h\nu - E^{CutOff} + E_F$, where E^{CutOff} is the secondary electron cutoff, E_F is the Fermi level if these magnitudes are considered on a kinetic energy scale. The Fermi level was determined based on an analysis of the energy spectrum of the valence band. To enable XPS investigation, the samples were created on a conducting C(B) film. This way, charge storage was prevented in the sample. The zero-value point of the binding energy scale corresponded to the Fermi level. In this case, the E^{CutOff} value marked on the kinetic energy scale coincides with φ.

The electrical properties of C(B) films on sapphire substrates were studied by a four-contact method in the van der Pauw geometry; Hall-effect measurements were performed at room temperature. A magnetic field, varying from 0 to 1 T, was used for Hall-effect characterization. Metallic contacts to the sample with a circular mesa were formed from an InSn alloy, and the linearity of the volt–ampere characteristics of all contacts was monitored. During the measurements, the direction of the current was switched to eliminate the effects of thermoelectric power. The resistivity was calculated by averaging the values from all pairs of contacts.

To study the photoelectrocatalytical properties of MoS$_x$/WSe$_2$/C(B)/Al$_2$O$_3$ samples, we irradiated these samples with 100 W Xe lamps in an 0.5 M H$_2$SO$_4$ aqueous solution. The light intensity was 100 mW/cm^2. A three-electrode configuration was used to determine the photo-assisted current in an electric circuit with modified cathodes. The polarization curves were measured using linear sweep voltammetry (LSV) with a change in the applied potential from -100 to 400 mV and a scan rate of 2 mV/s. When measuring LSV curves and the time evolution of the photocurrent, the light source was turned on and off. For chronoamperometry measurements, the potential of the photocathode was 0 V (relative to the reversible hydrogen electrode, RHE).

3. Results

3.1. On-Axis PLD of C(B) Films

Figure 2a,b show the morphology of a C(B) film formed on sapphire by traditional PLD. Detached rounded particles were observed on the smooth surface of C(B) films. The particle size ranged from 0.1 to 0.5 µm. This morphology is attributed to the deposition of B-rich particles [37]. The Raman spectrum of a C(B) film has two broad peaks at 1343 and 1545 cm^{-1} (Figure 2c), which correspond to the peaks marked D and G. The peaks are shifted to the lower wavenumber relative to the peaks associated with graphite (1360 cm^{-1} and 1580 cm^{-1} respectively). This shift meant that the films had a graphite-like local packing, which contained B atoms and some C atoms with sp^3-bonding in sp^2-matrix. A more detailed analysis of the Raman spectra of C(B) films obtained by PLD can be found in [37,41].

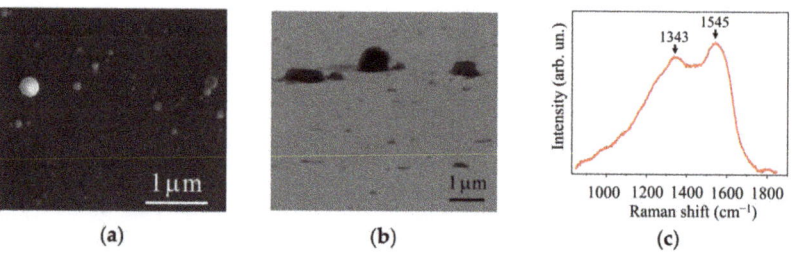

Figure 2. SEM images obtained at (**a**) normal and (**b**) 45° angles to the surface; (**c**) Raman spectrum of a C(B) film prepared on sapphire substrate by on-axis PLD in a vacuum.

The C(B) films had a specific resistance of ~1.5 mΩ·cm and p-type conductivity. At room temperature, the carrier concentration and mobility were 4.4×10^{19} cm^{-2} and 180 cm^2/V·s, respectively. The low resistance to the current flow in C(B) films enabled their use as a rear contact to the MoS$_{x\sim4}$/WSe$_2$ heterostructure, and p-type conductivity made it possible for holes formed upon illumination to move from the WSe$_2$ film to the external electric circuit.

3.2. Off-Axis PLD of WSe$_2$ Films

Figure 3 shows the morphology of the WSe$_2$ film obtained by off-axis PLD on the surface of C(B) film. The WSe$_2$ film covers the surface of the C(B) film with a continuous layer and the WSe$_2$ film had a nanocrystal structure consisted of petal-like crystals with random orientation relative to the film surface. The linear sizes of WSe$_2$ crystals reached 1 µm, whereas the thickness of the nanopetals did not exceed 50 nm.

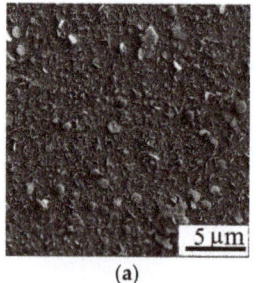 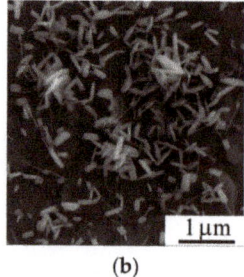

(**a**) (**b**)

Figure 3. (**a**,**b**) SEM images of the surface of the WSe$_2$ film (two magnifications) obtained by off-axis PLD on the surface of C(B)/Al$_2$O$_3$ sample.

A structural investigation by MRS and TEM/MD techniques demonstrated that the WSe$_2$ film had a crystal structure. An MRS spectrum (Figure 4a) only had peaks characteristic of the 2H-WSe$_2$ phase. The peaks associated with the vibrational modes E_{2g}^1 and A_{1g} coincided because the shift between them was approximately 3 cm^{-1} [42,43]. The narrow half-height width of the peak (3 cm^{-1}) points to the suitable quality of the crystal structure. A high-resolution TEM and SAED analysis of a single WSe$_2$ petal showed that it consisted of several nanocrystals with a hexagonal lattice of the 2H-WSe$_2$ phase (Figure 4b). The nanocrystal size was ~10 nm. Although the nanocrystals were oriented randomly relative to the c-axis, the basal plane of all the nanocrystals was parallel to the petal surface.

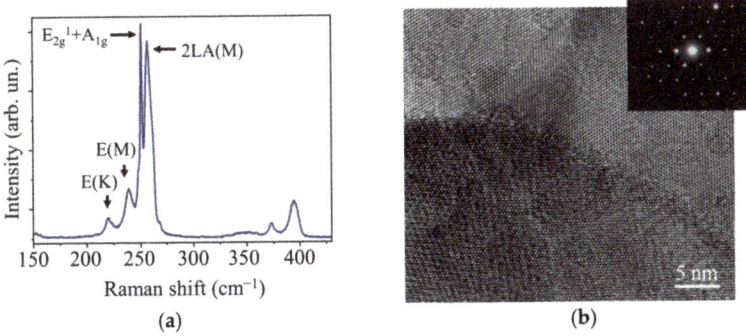

Figure 4. (a) Raman spectra and (b) HRTEM and SAED patterns of the WSe$_2$ film obtained by off-axis PLD.

Figure 5 shows part of the XPS spectra for the surface of WSe$_2$ film deposited on the surface of C(B) film. The W4f spectrum was well described by a doublet in which the W4f$_{7/2}$ and W4f$_{5/2}$ peaks had binding energies of 32.24 and 34.47 eV, respectively, which are characteristic of WSe$_2$. The Se3d spectrum was described by a doublet whose Se3d$_{5/2}$ and Se3d$_{3/2}$ peaks were at 54.50 and 55.37 eV, respectively. The XPS spectra indicated effective chemical interaction between Se and W during off-axis PLD [13,43,44].

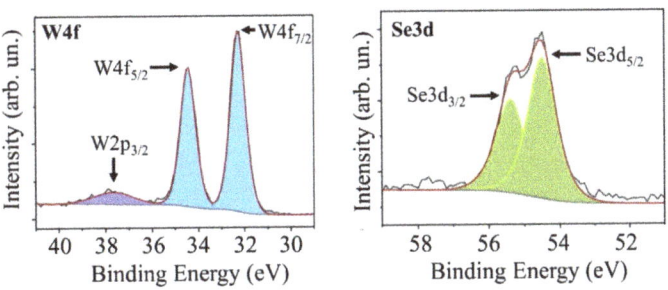

Figure 5. Core level XPS W4f and Se3d spectra of the WSe$_2$ film obtained by off-axis PLD.

An analysis of the energy spectrum of secondary electrons and the valence band showed that the work function for WSe$_2$ electrons was 4.9 eV (Figure 6a). The Fermi level was close to the bottom of the band gap 0.25 eV away from the upper edge of the valence band (Figure 6b). An investigation of the WSe$_2$ film optical properties demonstrated that the film had an absorption spectrum characteristic of WSe$_2$; the band gap width was 1.4 eV (Figure 7). A study of the band structure of the WSe$_2$ film proved that it had p-type conductivity typical of this compound [45,46].

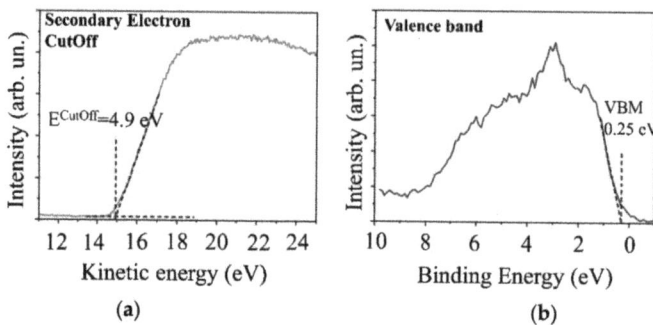

Figure 6. (a) XPS spectrum of secondary electrons cutoff and (b) evaluation of the valence band edge position for a WSe$_2$ film obtained by off-axis PLD on the surface of C(B) film.

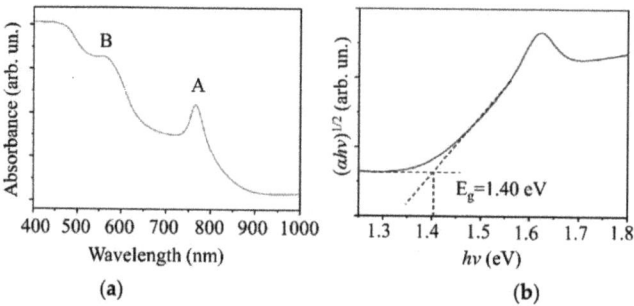

Figure 7. (a) Optical absorption spectra and (b) Tauc plots for the WSe$_2$ film on the sapphire substrate. The A and B peaks are explained by excitonic absorption.

3.3. On-Axis Reactive PLD of MoS$_{x\sim4}$ Film

Figure 8 shows an SEM image of the surface of WSe$_2$/C(B)/Al$_2$O$_3$ sample after MoS$_x$ film deposition by on-axis reactive PLD. MoS$_x$ film deposition did not cause a substantial change in the morphology of the sample surface. The principal difference between SEM images of the WSe$_2$ film before (Figure 3) and after MoS$_x$ film deposition (Figure 8) was that the sides of the WSe$_2$ nanocrystal petals lost their sharpness when coated by a thin MoS film, which is a porous structure. Mapping element distribution in the sample surface suggested that the MoS$_x$ film had a sufficiently homogeneous distribution over the sample surface (Figure 9). During PLD, the collision of Mo atoms with H$_2$S molecules ensured their scattering at different angles. As a result, a MoS$_x$ film could be formed even on those WSe$_2$ nanopetals that were oriented perpendicular to the surface of the substrate.

Figure 8. (a,b) SEM images (two magnifications) of the $MoS_x/WSe_2/C(B)/Al_2O_3$ sample.

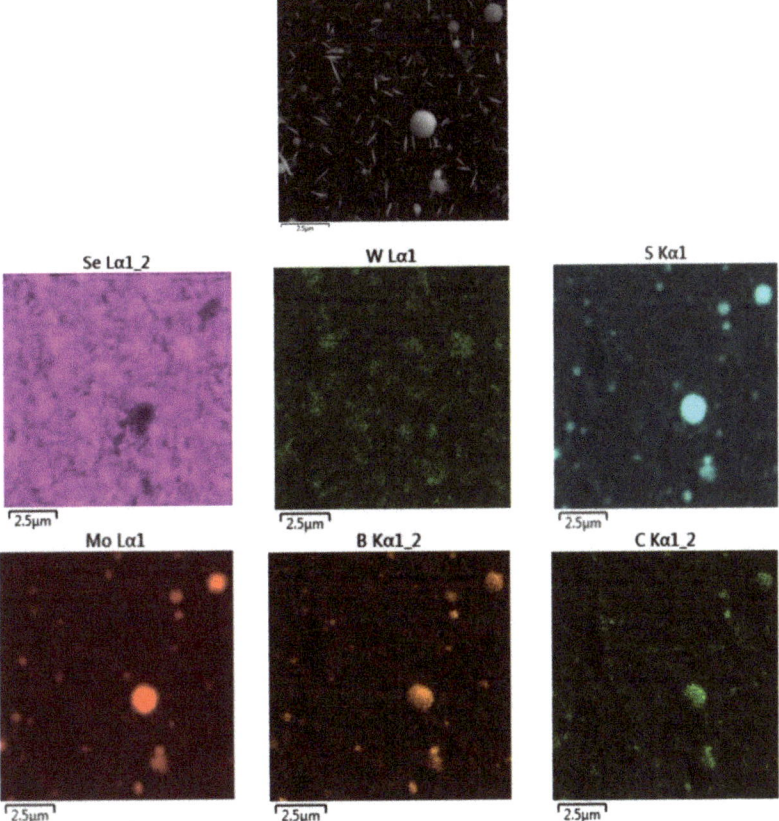

Figure 9. SEM image (top gray) and EDS maps (colored) of element distribution on the surface of the $MoS_x/WSe_2/C(B)/Al_2O_3$ sample. Intensity of different colors indicates where the corresponding elements (Se, W, S, Mo, B, and C) are most abundant. The presence of submicron rounded particles is explained by B-rich particle deposition during C(B) film formation.

Figure 10 shows the results of XPS investigation of a quite thick MoS_x film obtained by on-axis RPLD. An analysis of the chemical state of elements showed (Figure 10a,b) that core level XPS Mo3d spectrum was well described by a doublet corresponding to the Mo^{4+}

state. The bonding energy of the peak Mo3d$_{5/2}$ was 229.24 V, accounted for by chemical bonds with S atoms [26,34]. Molybdenum oxides (Mo^{6+}) or metallic ^{0} were not observed. The Mo3d$_{5/2}$ peak partially overlapped with the S2s peak. The S2s peak consisted of singular peaks whose position correlated with that of doublets in the S2p spectrum. When analyzing the S2p peak, we used the traditional approach, i.e., we identified the states of sulfur with high and low binding energy (HBE and LBE, respectively) [34,35]. The LBE doublet was associated with single S^{2-} atoms (in MoS$_2$-like clusters) and a terminal (S$_2^{2-}$)$_{tr}$ ligand (in Mo$_3$S$_{13}$/Mo$_3$S$_{12}$ clusters). The doublet had S2p$_{3/2}$ and S2p$_{1/2}$ peaks with binding energies of 162.04 and 163.35 eV, respectively. The HBE doublet had S2p$_{3/2}$ and S2p$_{1/2}$ peaks, whose binding energies were 163.28 and 164.50 eV. This doublet is usually attributed to apical S^{2-} and bridging (S$_2^{2-}$)$_{br}$ ligands in Mo$_3$S$_{13}$/Mo$_3$S$_{12}$ clusters. An XPS studies-based calculation of S/Mo atomic concentration ratios for this film confirmed that $x \sim 4.0$. Measuring the valence band spectrum showed that the Fermi level was 0.4 eV away from the bottom of the band gap (Figure 10c).

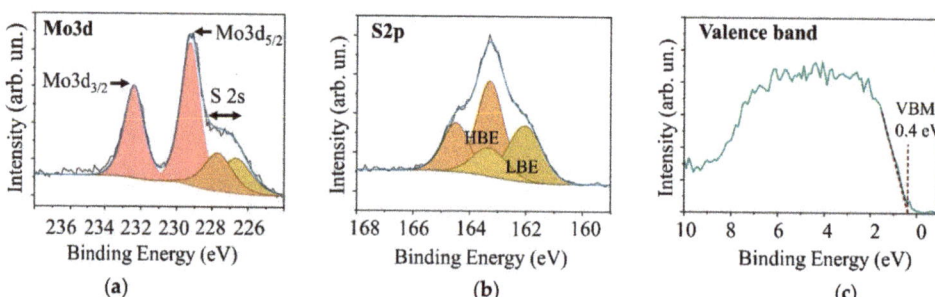

Figure 10. XPS spectra of (a,b) core level Mo3d and S2p and (c) the valence band of a relatively thick MoS$_x$ film obtained by on-axis RPLD.

Figure 11a shows the spectrum of optical absorption for the MoS$_{x \sim 4}$ film. Figure 11b demonstrates a Tauc plot calculated for that spectrum. The optical properties of MoS$_{x \sim 4}$ films are very similar to those of WSe$_2$. This similarity sets a limit on the thickness of the MoS$_x$ film in a MoS$_x$/WSe$_2$ heterostructure since both films absorbed light most efficiently at wavelengths below 500 nm. The width of the band gap in a MoS$_x$ film was 1.55 eV. A Fermi level in the lower part of the band gap indicated p-type conductivity in the MoS$_{x \sim 4}$ film.

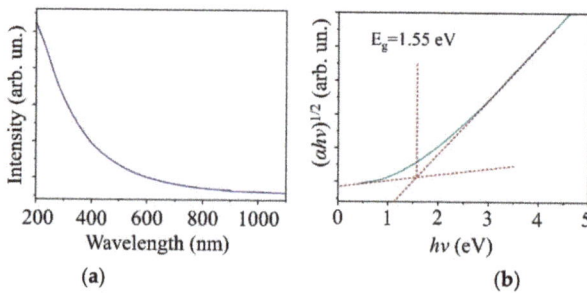

Figure 11. (a) Optical absorption spectra and (b) Tauc plots for the MoS$_{x \sim 4}$ film deposited on sapphire substrate.

The local packing of atoms in the MoS$_{x \sim 4}$ film was investigated by MRS. It can be seen in Figure 12 that the Raman spectrum of the film consists of a set of broadened strips, whose position correlates well with that of the bands in the Raman spectrum of a catalytic molybdenum sulfide film obtained by chemical synthesis in a solution [13] and by reactive magnetron sputtering [47]. The spectrum had two clear broadened peaks at ~525 and ~550 cm^{-1}, which were accounted for by the vibrational modes ν(S-S)$_{tr}$ and ν(S-S)$_{br}$ respectively in Mo$_3$S$_{13}$/Mo$_3$S$_{12}$ clusters. The peak at ~450 cm^{-1} is explained by the vibrations of apical S in Mo$_3$S$_{13}$ clusters. A broad band in the range 250–400 cm^{-1} is characteristic of an amorphous featureless structure of MoS$_x$. Thus, the selected regime of on-axis RPLD made it possible to obtain thin layers of an amorphous molybdenum sulfide containing Mo$_3$S$_{13}$/Mo$_3$S$_{12}$ clusters on the surface of a nanostructured WSe$_2$ film. The high electrocatalytic activity of such an amorphous molybdenum sulfide could contribute to a photo-assisted HER if the flux of nonequilibrium carriers (electrons) through the interface with WSe$_2$ was sufficient.

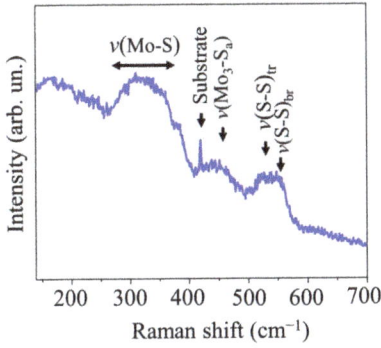

Figure 12. Raman spectra of MoS$_{x\sim4}$ film obtained by on-axis RPLD on a sapphire substrate.

3.4. Photoelectrocatalytic Properties of the MoS$_{x\sim4}$/WSe$_2$/C(B)/Al$_2$O$_3$ Cathode

Figure 13 shows the results of an investigation of photoelectrocatalytic properties of various heterostructure based on laser-deposited MoS$_{x\sim4}$, WSe$_2$, and C(B) films on a sapphire substrate. During photo-assisted HER, the MoS$_{x\sim4}$/WSe$_2$/C(B) materials combination had the most suitable properties (Figure 13a). A luminous flux caused the photo-current density to increase to ~3 A/cm^2 at a voltage of 0 V(RHE). The photocurrent magnitude was superimposed with the relatively high dark current raised due to transient effects [43]. The onset potential reached 400 mV (RHE). The heterojunction between MoS$_{x\sim4}$ and WSe$_2$ films was largely responsible for an efficient photo-assisted HER in this photocathode. Cathodes with a single semiconductive layer (MoS$_{x\sim4}$ or WSe$_2$) on the C(B) layer were associated with very low efficiency of photo-assisted HER (Figure 13b).

In the study of the temporal stability of the MoS$_{x\sim4}$/WSe$_2$/C(B)/Al$_2$O$_3$ photocathode, the current density was found to rapidly decrease by 20% in 20 min under chopped illumination. After a period of decline, the current density remained relatively stable for two hours. Longer tests of the temporal stability of this photocathode were not performed.

Table 1 contains collected data for comparison of the main parameters of metal chalcogenide-based photocathodes that characterize their performance in photo-assisted HER. It can be seen that the MoS$_{x\sim4}$/WSe$_2$/C(B)/Al$_2$O$_3$ photocathode created by pulsed laser deposition is not inferior in general in photo-assisted HER to the performance of photocathodes which were prepared by the methods of wet/dry chemical synthesis, exfoliation, spin coating, etc. Next, we will discuss the factors that should be overcome to enhance the photo-assisted HER efficiency of the MoS$_{x\sim4}$/WSe$_2$/C(B)/Al$_2$O$_3$ photocathode.

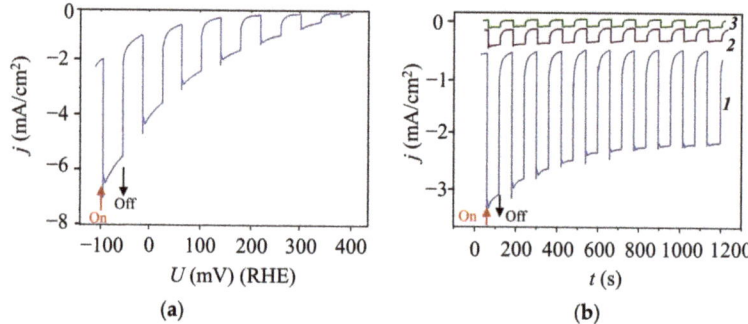

Figure 13. (a) Chopped LSV curve for the MoS$_{x\sim4}$/WSe$_2$/C(B)/Al$_2$O$_3$ photocathode in 0.5 M H$_2$SO$_4$ upon illumination; (b) Chopped photocurrent density versus time for MoS$_{x\sim4}$/WSe$_2$/C(B)/Al$_2$O$_3$ (curve **1**), MoS$_{x\sim4}$/C(B)/Al$_2$O$_3$ (curve **2**) and WSe$_2$/C(B)/Al$_2$O$_3$ (curve **3**) photocathodes at 0 V (RHE) in 0.5 M H$_2$SO$_4$ upon illumination.

Table 1. Comparison of photo-assisted HER performances for metal chalcogenide-based photocathodes with heterojunction structure.

Hetero-Structures	Rear Contact/Support	Preparation Methods	U_{onset}, mV (RHE)	Photocurrent at $U = 0$, mA/cm^2	Light Intensity, mW/cm^2	Ref.
WSe$_2$(Pt)	TiN:O/SiO$_2$/Si	aSLcS process *1	~500	≤1	100	[48]
(NH4)$_2$Mo$_3$S$_{13}$/WSe$_2$	TiN:O/quarts glass	Spin coating/aSLcS	~250	5.6	100	[13]
MoS$_x$O$_y$/2D-WSe$_2$	F:SnO$_2$/glass	SDCI *2/drop casting	~300	2.0	100	[49]
Mo$_x$S$_y$/WSe2	rGO/F:SnO$_2$/glass	Drop casting/successive dip coating	~0.2	~3–4	100	[50]
WSe$_2$-PANI (Polyaniline) nanohybrid		Vapor transport technique	280	~20	30	[51]
WSe$_2$(Pt-Cu)	F:SnO$_2$/glass	Exfoliation/spin-coating	~350	~4	100	[52]
Pt/(NH$_4$)$_2$MoS$_4$/WSe$_2$	TiN:O/glass	aSLcS/spin coating	~200	~5	100	[14]
MoS$_2$/WSe$_2$	F:SnO$_2$/glass	mechanical exfoliation/chemical vapor deposition	800 (SCE)	0.4	100	[53]
p-WSe$_2$/FePt	Metallic tungsten substrate	Chemical vapor transport	200	4	100	[54]
MoS$_4$/WSe$_2$	C(B)/Al$_2$O$_3$	RPLD/PLD	400	3	100	This work

*1 the amorphous solid–liquid–crystalline solid process with Pd promoter. *2 selective dip coating impregnation.

4. Discussion

XPS studies of MoS$_{x\sim4}$ and WSe$_2$ layers obtained by PLD showed that they had p-type conductivity. Such a combination of the electrophysical properties of contacting semiconductors creates a situation when the efficiency of photo-assisted HER processes largely depends on the structure of energy bands at the MoS$_{x\sim4}$/WSe$_2$ interface. Figure 14 shows band alignment at the interface which was determined through a comprehensive study of the films by XPS and optical methods. The conductive band offset (CBO) value was calculated using the formula:

$$CBO = VBO + E_g(WSe_2) - E_g(MoS_{x\sim4}).$$

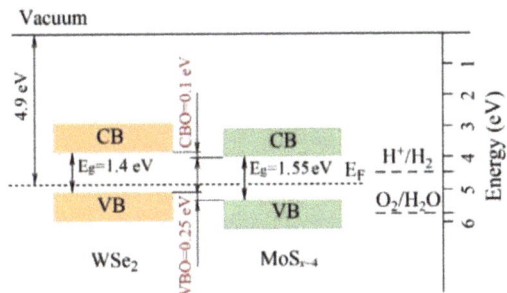

Figure 14. Band alignment diagram for the MoS $_{x\sim4}$/WSe$_2$ heterojunction system obtained by PLD.

The obtained CB value equaled 0.1 eV. Thus, the band alignment was of type II, which is associated with the most efficient separation of photo-generated electron-hole pairs. In this case, electrons will move into the MoS$_x$ layer from the WSe$_2$ and participate in the hydrogen evolution reaction, whilst holes will move from MoS$_x$ into the WSe$_2$ layer. From WSe$_2$, holes will migrate into the C(B) rear contact and further into the external electric circuit.

An additional study of MoS$_{x\sim4}$/WSe$_2$/C(B)/Al$_2$O$_3$ samples by electrochemical impedance spectroscopy (EIS) demonstrated that the C(B) contact layer did not ensure a sufficiently low resistance to the flow of current. The value of equivalent series resistance (R_s), which was extracted from EIS data, achieved 30 Ω. When a glassy carbon conducting substrate was used to create a MoS$_{x\sim4}$/WSe$_2$/GC photocathode, R_s did not exceed 4 Ω, and the density of the photo-assisted HER current increased. Therefore, to increase the efficiency of a photo-assisted HER when using a MoS $_{x\sim4}$/WSe$_2$ heterojunction system, it is recommended to choose a rear contact with an electrical resistance lower than that of the C(B) film. Further work may focus on the effect of the B concentration on the electrical properties of such films.

The analysis of the optical characteristics of the MoS$_{x\sim4}$ and WSe$_2$ films showed that these films absorb light rather efficiently. These nanomaterials are potentially active catalysts for the hydrogen evolution reaction. However, these factors did not provide effective photo-assisted HER in the MoS$_{x\sim4}$/C(B)/Al$_2$O$_3$ and WSe$_2$/C(B)/Al$_2$O$_3$ samples. Additional experiments with thicker MoS$_{x\sim4}$ and WSe$_2$ films did not reveal significant changes in the efficiency of photo-assisted HER. This indicated that after the generation of electron-hole pairs under a light flux, electrons and holes could rapidly recombine in the bulk of the films. The formation of a heterojunction turned out to be the most important factor contributing to an increase in the photocurrent. At the interface of the MoS$_{x\sim4}$ and WSe$_2$ films, not only the processes of separation of nonequilibrium electrons and holes due to the specificity of the energy bands alignment could occur, but also recombination processes can be expected. The recombination processes will facilitate to photo-assisted HER if electrons from WSe$_2$ and holes from MoS$_{x\sim4}$ actively participated in the recombination process (Z-schema) [35]. However, one cannot exclude the recombination at this interface of electrons and holes generated by the light flux in the WSe$_2$ film. In addition, the small size of the crystalline domains in the WSe$_2$ film and their random orientation resulted in a high density of edge states. This should lead to a decrease in the efficiency of charge separation since such edge states serve as recombination centers [13]. Insufficiently large values of CBO and VBO for the MoS$_{x\sim4}$/WSe$_2$ heterojunction could also be the reason limiting the efficiency of photo-assisted HER in our samples.

Another factor that could reduce the efficiency of a photo-assisted HER with a MoS$_{x\sim4}$/WSe$_2$/C(B)/Al$_2$O$_3$ photocathode is modification of the MoS$_{x\sim4}$/WSe$_2$ interface under the influence of hydrogen sulfide activated by laser-induced plasma. Figure 15 shows W4f and Mo3d XPS spectra measured for a very thin MoS$_x$ film formed by on-axis RPLD on the surface of the WSe$_2$ layer. A comparison of these spectra with those of pristine WSe$_2$ and MoS$_{x\sim4}$ (Figures 5 and 10) demonstrated that the chemical state of W has practically not changed after the deposition of MoS$_x$ film. The Mo3d spectrum shifted by 0.37 eV towards greater bonding energies, whilst the S2s spectrum increased in intensity. These changes indicated that, at the initial stage of the MoS$_x$ film growth, sulfur could be effectively deposited on the WSe$_2$ as a result of H$_2$S molecules interacting with the WSe$_2$ surface. The plasma that formed in H$_2$S during the ablation of the target could activate the process. The introduction of S atoms into the WSe$_2$ crystal lattice can distort the latter and thus cause the formation of new energy levels in the WSe$_2$ band gap. At the same time, energy bands will bend in the contact area. Band bending may cause a bonding energy shift for the Mo3d$_{5/2}$ peak and increase the width at the half maximum of the peak from 1.4 to 1.8 eV. At these energy levels, effective recombination of electrons and holes formed upon illumination may occur.

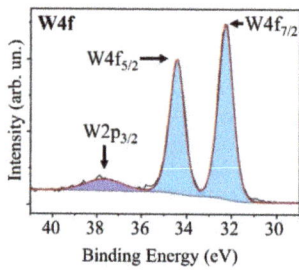
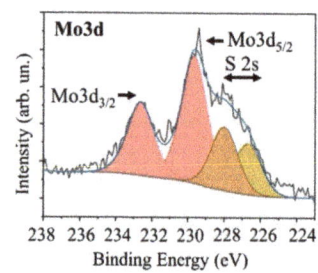

Figure 15. XPS W4f and Mo3d (overlapped with S2s) spectra for a very thin MoS$_x$ film obtained by on-axis reactive PLD on the surface of WSe$_2$ layer.

To change the conditions under which the MoS$_{x\sim4}$/WSe$_2$ interface is formed, one can employ a different technique for the deposition of a molybdenum sulfide film—one that prevents the influence of plasma-activated H$_2$S gas. Fominski et al. [55] and Giuffredi et al. [32] demonstrate that the pulsed laser ablation of a MoS$_2$ target in a buffer gas enables the formation of MoS$_x$ films with an increased concentration of sulfur ($x \geq 3$). These films have an extremely high electrocatalytic HER activity. The area of MoS$_x$ film deposition (i.e., WSe$_2$ nanopetals) can be oriented randomly to the axis of the plume expansion [56]. If this technique for MoS$_x$ film deposition is applied, the deposition of Mo and S atom flux on the interface with the WSe$_2$ film occurs almost simultaneously. This contributes to the formation of Mo-S chemical bonds in the growing film. The energy of atoms deposited during the ablation of the MoS$_2$ target in the on-axis PLD configuration is much lower than during the ablation of metallic Mo in the on-axis RPLD configuration. As a rule, the ablation of metals occurs in the conditions of effective laser plume ionization under the influence of more powerful laser pulses. This factor can also impact chemical processes at the MoS$_x$/WSe$_2$ interface.

The regulation of the MoS$_x$/WSe$_2$ interface formation is not the only factor that affects the efficiency of photo-assisted HER. Another one is the texture of the WSe$_2$ layer [10,13]. WSe$_2$ petals sitting along the substrate surface minimize the impact of edge states on the recombination on nonequilibrium carriers. Yet, the orthogonal orientation of the petals increases the area of the surface involved in catalysis. The negative effect of edge states can be reduced through their passivation by a MoS$_x$ catalyst. We carried out additional studies to obtain WSe$_2$ by off-axis PLD at varying buffer gas pressures. This factor did not have a marked effect on the texture of WSe$_2$ films. When this WSe$_2$ formation technique is used, other parameters of off-axis PLD may vary as well. These are laser fluence, the laser plume incidence angle, deposition temperature, etc. Co-deposition with some metals (for example, Pd [13]) will also affect the growth of WSe$_2$ films. The optimization of regimes for obtaining WSe$_2$ films with a targeted structure by laser-based methods is a central condition for creating HER photocatalysts with suitable characteristics. Achieving the latter requires further research into the MoS$_x$/WSe$_2$ heterojunction system.

5. Conclusions

Using different PLD configurations makes it possible to fully form a HER catalyst (in one production vessel) on a dielectric substrate (sapphire). A robust rear contact (conducting layer) was obtained using B-doped amorphous carbon by traditional on-axis PLD. A nanostructured WSe$_2$ layer was grown on the C(B) contact layer. The WSe$_2$ layer consisted of differently oriented nano-petals, which had a nearly perfect 2H-WSe$_2$ crystal lattice. To obtain a WSe$_2$ layer, off-axis PLD was performed in a buffer gas. A catalytic MoS$_{x\sim4}$ layer was created on the surface of WSe$_2$ petals by on-axis reactive PLD from Mo target in H$_2$S gas. The temperature of functional layer formation for a MoS$_{x\sim4}$/WSe$_2$/C(B)/Al$_2$O$_3$ photocathode ranged between 22–700 °C.

The MoS$_{x\sim4}$/WSe$_2$/C(B)/Al$_2$O$_3$ photocathode obtained by laser-based processes has the following characteristics as regards HER in 0.5M H$_2$SO$_4$ acid solution during light irradiation with an intensity of 100 mW/cm^2: the current density at 0 V (RHE) is ~3 A/cm^2; the onset potential reaches 400 mV (RHE). Given that these photocathodes are made from relatively cheap materials commonly found in nature, these are suitable characteristics. The performance of MoS$_x$/WSe$_2$ heterojunction system for photo-assisted water splitting for hydrogen production can be substantially increased by enhancing the composition of the photocatalyst (i.e., employing a different rear contact) and optimizing PLD regimes for creating functional semiconductor layers.

Author Contributions: Conceptualization, V.F. and R.R.; methodology, M.D.; PLD of the films, D.F.; investigation, XPS studies, O.R.; investigation, optical properties, S.N., V.V., and N.D.; writing—original draft preparation, V.F. and R.R.; writing—review and editing, V.F. All authors have read and agreed to the published version of the manuscript.

Funding: This research was funded by the Russian Science Foundation, grant number 19-19-00081.

Data Availability Statement: Not applicable.

Acknowledgments: Sample characterization by optical spectroscopy has been done in Center for Photonics and 2D Materials, MIPT. V.V., S.N., and N.D. gratefully acknowledge financial support from the Ministry of Science and Higher Education of the Russian Federation (No. 0714-2020-0002).

Conflicts of Interest: The authors declare no conflict of interest.

References

1. Wang, F.; Shifa, T.A.; Zhan, X.; Huang, Y.; Liu, K.; Cheng, Z.; Jiang, C.; He, J. Recent advances in transition-metal dichalcogenide based nanomaterials for water splitting. *Nanoscale* **2015**, *7*, 19764–19788. [CrossRef]
2. Zhang, Q.; Wang, W.; Zhang, J.; Zhu, X.; Zhang, Q.; Zhang, Y.; Ren, Z.; Song, S.; Wang, J.; Ying, Z.; et al. Highly Efficient Photocatalytic Hydrogen Evolution by ReS$_2$ via a Two-Electron Catalytic Reaction. *Adv. Mater.* **2018**, *30*, 1707123. [CrossRef]
3. Andoshe, D.; Jeon, J.; Kim, S.; Jang, H. Two-Dimensional Transition Metal Dichalcogenide Nanomaterials for Solar Water Splitting. *Electron. Mater. Lett.* **2015**, *11*, 3. [CrossRef]
4. Wang, Q.; Kalantar-Zadeh, K.; Kis, A.; Coleman, J.; Strano, M. Electronics and optoelectronics of two-dimensional transition metal dichalcogenides. *Nat. Nanotechnol.* **2012**, *7*, 699–712. [CrossRef]
5. Lu, Q.; Yu, Y.; Ma, Q.; Chen, B.; Zhang, H. 2D Transition-Metal-Dichalcogenide-Nanosheet-Based Composites for Photocatalytic and Electrocatalytic Hydrogen Evolution Reactions. *Adv. Mater.* **2016**, *28*, 1917–1933. [CrossRef]
6. Huang, X.; Zeng, Z.; Zhang, H. Metal dichalcogenide nanosheets: Preparation, properties and applications. *Chem. Soc. Rev.* **2013**, *42*, 1934–1946. [CrossRef]
7. Laursen, A.; Kegnæs, S.; Dahl, S.; Chorkendorff, I. Molybdenum sulfides-efficient and viable materials for electro- and photoelectrocatalytic hydrogen evolution. *Energy Environ. Sci.* **2012**, *5*, 557. [CrossRef]
8. Luo, Z.; Wang, T.; Gong, J. Single-crystal silicon-based electrodes for unbiased solar water splitting: Current status and prospects. *Chem. Soc. Rev.* **2018**, *48*, 2158–2181. [CrossRef]
9. Lin, H.; Li, S.; Yang, G.; Kai Zhang, K.; Tang, D.; Su, Y.; Li, Y.; Luo, S.; Chang, K.; Ye, J. In Situ Assembly of MoS$_x$ Thin-Film through Self-Reduction on p-Si for Drastic Enhancement of Photoelectrochemical Hydrogen Evolution. *Adv. Funct. Mater.* **2020**, 2007071. [CrossRef]

10. McKone, J.; Adam, P.; Pieterick, A.; Gray, H.; Nathan, S.; Lewis, N. Hydrogen Evolution from Pt/Ru-Coated p-Type WSe_2 Photocathodes. *J. Am. Chem. Soc.* **2013**, *135*, 223–231. [CrossRef]
11. Li, C.; Cao, Q.; Wang, F.; Xiao, Y.; Li, Y.; Delaunay, J.-J.; Zhu, H. Engineering graphene and TMDs based van der Waals heterostructures for photovoltaic and photoelectrochemical solar energy conversion. *Chem. Soc. Rev.* **2018**, *47*, 4981. [CrossRef]
12. Zhong, S.; Xi, Y.; Wu, S.; Liu, Q.; Zhao, L.; Bai, S. Hybrid cocatalysts in semiconductor based photocatalysis and photoelectrocatalysis. *J. Mater. Chem. A* **2020**, *8*, 14863–14894. [CrossRef]
13. Bozheyev, F.; Xi, F.; Plate, P.; Dittrich, T.; Fiechter, S.; Ellmer, K. Efficient charge transfer at a homogeneously distributed $(NH_4)_2Mo_3S_{13}/WSe_2$ heterojunction for solar hydrogen evolution. *J. Mater. Chem. A* **2019**, *7*, 10769–10780. [CrossRef]
14. Bozheyev, F.; Xi, F.; Ahmet, I.; Hohn, C.; Ellmer, K. Evaluation of Pt, Rh, SnO_2, $(NH_4)_2Mo_3S_{13}$, $BaSO_4$ protection coatings on WSe_2 photocathodes for solar hydrogen evolution. *Int. J. Hydrog. Energy* **2020**, *45*, 19112–19120. [CrossRef]
15. Zhang, W.; Chiu, M.-H.; Chen, C.-H.; Chen, W.; Li, L.-J.; Thye, A.; Wee, S. Role of Metal Contacts in High Performance Phototransistors Based on WSe_2 Monolayers. *Am. Chem. Soc.* **2014**, *8*, 8653–8661. [CrossRef] [PubMed]
16. Chaudhary, S.; Umar, A.; Mehta, S.K. Selenium nanomaterials: An overview of recent developments in synthesis, properties and potential applications. *Prog. Mater. Sci.* **2016**, *83*, 270–329. [CrossRef]
17. Li, H.; Zou, J.; Xie, S.; Leng, X.; Gao, D.; Yang, H.; Mao, X. WSe_2 nanofilms grown on graphite as efficient electrodes for hydrogen evolution reactions. *J. Alloys Compd.* **2017**, *725*, 884–890. [CrossRef]
18. Fominski, V.Y.; Grigoriev, S.N.; Romanov, R.I.; Volosova, M.A.; Grunin, A.I.; Teterina, G.D. The Formation of a Hybrid Structure from Tungsten Selenide and Oxide Plates for a Hydrogen-Evolution Electrocatalyst. *Tech. Phys. Lett.* **2016**, *42*, 553–556. [CrossRef]
19. Zheng, Z.; Zhang, T.; Yao, J.; Zhang, Y.; Xu, J.; Yang, G. Flexible, transparent and ultra-broadband photodetector based on large-area WSe_2 film for wearable devices. *Nanotechnology* **2016**, *27*, 225501. [CrossRef]
20. Grigoriev, S.N.; Fominski, V.Y.; Nevolin, V.N.; Romanov, R.I.; Volosova, M.A.; Irzhak, A.V. Formation of Thin Catalytic WSe_x Layer on Graphite Electrodes for Activation of Hydrogen Evolution Reaction in Aqueous Acid. *Inorg. Mater. Appl. Res.* **2016**, *7*, 2–285. [CrossRef]
21. Seo, S.; Choi, H.; Kim, S.-Y.; Lee, J.; Kim, K.; Yoon, S.; Lee, B.; Lee, S. Growth of Centimeter-Scale Monolayer and Few-Layer WSe_2 Thin Films on SiO_2/Si Substrate via Pulsed Laser Deposition. *Adv. Mater. Interfaces* **2018**, *5*, 1800524. [CrossRef]
22. Grigoriev, S.N.; Fominski, V.Y.; Romanov, R.I.; Gnedovets, A.G.; Volosova, M.A. Shadow masked pulsed laser deposition of WSe_x films: Experiment and modeling. *Appl. Surf. Sci.* **2013**, *282*, 607–614. [CrossRef]
23. Fominski, V.Y.; Grigoriev, S.N.; Gnedovets, A.G.; Romanov, R.I.; Volosova, M.A. Experimental study and modelling of laser plasma ion implantation for $WSe_x/^{57}Fe$ interface modification. *Appl. Surf. Sci.* **2013**, *276*, 242–248. [CrossRef]
24. Fominski, V.Y.; Grigoriev, S.N.; Romanov, R.I.; Volosova, M.A.; Demin, M.V. Chemical composition, structure and light reflectance of W-Se and W-Se-C films prepared by pulsed laser deposition in rare and reactive buffer gases. *Vacuum* **2015**, *119*, 19–29. [CrossRef]
25. Fominski, V.Y.; Grigoriev, S.N.; Gnedovets, A.G.; Romanov, R.I. On the Mechanism of Encapsulated Particle Formation during Pulsed Laser Deposition of WSe_x Thin-Film Coatings. *Tech. Phys. Let.* **2013**, *39*, 312–315. [CrossRef]
26. Fominski, V.Y.; Markeev, A.M.; Nevolin, V.N.; Prokopenko, V.B.; Vrublevski, A.R. Pulsed laser deposition of MoS_x films in a buffer gas atmosphere. *Thin Solid Films* **1994**, *248*, 240–246. [CrossRef]
27. Fominski, V.Y.; Nevolin, V.N.; Romanov, R.I.; Smurov, I. Ion-assisted deposition of MoS_x films from laser-generated plume under pulsed electric field. *J. Appl. Phys.* **2001**, *89*, 1449–1457. [CrossRef]
28. Fominski, V.; Demin, M.; Fominski, D.; Romanov, R.; Goikhman, A.; Maksimova, K. Comparative study of the structure, composition, and electrocatalytic performance of hydrogen evolution in $MoS_{x\sim2+\delta}/Mo$ and $MoS_{x\sim3+\delta}$ films obtained by pulsed laser deposition. *Nanomaterials* **2020**, *10*, 201. [CrossRef]
29. Li, B.; Jiang, L.; Li, X.; Cheng, Z.; Ran, P.; Zuo, P.; Qu, L.; Zhang, J.; Lu, Y. Controllable Synthesis of Nanosized Amorphous MoS_x Using Temporally Shaped Femtosecond Laser for Highly Efficient Electrochemical Hydrogen Production. *Adv. Funct. Mater.* **2019**, *29*, 1806229. [CrossRef]
30. Li, Y.; Yu, Y.; Huang, Y.; Nielsen, R.; William, A.; Goddard, W.A., III; Li, Y.; Cao, L. Engineering the Composition and Crystallinity of Molybdenum Sulfide for High-Performance Electrocatalytic Hydrogen Evolution. *ACS Catal.* **2015**, *5*, 448–455. [CrossRef]
31. Ding, R.; Wang, M.; Wang, X.; Wang, H.; Wang, L.; Mu, Y.; Lv, B. N-Doped amorphous MoS_x for the hydrogen evolution reaction. *Nanoscale* **2019**, *11*, 11217–11226. [CrossRef]
32. Giuffredi, G.; Mezzetti, A.; Perego, A.; Mazzolini, P.; Prato, M.; Fumagalli, F.; Lin, Y.-C.; Liu, C.; Ivanov, I.; Belianinov, A.; et al. Non-Equilibrium Synthesis of Highly Active Nanostructured, Oxygen-Incorporated Amorphous Molybdenum Sulfide HER Electrocatalyst. *Small* **2020**, *2004047*. [CrossRef]
33. Wang, R.; Sun, P.; Wang, H.; Wang, X. Pulsed laser deposition of amorphous molybdenum disulfide films for efficient hydrogen evolution reaction. *Electrochim. Acta* **2017**, *258*, 876–882. [CrossRef]
34. Fominski, V.Y.; Romanov, R.I.; Fominski, D.V.; Shelyakov, A.V. Regulated growth of quasi-amorphous MoS_x thin-film hydrogen evolution catalysts by pulsed laser deposition of Mo in reactive H_2S gas. *Thin Solid Films* **2017**, *642*, 58–68. [CrossRef]
35. Fominski, V.; Romanov, R.; Fominski, D.; Soloviev, A.; Rubinkovskaya, O.; Demin, M.; Maksimova, K.; Shvets, P.; Goikhman, A. Performance and Mechanism of Photoelectrocatalytic Activity of MoS_x/WO_3 Heterostructures Obtained by Reactive Pulsed Laser Deposition for Water Splitting. *Nanomaterials* **2020**, *10*, 871. [CrossRef]

36. Yang, X.; Liu, W.; Bastiani, M.; Allen, T.; Kang, J.; Xu, H.; Aydin, E.; Xu, L.; Bi, Q.; Dang, H. Dual-Function Electron-Conductive, Hole-Blocking Titanium Nitride Contacts for Efficient Silicon Solar Cells. *Joule* **2019**, *3*, 1314–1327. [CrossRef]
37. Fominski, V.Y.; Romanov, R.I.; Vasil'evskii, I.S.; Safonov, D.A.; Soloviev, A.A.; Zinin, P.V.; Bulatov, K.M.; Filonenko, V.P. Structural, electrical and mechanical properties of BC films prepared by pulsed laser deposition from mixed and dual boron-diamond/graphite targets. *Diam. Relat. Mater.* **2019**, *92*, 266–277. [CrossRef]
38. Fominski, V.; Demin, M.; Nevolin, V.; Fominski, D.; Romanov, R.; Gritskevich, M.; Smirnov, N. Reactive Pulsed Laser Deposition of Clustered-Type MoS_x ($x \sim 2, 3$, and 4) Films and Their Solid Lubricant Properties at Low Temperature. *Nanomaterials* **2020**, *10*, 653. [CrossRef]
39. Chiu, M.; Zhang, C.; Shiu, H.; Chuu, C.; Chen, C.; Chang, C.S.; Chen, C.; Chou, M.; Shih, C.; Li, L. Determination of band alignment in the single-layer MoS_2/WSe_2 heterojunction. *Nat. Commun.* **2015**, *6*, 1–6. [CrossRef]
40. Xing, S.; Zhao, G.; Wang, J.; Xu, Y.; Ma, Z.; Li, X.; Yang, W.; Liu, G.; Yang, J. Band alignment of wo-dimensional $h-BN/MoS_2$ van der Waals heterojunction measured by X-ray photoelectron spectroscopy. *J. Alloys Compd.* **2020**, *834*, 155108. [CrossRef]
41. Fominski, V.Y.; Romanov, R.I.; Vasil'evskii, I.S.; Safonov, D.A.; Soloviev, A.A.; Ivanov, A.A.; Zinin, P.V.; Krasnoborodko, S.Y.; Vysokikh, Y.E.; Filonenko, V.P. Pulsed laser modification of layered B-C and mixed BC_x films on sapphire substrate. *Diam. Relat. Mater.* **2021**, *114*, 108336. [CrossRef]
42. Luo, X.; Zhao, Y.; Zhang, J.; Toh, M.; Kloc, C.; Xiong, Q.; Quek, S.Y. Effect of lower symmetry and dimensionality on Raman spectra in two-dimensional WSe_2. *Phys. Rev. B* **2013**, *88*, 195313. [CrossRef]
43. Yu, X.; Prévot, M.S.; Guijarro, N.; Sivula, K. Self-assembled 2D WSe_2 thin films for photoelectrochemical hydrogen production. *Nat. Commun.* **2015**, *6*, 1–8. [CrossRef] [PubMed]
44. Boscher, N.D.; Carmalt, C.J.; Parkin, I.P. Atmospheric pressure chemical vapor deposition of WSe_2 thin films on glass-highly hydrophobic sticky surfaces. *J. Mater. Chem.* **2006**, *16*, 122–127. [CrossRef]
45. Lee, C.; Lee, G.; Van der Zande, A.M.; Chen, W.; Li, Y.; Han, M.; Cui, X.; Arefe, G.; Nuckolls, C.; Heinz, T.F.; et al. Atomically thin p–n junctions with van der Waals heterointerfaces. *Nat. Nanotechnol.* **2014**, *9*, 676–681. [CrossRef]
46. Doan, M.; Jin, Y.; Adhikari, S.; Lee, S.; Zhao, J.; Lim, S.C.; Lee, Y.H. Charge Transport in $MoS2/WSe_2$ van der Waals Heterostructure with Tunable Inversion Layer. *ACS Nano* **2017**, *11*, 3832–3840. [CrossRef]
47. Xi, F.; Bogdanoff, P.; Harbauer, K.; Plate, P.; Höhn, C.; Rappich, J.; Wang, B.; Han, X.; Van de Krol, R.; Fiechter, S. Structural transformation identification of sputtered amorphous MoS_x as efficient hydrogen evolving catalyst during electrochemical activation. *ACS Catal.* **2019**, *9*, 2368–2380. [CrossRef]
48. Bozheyev, F.; Rengacharid, .; Berglunde, S.; Abou-Rase, D.; Ellmere, K. Passivation of recombination active $PdSe_x$ centers in (001)-textured photoactive WSe_2 films. *Mat. Sci. Semicon. Proc.* **2019**, *93*, 284–289. [CrossRef]
49. Barbosa, J.B.; Taberna, P.L.; Bourdon, V.; Gerber, I.C.; Poteau, R.; Balocchi, A.; Marie, X.; Esvan, J.; Puech, P.; Barnabé, A.; et al. Mo thio and oxo-thio molecular complexes film as self-healing catalyst for photocatalytic hydrogen evolution on 2D materials. *Appl. Catal. B* **2020**, *278*, 119288. [CrossRef]
50. Taberna, P.L.; Barbosa, J.B.; Balocchi, A.; Gerber, K.U.; Barnabe, A.; Marie, X.; Chane-Ching, J.Y. Patch-like, Two Dimensional WSe_2-Based Hetero-structures Activated by a Healing Catalyst for H_2 Photocatalytic Generation. *Chem. Eng. J.* **2021**, 130433. [CrossRef]
51. Kannichankandy, D.; Pataniya, P.M.; Sumesh, C.K.; Solanki, G.K.; Pathak, V.M. WSe_2-PANI nanohybrid structure as efficient electrocatalyst for photo-enhanced hydrogen evolution reaction. *J. Alloys Compd.* **2021**, *876*, 160179. [CrossRef]
52. Yu, X.; Guijarro, N.; Johnson, M.; Sivula, K. Defect mitigation of Solution-Processed 2D WSe_2 Nano-flakes for Solar-to Hydrogen Conversion. *Nano Lett.* **2018**, *18*, 215–222. [CrossRef]
53. Si, K.; Ma, J.; Lu, C.; Zhou, Y.; He, C.; DanYang, D.; Wang, X.; Xu, X. A two-dimensional MoS_2/WSe_2 van der Waals heterostructure for enhanced photoelectric performance. *Appl. Surf. Sci.* **2020**, *507*, 145082. [CrossRef]
54. Zheng, X.; Zhang, G.; Xu, X.; Liu, L.; Zhang, J.; Xu, Q. Synergistic effect of mechanical strain and interfacial-chemical interaction for stable $1T-WSe_2$ by carbon nanotube and cobalt. *Appl. Surf. Sci.* **2019**, *496*, 143694. [CrossRef]
55. Fominski, V.Y.; Romanov, R.I.; Fominski, D.V.; Dzhumaev, P.S.; Troyan, I.A. Normal and grazing incidence pulsed laser deposition of nanostructured MoS_x hydrogen evolution catalysts from a MoS_2 target. *Opt. Laser Technol.* **2018**, *102*, 74–84. [CrossRef]
56. Nevolin, V.N.; Fominski, D.V.; Romanov, R.I.; Esin, M.I.; Fominski, V.Y.; Kartsev, P.F. Selection of pulsed laser deposition conditions for preparation of perfect thin-film MoS_x hydrogen evolution catalysts. *J. Phys. Conf. Ser.* **2019**, *1*, 1238. [CrossRef]

Article

Synthesis of Air-Stable Cu Nanoparticles Using Laser Reduction in Liquid

Ashish Nag, Laysa Mariela Frias Batista and Katharine Moore Tibbetts *

Department of Chemistry, Virginia Commonwealth University, Richmond, VA 23284, USA; nagab@vcu.edu (A.N.); friasbatistlm@vcu.edu (L.M.F.B.)
* Correspondence: kmtibbetts@vcu.edu

Abstract: We report the synthesis of air-stable Cu nanoparticles (NPs) using the bottom-up laser reduction in liquid method. Precursor solutions of copper acetlyacetonate in a mixture of methanol and isopropyl alcohol were irradiated with femtosecond laser pulses to produce Cu NPs. The Cu NPs were left at ambient conditions and analyzed at different ages up to seven days. TEM analysis indicates a broad size distribution of spherical NPs surrounded by a carbon matrix, with the majority of the NPs less than 10 nm and small numbers of large particles up to ~100 nm in diameter. XRD collected over seven days confirmed the presence of fcc-Cu NPs, with some amorphous Cu_2O, indicating the stability of the zero-valent Cu phase. Raman, FTIR, and XPS data for oxygen and carbon regions put together indicated the presence of a graphite oxide-like carbon matrix with oxygen functional groups that developed within the first 24 h after synthesis. The Cu NPs were highly active towards the model catalytic reaction of *para*-nitrophenol reduction in the presence of $NaBH_4$.

Keywords: laser synthesis; laser reduction in liquid; copper nanoparticles; para-nitrophenol

Citation: Nag, A.; Frias Batista, L.M.; Tibbetts, K.M. Synthesis of Air-Stable Cu Nanoparticles Using Femtosecond Laser Reduction in Liquid. *Nanomaterials* **2021**, *11*, 814. https://doi.org/10.3390/nano11030814

Academic Editor: Mohamed Boutinguiza

Received: 15 March 2021
Accepted: 21 March 2021
Published: 23 March 2021

Publisher's Note: MDPI stays neutral with regard to jurisdictional claims in published maps and institutional affiliations.

Copyright: © 2021 by the authors. Licensee MDPI, Basel, Switzerland. This article is an open access article distributed under the terms and conditions of the Creative Commons Attribution (CC BY) license (https://creativecommons.org/licenses/by/4.0/).

1. Introduction

Laser synthesis techniques have emerged over the last decade as reliable methods for producing pure nanomaterials (NMs) [1–3]. Laser synthesis has several advantages over traditional chemical synthesis methods including: facile generation of metastable phases and bonding environments, rapid conversion of precursors to NM products, and avoidance of capping ligands. For instance, alloys of immiscible metals can be formed [4], reactions can be completed in seconds [5], and grams per hour synthesis yields can be attained with high-repetition rate lasers [6]. Moreover, laser synthesis techniques are considered 'green' because they do not employ toxic chemical reducing agents or surfactants and produce little chemical waste [7].

The lack of otherwise required capping ligands results in high purity nanoparticles that are important to several applications. Biofunctional assemblies can be produced by functionalizing laser-generated nanoparticles (NPs) for in-vitro applications [8]. The absence of ligands makes more active sites available for catalysis, resulting in laser-synthesized NPs often having higher activities than their conventional counterparts [9–11]. Higher purity also makes laser-synthesized NPs attractive candidates for other biomedical applications and as references for modeling chemical reactions [12–15]. In addition, photoluminescence can be introduced to the NPs by in-situ generation of carbon shells through laser-induced decomposition of organic solvents [16,17].

Cu NPs in particular are of high interest both due to the natural abundance of copper and their myriad applications in catalysis, electronics, and biology [18–22]. However, conventional wet-chemical methods used to synthesize Cu-based NPs require toxic solvents, reducing agents, or both [23–25]. Hence, greener synthesis routes to Cu NPs are of primary importance. To address this need, pulsed laser ablation in liquid (PLAL) has been widely employed to generate Cu NPs [26–30]. In these syntheses, copper oxides are a major product when ablation is conducted in water [26–29], and Cu^0 phases are only stable when the ablation liquid contains an organic solvent, such as acetone, methanol, or ethanol [28–30].

The *top-down* PLAL method is by far the most common laser synthesis technique used to produce colloidal metal NPs [1]. PLAL involves focusing of laser beam on the surface of a solid or powdered target immersed in a solvent, causing the removal of target material from the surface and its coalescence into colloidal NPs. However, NP products from PLAL in many cases exhibit bimodal size distributions due to the ejection of both small clusters and large droplets from the surface [31–33]. An alternative method is *bottom-up* laser reduction in liquid (LRL), which involves focusing picosecond (ps, 10^{-12} s) or femtosecond (fs, 10^{-15} s) laser pulses into a solution of molecular precursors to generate a dense plasma containing electrons that reduce metal ions to colloidal NPs. LRL can enable superior control over Au NP sizes in a single step when the chemistry of the precursor solution and laser irradiation conditions are carefully controlled [34,35].

A current limitation to the wide use of LRL is that the vast majority of studies reported to date focus on the easily reduced noble metals Au [34–38], Ag [39–41], Pt [42], and their alloys [43–45] because these metals are resistant to oxidation. Unlike in PLAL where the zero-valent metal is present in the initial target, metal NP formation in LRL requires chemical reactions between the molecular precursor(s) and the reactive species in LRL plasma. In aqueous solution, the major species are hydrated electrons (e_{aq}^-) and hydroxyl radicals (OH·) [38]. Whereas e_{aq}^- are exceptionally strong reducing agents towards metal ions, OH· radicals and their recombination product H_2O_2 can back-oxidize zero-valent metal atoms. Although Au is resistant to back-oxidation and $HAuCl_4$ precursor can be reduced by H_2O_2 [37], effective production of Ag-containing NPs by LRL requires the addition of OH· scavengers, such as ammonia or isopropyl alcohol, to prevent back-oxidation of Ag [40,41,45]. Moreover, the few LRL studies on the non-noble metal Fe report production of only Fe oxides [46,47].

The detrimental effects of reactive oxygen species on Cu NP synthesis have been widely known since Dhas et al. [48] reported that OH· and H_2O_2 formed during sonochemical synthesis lead to the formation of copper oxides. In PLAL of Cu metal in water, Cu oxides are formed due to the presence of OH· radicals in the laser plasma from water, dissolved O_2, or both [28]. Oxide formation during PLAL can be hindered through the use of organic solvents [27–30], which also can result in the formation of a protective carbon shell around the Cu NPs [27,28]. On the basis of these results, we designed an air-free LRL synthesis route to Cu NPs using an isopropyl alcohol/methanol solvent mixture. We report, to the best of our knowledge, the first LRL synthesis of air-stable Cu NPs and their catalytic activity towards reduction of *para*-nitrophenol in the presence of $NaBH_4$.

2. Materials and Methods

2.1. Materials

Copper acetylacetonate ($Cu(acac)_2$, Acros Organics, Fair Lawn, NJ, USA), isopropyl alcohol (IPA, Fisher Scientific, Waltham, MA, USA), methanol (MeOH, Fisher Scientific, Waltham, MA, USA), sodium borohydride ($NaBH_4$, Acros Organics, Fair Lawn, NJ, USA), and *para*-nitrophenol (PNP, Acros Organics, Fair Lawn, NJ, USA) were used as received.

2.2. Cu NP synthesis

A working solution of 0.8 mM $Cu(acac)_2$ in 25%/75% (v/v) MeOH/IPA solvent was purged with nitrogen for 20 min before 3.0 mL of the solution was transferred to a nitrogen-filled 10 × 10 × 40 mm quartz fluorimeter cuvette. The solution was then irradiated with laser pulses to yield Cu NPs. Upon formation, Cu NPs were exposed to air and analyzed at different time intervals (0 h, 3 h, 24 h, and 7 days) to study the aging of NPs. By 24 h after synthesis, all the Cu NPs are precipitated out of the solution but can be redispersed via brief sonication.

2.3. Instrumentation

2.3.1. Laser Synthesis

The experimental setup is described elsewhere [38]. Briefly, samples were irradiated using a commercial titanium-sapphire chirped-pulse amplifier (Astrella, Coherent, Inc., Santa Clara, CA, USA), delivering 30 fs pulses, with the bandwidth centered at 800 nm and a repetition rate of 1 kHz; 2 mJ laser pulses were focused using a $f = 5$ cm aspheric lens into the center of the sample cuvette to produce a peak irradiance of 5×10^{16} Wcm^{-2}. Details for calculating peak irradiance can be found in Ref. [38]. The sealed cuvettes containing Cu(acac)$_2$ solution were irradiated for 10 min.

2.3.2. UV-Vis Spectroscopy

Conversion of precursors to Cu NPs was monitored using a home-built in-situ UV-vis spectrophotometer described in Ref. [38] and the irradiation was stopped when no further growth in Cu surface plasmon resonance (SPR) was observed. The catalytic performance of Cu NPs was tested for the reduction of PNP with NaBH$_4$ (see Section 2.4) by employing a second home-built in-situ UV-vis spectrophotometer described in Ref. [49]. Finally, to observe the effect of aging of Cu NPs, absorbance data was recorded using Agilent 8453 UV-vis spectrophotometer at times up to 7 days after synthesis.

2.3.3. Transmission Electron Microscopy (TEM)

Cu NPs were visualized using TEM (JEOL JEM-1230 TEM) at 120 kV. A diluted solution of colloidal Cu NPs was drop-casted onto carbon-coated grids (Structure Probe, Inc., West Chester, PA, USA) and left to dry for 24 h or longer. Average sizes and size distributions were measured using ImageJ software. At least 350 particles from images of three separate areas of a TEM grid were evaluated.

2.3.4. X-ray Photoelectron Spectroscopy (XPS)

XP spectra were collected on a PHI VersaProbe III Scanning XPS Microprobe with a monochromated Al Kα X-ray source (1486.6 eV), with a typical resolution of 0.4–0.5 eV. Survey scans and high resolution scans were collected with pass energies of 280 eV and 26 eV, respectively. Charge neutralization was done by running an ion gun and a flood gun during sample analysis. The measurement spot diameter was 200 µm with take off angle of 90° and the detector at 45°. Spectral analysis was carried out using PHI Multipak XPS software with 70% Gaussian/Lorentzian convolution to fit each spectral peak. Samples were prepared by drop-casting Cu NPs on a gold-sputtered silicon wafer, followed by drying under vacuum for at least 24 h. All spectra were corrected using a Au4f peak shift to center at 84.0 eV.

2.3.5. FTIR Spectroscopy

A Thermo Scientific Nicolet iS50 FTIR spectrometer equipped with a mid- and far-IR-capable diamond ATR was used to record FTIR spectra. All spectra were obtained using 32 scans in the range from 4000 to 400 cm^{-1} with 4 cm^{-1} resolution. Cu NPs were directly drop-casted on to the diamond crystal.

2.3.6. X-ray Diffraction (XRD)

XRD data was collected on a Panalytical Empyrean Diffractometer with CuKα radiation (0.15418 nm) at 40 kV and 45 mA, with scanning angle (2θ) of 30–80° and a gonio focusing geometry. Sample preparation involved drop-casting of Cu NPs on a low-background silicon substrate, followed by drying under vacuum for at least 24 h.

2.3.7. Raman Spectroscopy

Raman spectra were recorded on a Thermo Scientific DXR3 SmartRaman Spectrometer involving a 532 nm excitation laser. All spectra obtained were averaged over 32 scans in the range of 50 to 3500 cm^{-1} with 5 cm^{-1} resolution. Samples were prepared by drop-casting Cu NPs on a silicon wafer, followed by drying under vacuum for at least 24 h.

2.4. Catalytic Reduction of Para-Nitrophenol (PNP)

PNP reduction reactions were carried out in a home-built in situ UV-vis spectrometer setup described elsewhere [35] at different time intervals after laser synthesis (0 h, 3 h, 24 h, and 7 days). Spectra were recorded every 2.4 s using LabVIEW software (National Instruments). A solution containing a final concentration of 0.1 mM PNP and 10 mM of freshly prepared NaBH$_4$ was prepared in a 10 × 10 × 40 mm quartz fluorimeter cuvette with a magnetic stir bar, resulting in the formation of p-nitrophenolate ion with UV-vis absorbance at 400 nm. Prior to the addition of the catalyst, this peak was observed for 24 s to confirm no reaction occurred in the absence of a catalyst. After this period, 300 μL of the Cu NP catalyst was added, triggering the reduction of PNP. Data collection was terminated when the absorbance at 400 nm (p-nitrophenolate ion) had disappeared.

3. Results

3.1. Physical Characterization

Figure 1 shows TEM images and size distributions for Cu NPs samples obtained 0 h (a), 3 h (b), and 24 h (c) after laser synthesis. In all images, small <10 nm NPs surrounding significantly larger NPs are observed. Fitting the size distributions to a log-normal function primarily captured the small NPs, producing mean diameters of 3.30 nm (PDI = 0.16), 4.54 nm (PDI = 0.19), and 4.57 nm (PDI = 0.16) for 0 h, 3 h, and 24 h samples, respectively. Despite the large numbers of small NPs, they make up a tiny fraction of the overall mass of Cu NPs, as evident in the mass-weighted size distributions. Nevertheless, the contribution of NPs smaller than 10 nm to the overall mass slightly increased from 0.16 % at 0 h to 0.34 % at 3 h and 24 h. The slight size increase in the small NPs after the initial synthesis was also observed for Fe oxide NPs synthesized by LRL from Fe(acac)$_3$ in hexane [47], although, in that work, no large NPs were observed. A carbon shell around clusters of Cu NPs is also visible in Figure 1, particularly for the 3 h and 24 h samples. Similar carbon shells have been observed for Cu NPs produced by PLAL in acetone [28] and methanol [27].

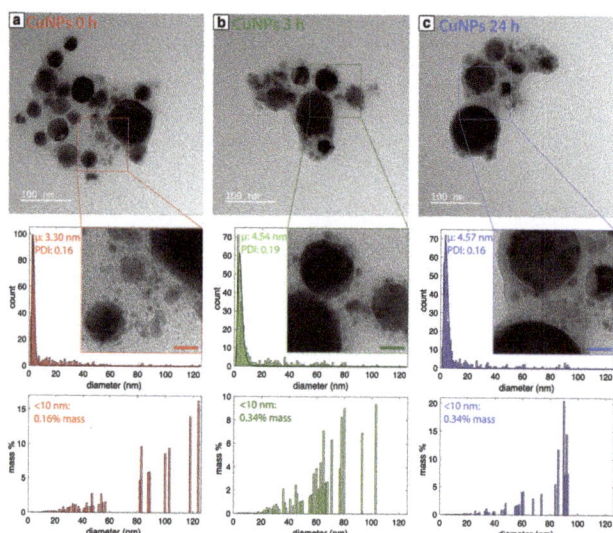

Figure 1. TEM images and corresponding number-weighted and mass-weighted size distributions for samples analyzed (**a**) 0 h, (**b**) 3 h, and (**c**) 24 h after synthesis. Scale bars in magnified regions are 20 nm.

Figure 2 shows the XRD spectrum of the 24 h sample. Intense diffraction peaks observed at 43.35°, 50.47°, and 74.18° were indexed to (111), (200), and (220) planes of fcc-Cu (JCPDS no. 01-085-1326), respectively. A broad feature centered at 36.85° indexed to cubic-Cu$_2$O(111) (JCPDS no. 00-005-0667) was also observed. Using the Scherrer equation, the crystallite sizes were calculated to be 29.89 nm for the Cu(111) [50]. Thus, we can infer that the large NPs observed in Figure 1 are Cu metal NPs, whereas Cu$_2$O exists as an amorphous thin layer around the Cu NPs, as the smaller NPs, or both.

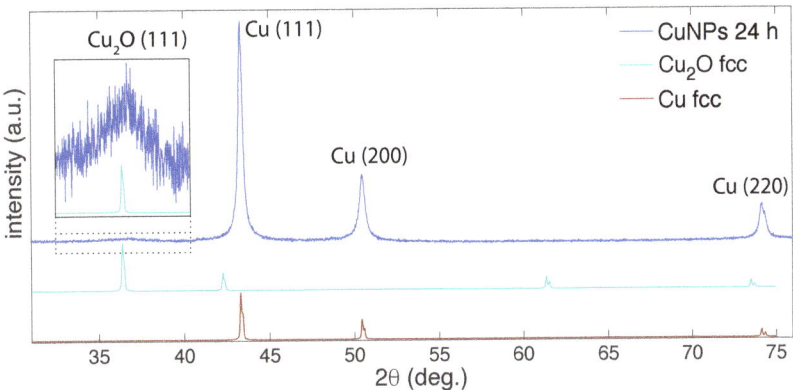

Figure 2. XRD spectrum of 24 h sample (blue), with references for Cu$_2$O (cyan) and Cu (dark red). Inset magnifies a broad peak observed in the spectra for 24 h sample indexed to Cu$_2$O (111).

Figure 3 shows XP spectra obtained for the 24 h sample. Two peaks in the Cu2p$_{3/2}$ region (Figure 3a) at 933.0 eV (cyan) and 934.2 eV (violet) were assigned to either Cu^0 or Cu$^+$ and Cu^{2+}, respectively [51]. To distinguish between Cu^0 and Cu$^+$, the CuLMM region was analyzed (Figure 3b) and resulted in two peaks at 568.4 eV (blue) and 570.6 eV (magenta) assigned to Cu^0 and Cu$^+$, respectively [51]. The presence of both Cu^0 and Cu$^+$ in the XP spectra is consistent with the XRD results (Figure 2). In the O1s region (Figure 3c), the peaks at 530.8 eV (magenta), 531.7 eV (light blue), 532.2 eV (light green), and 533.0 eV (dark green) were assigned to Cu–O, C=O, –OH and C–O species, respectively [52,53]. In the C1s region (Figure 3d), peaks at 284.4 eV (red), 284.8 eV (orange), 286.1 eV (dark green), and 288.4 eV (light blue) are assigned to C=C, C–C, C–O, and C=O species, respectively [53]. This collection of carbon species is associated with the carbon shell observed around the Cu NPs (Figure 1).

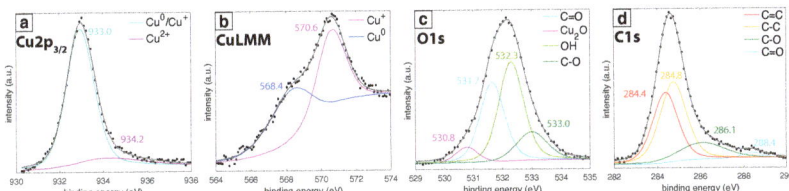

Figure 3. XPS spectra of (**a**) Cu2p$_{3/2}$, (**b**) CuLMM, (**c**) O1s and (**d**) C1s regions with fitted peaks for 24 h samples.

Spectral characterization of the Cu NPs (Figure 4) confirms the chemical species assigned in Figures 2 and 3. Figure 4a shows the UV-vis absorbance data collected at 0 h, 3 h, and 24 h after synthesis. An absorbance peak at 574 nm assigned to the Cu surface plasmon resonance (SPR) was observed at 0 h, followed by broadening and red-shifting to 586 nm and 596 nm at 3 h and 24 h, respectively. These changes in the absorbance peaks are associated with the oxidation of Cu to Cu$_2$O over time [54] and likely indicate the formation of a Cu_2O shell around the large Cu NPs on the basis of the XRD spectrum (Figure 2). Raman spectra shown in Figure 4b closely resemble graphite oxide spectra [55]. The intense G band at 1600 cm^{-1} and D band at 1350 cm^{-1} correspond to the in-plane vibrational modes of sp^2-hybridized carbon atoms and structural disorder due to functionalization, respectively [56]. Moreover, the weak and broad 2D and D + G bands at 2680 cm^{-1} and 2910 cm^{-1} arise from disorder due to formation of oxygen functional groups [55]. The presence of oxygen functional groups is confirmed by the FTIR spectra (Figure 4c) with peaks indicating C=O stretch of COOH groups at 1722 and 1696 cm^{-1}, COO^- stretch at 1415, 1522, 1575 cm^{-1}, O–H deformations of C–OH groups at 1356 cm^{-1}, C–O stretch of epoxide groups at 1261 cm^{-1}, C–O stretch at 1000–1100 cm^{-1}, and C–H bend at 800 cm^{-1} [56,57], although some of these bands may be overlapped with vibrational modes of other functional groups. Notably, these IR peaks grow in intensity over the course of 24 h after synthesis, suggesting that the carbon shell formation occurs mostly after laser irradiation is terminated. Overall, these data and the XP spectra for C1s and O1s (Figure 3a,b) indicate that the carbon shell around the Cu NPs consists of disordered graphite oxide-like structures that contain multiple different functional groups.

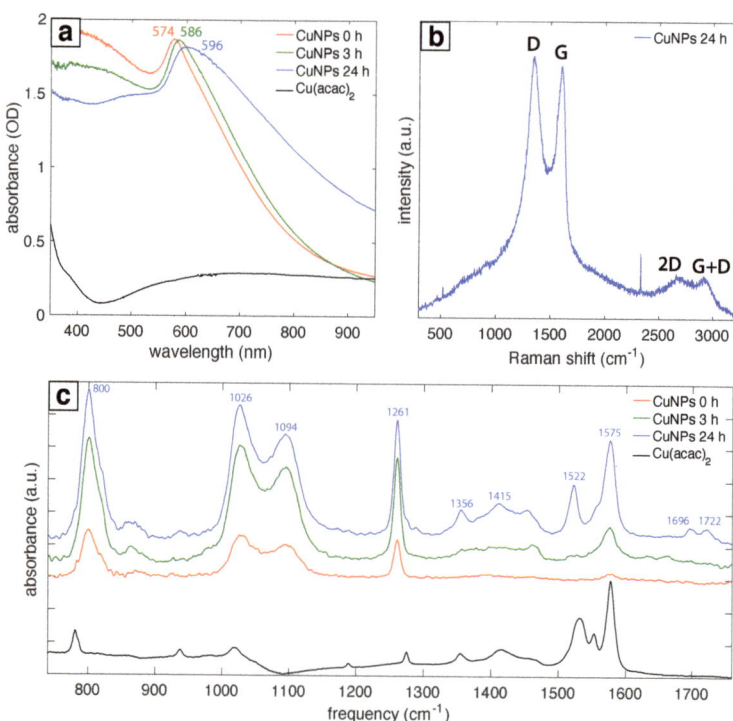

Figure 4. (a) UV-vis absorbance spectra for 0 h, 3 h, 24 h samples and precursor. (b) Raman spectra for 24 h samples. (c) FTIR spectra for 0 h, 3 h, 24 h samples and precursor.

3.2. Catalytic Activity of Cu NPs

Reduction of PNP by $NaBH_4$ is a commonly used model reaction to test catalytic activity of metal NPs by employing UV-vis spectroscopy to monitor the decrease in absorbance of the *p*-nitrophenolate ion at 400 nm, which allows for convenient determination of pseudo-first-order rate constants when excess $NaBH_4$ is present [58,59]. Figure 5a shows representative kinetic curves obtained from the ratio of the natural log of the 400 nm absorbance feature at time t, $A(t)$, to the initial absorbance, $A(0)$, as a function of reaction time t for the 0 h, 3 h, and 24 h samples. The slopes obtained from the linear regions of the curves (shown in black) give the apparent rate constants (k_{app}). The values of k_{app} were averaged over three different samples for all three post-synthesis times and then converted to the mass-specific rate constants (k) of 1084, 1479, and 1927 s^{-1}g^{-1} for 0 h, 3 h, and 24 h samples (Figure 5b). The specific rate at 3 h is comparable to the rate of 1490 s^{-1}g^{-1} for Cu NPs synthesized by PLAL in a water-ethanol mixture [30]. Although the rate constant is higher at 24 h, this increase could be due to some evaporation of the solvent over time, resulting in concentration of the Cu NPs and an artificially high measured rate.

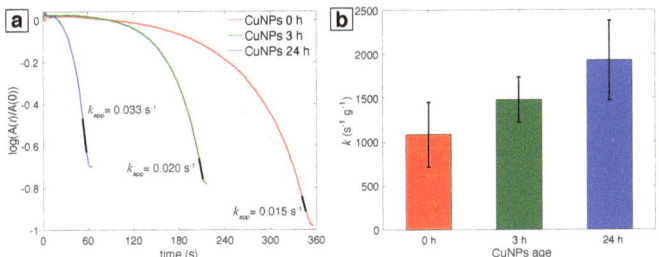

Figure 5. (a) Time dependence of the absorbance of p-nitrophenolate ions at 400 nm for a representative sample at 0 h, 3 h, and 24 h; black portion of line is where k_{app} (s^{-1}) is extracted. (b) Specific rate constants (s^{-1}g^{-1}) with error bars generated over three runs for 0 h, 3 h, 24 h samples each.

3.3. Stability of Cu NPs

Figure 6 compares the UV-vis absorbance, XRD spectrum, and PNP kinetics for 1 day (24 h) and 7 day samples. The Cu SPR peak observed for 7 day sample was only slightly red shifted compared to 1 day sample (Figure 6a), indicating no significant oxidation of Cu NPs. No changes were observed in the XRD peaks for 7 day samples (Figure 6b) compared 1 day samples (Figure 2), further confirming no or insignificant oxidation of Cu NPs. Accordingly, similar rate constants were obtained for 1 and 7 day samples for the PNP reduction reaction (Figure 6c). Collectively, these data indicate that the LRL Cu NPs are highly stable to oxidation.

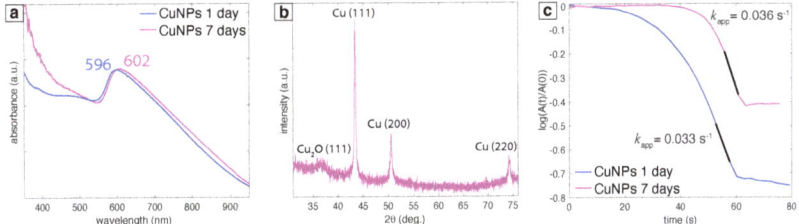

Figure 6. (a) UV-vis absorbance spectra measured at 1 day (blue) and 7 days (magenta). (b) XRD spectra for 7 day sample. (c) Time dependence of the absorbance of p-nitrophenolate ions at 400 nm for 1 day (blue) and 7 days (bright pink) samples; black portion of line is where k_{app} (s^{-1}) is extracted.

4. Discussion

The production of Cu metal NPs using LRL requires both an organic solvent and air-free conditions, as no laser-induced conversion of Cu(acac)$_2$ precursor was observed when the solvent was water or when the precursor in IPA/methanol mixture was irradiated under ambient conditions. This result underscores the importance of minimizing reactive oxygen species formation during LRL, which was also needed to effectively control the sizes of Au NPs [35] and enable formation of Ag-containing NPs [41,44,45]. The important role of reactive oxygen species has also been noted in PLAL synthesis of Cu NPs, where air removal completely eliminated formation of the highly oxidized CuO phase during PLAL in water [28]. Collectively, these results highlight similarities in the chemical reaction pathways induced during both PLAL and LRL that can lead to metal NP oxidation, as well as the potential for the same strategies to mitigate oxide formation.

The formation of substantial quantities of sub-10 nm Cu NPs in our LRL synthesis is similar to previous LRL results for Au and Ag NPs using well-controlled solvent chemistry [35,45], whereas the large Cu NPs up to ~100 nm in diameter resemble the large Fe oxide NPs obtained from LRL of ferrocene in hexane [46]. The formation of two distinct size distributions of Cu NPs strongly suggests the participation of two different reaction mechanisms [33], although specific identification of these mechanisms is beyond the scope of this work. Nevertheless, on the basis of the PLAL literature demonstrating that substantial control over Cu NP size and morphology can be achieved by changing solvent mixtures [27–30], we anticipate that further exploration of different solvents will enable better control over Cu NP sizes using LRL. Finally, we note that the catalytic activity of the LRL Cu NPs to PNP reduction is comparable to that reported for PLAL Cu NPs [30], despite the presence of a thick carbon shell around our NPs. This result indicates that the carbon shell is sufficiently permeable to allow for catalytic reactions to take place and suggests that our LRL Cu NPs may have additional catalytic applications in areas, such as CO_2 reduction or cross-coupling.

The formation of a substantial carbon shell around the LRL Cu NPs after termination of laser irradiation strongly suggests that the carbon formation is not entirely attributable to direct laser-induced solvent decomposition, as with PLAL in organic solvents [16,17,27,28]. Moreover, the lack of carbon shell formation during LRL of $Fe(acac)_3$ in water [47] indicates that the acetylacetonate ligands from the precursor are not sufficient to induce carbon shell formation in LRL. On the basis of these results, we can speculate that the carbon shell is formed from catalytic activation on the Cu NP metal surfaces of long-lived solvent and ligand byproducts produced during laser irradiation. Cu metal is known to be a highly active catalyst in carbon-carbon cross-coupling reactions [19] and induces graphene growth from aliphatic alcohols under low-temperature CVD conditions [60]. Hence, we anticipate that the active bare Cu surfaces present immediately following laser irradiation catalyze the formation of the carbon shell. Although carbon shells are widely known to effectively protect transition metal NPs from oxidation [61–63], the evident permeability of the carbon around the LRL Cu NPs on the basis of their high catalytic activity suggests that a thin Cu_2O layer around the Cu NPs also contributes to their observed stability [61].

5. Conclusions

We have reported the first synthesis of air-stable Cu NPs using the laser reduction in liquid (LRL) approach. Both small (<10 nm) and large (up to 100 nm) spherical NPs were observed, which primarily consisted of Cu^0 metal on the basis of XRD and XPS analysis, although a Cu_2O shell around the large NPs may be present. The Cu NPs exhibit remarkable stability over 7 days on the basis of the lack of significant changes observed in the UV-vis absorbance and XRD features and the similar rate constants obtained for PNP reduction. The LRL Cu NPs compare favorably with PLAL-synthesized Cu NPs in terms of Cu^0 content, stability, and catalytic activity. The insights into LRL of Cu^{2+} gained in this manuscript can be extended to other metals for which oxidation of NPs is commonly observed during PLAL or LRL. Finally, further development of the synthesis conditions, such as conducting LRL in a flow setup, could increase Cu NPs yield to produce heterogeneous catalysts for applications, such as cross-coupling reactions and CO_2 reduction.

Author Contributions: K.M.T. and A.N. designed the research. A.N. performed the experiments and analyzed the data. L.M.F.B. performed experiments and processed the TEM data. A.N. and K.M.T. wrote the manuscript. All authors have read and agreed to the published version of the manuscript.

Funding: This work was supported by the American Chemical Society Petroleum Research Fund through Grant 57799-DNI10. L.M.F.B. acknowledges generous financial support from an Altria Graduate Research Fellowship.

Institutional Review Board Statement: Not applicable.

Informed Consent Statement: Not applicable.

Data Availability Statement: Data presented in this study are available on request from the corresponding author.

Acknowledgments: Microscopy was performed at the VCU Department of Anatomy and Neurobiology Microscopy Facility, supported by the Higher Education Equipment Trust Fund Grant No. 236160307. We would like to acknowledge the VCU Nanomaterials Core Characterization Facility and Chemistry Instrumentation Facility for additional characterization.

Conflicts of Interest: The authors declare no conflict of interest.

References

1. Zhang, D.; Gökce, B.; Barcikowski, S. Laser Synthesis and Processing of Colloids: Fundamentals and Applications. *Chem. Rev.* **2017**, *117*, 3990–4103. [CrossRef] [PubMed]
2. Reichenberger, S.; Marzun, G.; Muhler, M.; Barcikowski, S. Perspective of Surfactant-free Colloidal Nanoparticles in Heterogeneous Catalysis. *ChemCatChem* **2019**, *11*, 4489–4518. [CrossRef]
3. Amans, D.; Cai, W.; Barcikowski, S. Status and demand of research to bring laser generation of nanoparticles in liquids to maturity. *Appl. Surf. Sci.* **2019**, *488*, 445–454. [CrossRef]
4. Swiatkowska-Warkocka, Z.; Pyatenko, A.; Krok, F.; Jany, B.R.; Marszalek, M. Synthesis of new metastable nanoalloys of immiscible metals with a pulse laser technique. *Sci. Rep.* **2015**, *5*, 9849. [CrossRef]
5. Tangeysh, B.; Odhner, J.H.; Wang, Y.; Wayland, B.B.; Levis, R.J. Formation of Copper(I) Oxide- and Copper(I) Cyanide–Polyacetonitrile Nanocomposites through Strong-Field Laser Processing of Acetonitrile Solutions of Copper(II) Acetate Dimer. *J. Phys. Chem. A* **2019**, *123*, 6430–6438. [CrossRef]
6. Streubel, R.; Barcikowski, S.; Gökce, B. Continuous multigram nanoparticle synthesis by high-power, high-repetition-rate ultrafast laser ablation in liquids. *Opt. Lett.* **2016**, *41*, 1486–1489. [CrossRef]
7. Amendola, V.; Meneghetti, M. What controls the composition and the structure of nanomaterials generated by laser ablation in liquid solution? *Phys. Chem. Chem. Phys.* **2013**, *15*, 3027–3046. [CrossRef]
8. Mastrotto, F.; Caliceti, P.; Amendola, V.; Bersani, S.; Magnusson, J.P.; Meneghetti, M.; Mantovani, G.; Alexander, C.; Salmaso, S. Polymer control of ligand display on gold nanoparticles for multimodal switchable cell targeting. *Chem. Commun.* **2011**, *47*, 9846–9848. [CrossRef]
9. Li, S.; Zhang, J.; Kibria, M.G.; Mi, Z.; Chaker, M.; Ma, D.; Nechache, R.; Rosei, F. Remarkably enhanced photocatalytic activity of laser ablated Au nanoparticle decorated BiFeO3 nanowires under visible-light. *Chem. Commun.* **2013**, *49*, 5856–5858. [CrossRef]
10. Zhang, J.; Chen, G.; Chaker, M.; Rosei, F.; Ma, D. Gold nanoparticle decorated ceria nanotubes with significantly high catalytic activity for the reduction of nitrophenol and mechanism study. *Appl. Catal. B-Environ.* **2013**, *132-133*, 107–115. [CrossRef]
11. Blakemore, J.D.; Gray, H.B.; Winkler, J.R.; Müller, A.M. Co3O4 Nanoparticle Water-Oxidation Catalysts Made by Pulsed-Laser Ablation in Liquids. *ACS Catal.* **2013**, *3*, 2497–2500. [CrossRef]
12. Gu, S.; Kaiser, J.; Marzun, G.; Ott, A.; Lu, Y.; Ballauff, M.; Zaccone, A.; Barcikowski, S.; Wagener, P. Ligand-free Gold Nanoparticles as a Reference Material for Kinetic Modelling of Catalytic Reduction of 4-Nitrophenol. *Catal. Lett.* **2015**, *145*, 1105–1112. [CrossRef]
13. Longano, D.; Ditaranto, N.; Cioffi, N.; Di Niso, F.; Sibillano, T.; Ancona, A.; Conte, A.; Del Nobile, M.A.; Sabbatini, L.; Torsi, L. Analytical characterization of laser-generated copper nanoparticles for antibacterial composite food packaging. *Anal. Bioanal. Chem.* **2012**, *403*, 1179–1186. [CrossRef] [PubMed]
14. Naddeo, J.; Ratti, M.; O'Malley, S.; Griepenburg, J.; Bubb, D.; Klein, E. Antibacterial Properties of Nanoparticles: A Comparative Review of Chemically Synthesized and Laser-Generated Particles. *Adv. Sci. Eng. Med.* **2015**, *7*, 1044–1057. [CrossRef]
15. Petersen, S.; Barcikowski, S. In Situ Bioconjugation: Single Step Approach to Tailored Nanoparticle-Bioconjugates by Ultrashort Pulsed Laser Ablation. *Adv. Funct. Mater.* **2009**, *19*, 1167–1172. [CrossRef]
16. Zhang, H.; Liang, C.; Liu, J.; Tian, Z.; Shao, G. The formation of onion-like carbon-encapsulated cobalt carbide core/shell nanoparticles by the laser ablation of metallic cobalt in acetone. *Carbon* **2013**, *55*, 108–115. [CrossRef]
17. Yu-jin, K.; Ma, R.; Reddy, D.A.; Kim, T.K. Liquid-phase pulsed laser ablation synthesis of graphitized carbon-encapsulated palladium core–shell nanospheres for catalytic reduction of nitrobenzene to aniline. *Appl. Surf. Sci.* **2015**, *357*, 2112–2120. [CrossRef]
18. Kim, J.H.; Chung, Y.K. Copper nanoparticle-catalyzed cross-coupling of alkyl halides with Grignard reagents. *Chem. Commun.* **2013**, *49*, 11101–11103. [CrossRef] [PubMed]
19. Gawande, M.B.; Goswami, A.; Felpin, F.X.; Asefa, T.; Huang, X.; Silva, R.; Zou, X.; Zboril, R.; Varma, R.S. Cu and Cu-Based Nanoparticles: Synthesis and Applications in Catalysis. *Chem. Rev.* **2016**, *116*, 3722–3811. [CrossRef] [PubMed]
20. Kwon, Y.T.; Lee, Y.I.; Kim, S.; Lee, K.J.; Choa, Y.H. Full densification of inkjet-printed copper conductive tracks on a flexible substrate utilizing a hydrogen plasma sintering. *Appl. Surf. Sci.* **2017**, *396*, 1239–1244. [CrossRef]
21. Pugazhendhi, A.; Edison, T.N.J.I.; Karuppusamy, I.; Kathirvel, B. Inorganic nanoparticles: A potential cancer therapy for human welfare. *Int. J. Pharm.* **2018**, *539*, 104–111. [CrossRef]

22. Park, J.H.; Seo, J.; Kim, C.; Joe, D.J.; Lee, H.E.; Im, T.H.; Seok, J.Y.; Jeong, C.K.; Ma, B.S.; Park, H.K.; et al. Flash-Induced Stretchable Cu Conductor via Multiscale-Interfacial Couplings. *Adv. Sci.* **2018**, *5*, 1801146. [CrossRef]
23. Song, X.; Sun, S.; Zhang, W.; Yin, Z. A method for the synthesis of spherical copper nanoparticles in the organic phase. *J. Colloid Interface Sci.* **2004**, *273*, 463–469. [CrossRef]
24. Dharmadasa, R.; Jha, M.; Amos, D.A.; Druffel, T. Room Temperature Synthesis of a Copper Ink for the Intense Pulsed Light Sintering of Conductive Copper Films. *ACS Appl. Mater. Interfaces* **2013**, *5*, 13227–13234. PMID: 24283767. [CrossRef]
25. Mott, D.; Galkowski, J.; Wang, L.; Luo, J.; Zhong, C.J. Synthesis of Size-Controlled and Shaped Copper Nanoparticles. *Langmuir* **2007**, *23*, 5740–5745. PMID: 17407333. [CrossRef] [PubMed]
26. Swarnkar, R.K.; Singh, S.C.; Gopal, R. Effect of aging on copper nanoparticles synthesized by pulsed laser ablation in water: structural and optical characterizations. *Bull. Mater. Sci.* **2011**, *34*, 1363–1369. [CrossRef]
27. Fernández-Arias, M.; Boutinguiza, M.; Del Val, J.; Covarrubias, C.; Bastias, F.; Gómez, L.; Maureira, M.; Arias-González, F.; Riveiro, A.; Pou, J. Copper nanoparticles obtained by laser ablation in liquids as bactericidal agent for dental applications. *Appl. Surf. Sci.* **2020**, *507*, 145032. [CrossRef]
28. Marzun, G.; Levish, A.; Mackert, V.; Kallio, T.; Barcikowski, S.; Wagener, P. Laser synthesis, structure and chemical properties of colloidal nickel-molybdenum nanoparticles for the substitution of noble metals in heterogeneous catalysis. *J. Colloid Interface Sci.* **2017**, *489*, 57–67. [CrossRef] [PubMed]
29. Goncharova, D.A.; Kharlamova, T.S.; Lapin, I.N.; Svetlichnyi, V.A. Chemical and Morphological Evolution of Copper Nanoparticles Obtained by Pulsed Laser Ablation in Liquid. *J. Phys. Chem. C* **2019**, *123*, 21731–21742. [CrossRef]
30. Goncharova, D.A.; Kharlamova, T.S.; Reutova, O.A.; Svetlichnyi, V.A. Water-ethanol CuOx nanoparticle colloids prepared by laser ablation: Colloid stability and catalytic properties in nitrophenol hydrogenation. *Colloids Surf. A Physicochem. Eng. Asp.* **2021**, *613*, 126115. [CrossRef]
31. Rehbock, C.; Merk, V.; Gamrad, L.; Streubel, R.; Barcikowski, S. Size Control of Laser-Fabricated Surfactant-Free Gold Nanoparticles with Highly Diluted Electrolytes and their Subsequent Bioconjugation. *Phys. Chem. Chem. Phys.* **2013**, *15*, 3057–3067. [CrossRef]
32. Sylvestre, J.P.; Kabashin, A.V.; Sacher, E.; Meunier, M. Femtosecond laser ablation of gold in water: influence of the laser-produced plasma on the nanoparticle size distribution. *Appl. Phys. A* **2005**, *80*, 753–758. [CrossRef]
33. Shih, C.Y.; Streubel, R.; Heberle, J.; Letzel, A.; Shugaev, M.V.; Wu, C.; Schmidt, M.; Gökce, B.; Barcikowski, S.; Zhigilei, L.V. Two mechanisms of nanoparticle generation in picosecond laser ablation in liquids: the origin of the bimodal size distribution. *Nanoscale* **2018**, *10*, 6900–6910. [CrossRef]
34. Rodrigues, C.J.; Bobb, J.A.; John, M.G.; Fisenko, S.P.; El-Shall, M.S.; Tibbetts, K.M. Nucleation and growth of gold nanoparticles initiated by nanosecond and femtosecond laser irradiation of aqueous [AuCl4]-. *Phys. Chem. Chem. Phys.* **2018**, *20*, 28465–28475. [CrossRef]
35. Frias Batista, L.M.; Meader, V.K.; Romero, K.; Kunzler, K.; Kabir, F.; Bullock, A.; Tibbetts, K.M. Kinetic Control of $[AuCl_4]^-$ Photochemical Reduction and Gold Nanoparticle Size with Hydroxyl Radical Scavengers. *J. Phys.Chem. B* **2019**, *123*, 7204–7213. [CrossRef] [PubMed]
36. Nakamura, T.; Mochidzuki, Y.; Sato, S. Fabrication of Gold Nanoparticles in Intense Optical Field by Femtosecond Laser Irradiation of Aqueous Solution. *J. Mater. Res.* **2008**, *23*, 968–974. [CrossRef]
37. Tangeysh, B.; Moore Tibbetts, K.; Odhner, J.H.; Wayland, B.B.; Levis, R.J. Gold Nanoparticle Synthesis Using Spatially and Temporally Shaped Femtosecond Laser Pulses: Post-Irradiation Auto-Reduction of Aqueous $[AuCl_4]^-$. *J. Phys. Chem. C* **2013**, *117*, 18719–18727. [CrossRef]
38. Meader, V.K.; John, M.G.; Rodrigues, C.J.; Tibbetts, K.M. Roles of Free Electrons and H2O2 in the Optical Breakdown-Induced Photochemical Reduction of Aqueous [AuCl4]-. *J. Phys. Chem. A* **2017**, *121*, 6742–6754. [CrossRef] [PubMed]
39. Nakamura, T.; Magara, H.; Herbani, Y.; Sato, S. Fabrication of silver nanoparticles by highly intense laser irradiation of aqueous solution. *Appl. Phys. A* **2011**, *104*, 1021. [CrossRef]
40. Herbani, Y.; Nakamura, T.; Sato, S. Silver nanoparticle formation by femtosecond laser induced reduction of ammonia-containing AgNO 3 solution. *J. Phys. Conf. Ser.* **2017**, *817*, 012048. [CrossRef]
41. Meader, V.K.; John, M.G.; Frias Batista, L.M.; Ahsan, S.; Tibbetts, K.M. Radical Chemistry in a Femtosecond Laser Plasma: Photochemical Reduction of Ag^+ in Liquid Ammonia Solution. *Molecules* **2018**, *23*, 532. [CrossRef]
42. Nakamura, T.; Takasaki, K.; Ito, A.; Sato, S. Fabrication of platinum particles by intense, femtosecond laser pulse irradiation of aqueous solution. *Appl. Surf. Sci.* **2009**, *255*, 9630–9633. [CrossRef]
43. Herbani, Y.; Nakamura, T.; Sato, S. Synthesis of platinum-based binary and ternary alloy nanoparticles in an intense laser field. *J. Colloid Interface Sci.* **2012**, *375*, 78–87. [CrossRef]
44. Herbani, Y.; Nakamura, T.; Sato, S. Spectroscopic monitoring on irradiation-induced formation of AuAg alloy nanoparticles by femtosecond laser. In *AIP Conference Proceedings*; AIP Publishing LLC: Melville, NY, USA, 2016; Volume 1711, p. 030005. [CrossRef]
45. Nguyen, C.M.; Frias Batista, L.M.; John, M.G.; Rodrigues, C.J.; Tibbetts, K.M. Mechanism of Gold–Silver Alloy Nanoparticle Formation by Laser Coreduction of Gold and Silver Ions in Solution. *J. Phys. Chem. B* **2021**, *125*, 907–917. PMID: 33439650. [CrossRef]

46. Okamoto, T.; Nakamura, T.; Kihara, R.; Asahi, T.; Sakota, K.; Yatsuhashi, T. Synthesis of Bare Iron Nanoparticles from Ferrocene Hexane Solution by Femtosecond Laser Pulses. *ChemPhysChem* **2018**, *19*, 2480–2485. [CrossRef]
47. Okamoto, T.; Nakamura, T.; Tahara, Y.O.; Miyata, M.; Sakota, K.; Yatsuhashi, T. Effects of Ligand and Solvent on the Synthesis of Iron Oxide Nanoparticles from Fe(acac)3 Solution by Femtosecond Laser Irradiation. *Chem. Lett.* **2020**, *49*, 75–78. [CrossRef]
48. Dhas, N.A.; Raj, C.P.; Gedanken, A. Synthesis, Characterization, and Properties of Metallic Copper Nanoparticles. *Chem. Mater.* **1998**, *10*, 1446–1452. [CrossRef]
49. John, M.G.; Tibbetts, K.M. One-step femtosecond laser ablation synthesis of sub-3 nm gold nanoparticles stabilized by silica. *Appl. Surf. Sci.* **2019**, *475*, 1048–1057. [CrossRef]
50. Holzwarth, U.; Gibson, N. The Scherrer equation versus the 'Debye-Scherrer equation'. *Nat. Nanotechnol.* **2011**, *6*, 534. [CrossRef] [PubMed]
51. Ghodselahi, T.; Vesaghi, M.; Shafiekhani, A.; Baghizadeh, A.; Lameii, M. XPS study of the Cu@Cu2O core-shell nanoparticles. *Appl. Surf. Sci.* **2008**, *255*, 2730–2734. [CrossRef]
52. Wagner, C.; M, R.W.; Davis, L.E.; Moulder, J.F.; Muilenburg, G.E. *Handbook of X-ray Photoelectron Spectroscopy*; Physical Electronics Division; Perkin-Elmer Corp.: Maharashtra, India, 1979.
53. Yu, B.; Wang, X.; Qian, X.; Xing, W.; Yang, H.; Ma, L.; Lin, Y.; Jiang, S.; Song, L.; Hu, Y.; et al. Functionalized graphene oxide/phosphoramide oligomer hybrids flame retardant prepared via in situ polymerization for improving the fire safety of polypropylene. *RSC Adv.* **2014**, *4*, 31782–31794. [CrossRef]
54. Rice, K.P.; Walker, E.J.; Stoykovich, M.P.; Saunders, A.E. Solvent-Dependent Surface Plasmon Response and Oxidation of Copper Nanocrystals. *J. Phys. Chem. C* **2011**, *115*, 1793–1799. [CrossRef]
55. Tiwari, S.K.; Hatui, G.; Oraon, R.; De Adhikari, A.; Nayak, G.C. Mixing sequence driven controlled dispersion of graphene oxide in PC/PMMA blend nanocomposite and its effect on thermo-mechanical properties. *Curr. Appl. Phys.* **2017**, *17*, 1158–1168. [CrossRef]
56. Bobb, J.A.; Rodrigues, C.J.; El-Shall, M.S.; Tibbetts, K.M. Laser-assisted synthesis of gold–graphene oxide nanocomposites: effect of pulse duration. *Phys. Chem. Chem. Phys.* **2020**, *22*, 18294–18303. [CrossRef] [PubMed]
57. Baharuddin, A.A.; Ang, B.C.; Wong, Y.H. Self-assembly and electrical characteristics of 4-pentynoic acid functionalized Fe3O4-γ-Fe2O3 nanoparticles on SiO2/n-Si. *Appl. Surf. Sci.* **2017**, *423*, 236–244. [CrossRef]
58. Wunder, S.; Polzer, F.; Lu, Y.; Mei, Y.; Ballauff, M. Kinetic Analysis of Catalytic Reduction of 4-Nitrophenol by Metallic Nanoparticles Immobilized in Spherical Polyelectrolyte Brushes. *J. Phys. Chem. C* **2010**, *114*, 8814–8820. [CrossRef]
59. Aditya, T.; Pal, A.; Pal, T. Nitroarene reduction: a trusted model reaction to test nanoparticle catalysts. *Chem. Commun.* **2015**, *51*, 9410–9431. [CrossRef] [PubMed]
60. Guermoune, A.; Chari, T.; Popescu, F.; Sabri, S.S.; Guillemette, J.; Skulason, H.S.; Szkopek, T.; Siaj, M. Chemical vapor deposition synthesis of graphene on copper with methanol, ethanol, and propanol precursors. *Carbon* **2011**, *49*, 4204–4210. [CrossRef]
61. Galaburda, M.; Kovalska, E.; Hogan, B.T.; Baldycheva, A.; Nikolenko, A.; Dovbeshko, G.I.; Oranska, O.I.; Bogatyrov, V.M. Mechanochemical synthesis of carbon-stabilized Cu/C, Co/C and Ni/C nanocomposites with prolonged resistance to oxidation. *Sci. Rep.* **2019**, *9*, 17435. [CrossRef]
62. Stein, M.; Wieland, J.; Steurer, P.; Tölle, F.; Mülhaupt, R.; Breit, B. Iron Nanoparticles Supported on Chemically-Derived Graphene: Catalytic Hydrogenation with Magnetic Catalyst Separation. *Adv. Syn. Catal.* **2011**, *353*, 523–527. [CrossRef]
63. Tsang, S.C.; Caps, V.; Paraskevas, I.; Chadwick, D.; Thompsett, D. Magnetically Separable, Carbon-Supported Nanocatalysts for the Manufacture of Fine Chemicals. *Angew. Chem. Int. Ed.* **2004**, *43*, 5645–5649. [CrossRef] [PubMed]

Article

Fabrication of Hollow Channels Surrounded by Gold Nanoparticles in Hydrogel by Femtosecond Laser Irradiation

Izumi Takayama [1], Akito Katayama [1] and Mitsuhiro Terakawa [1,2,*]

1. School of Integrated Design Engineering, Keio University, 3-14-1, Hiyoshi, Kohoku-ku, Yokohama 223-8522, Japan; i.takayama@tera.elec.keio.ac.jp
2. Department of Electronics and Electrical Engineering, Keio University, 3-14-1, Hiyoshi, Kohoku-ku, Yokohama 223-8522, Japan
* Correspondence: terakawa@elec.keio.ac.jp

Received: 30 November 2020; Accepted: 16 December 2020; Published: 16 December 2020

Abstract: The fabrication of hollow channels surrounded by gold nanoparticles in poly(ethylene glycol) diacrylate (PEGDA) is demonstrated. The absorption spectra show that gold nanoparticles were formed at the periphery of the focus by reduction of gold ions. The microscope observation and Raman spectroscopy analyses indicate that the center of the channels were void of PEGDA, which can be attributed to the femtosecond laser-induced degradation of the hydrogel. Since both the hydrogel and gold nanoparticles are biocompatible, this technique of fabricating hollow channels surrounded by gold nanoparticles is promising for tissue engineering, drug screening, and lab-on-a-chip devices.

Keywords: femtosecond laser; PEGDA; multiphoton reduction; gold nanoparticles; hollow channel

---

1. Introduction

The fabrication of tissue in vitro is essential in advancing the development of tissue engineering and drug screening. A tissue scaffold is necessary for growing cells and to support the tissue. Cell adhesion, differentiation, and proliferation are highly dependent on cell interactions with the tissue scaffold; therefore, it is necessary to control the shape and properties of the tissue scaffold when fabricating a tissue that resembles biological tissue. The mechanical strength of the material of the tissue scaffold should be comparable to that of the tissue from which the cell originates [1]. In addition, a chemical composition and surface structure that has a high compatibility to cells are required for the tissue scaffold.

Hydrogels have been used as tissue scaffolds because of their high biocompatibility. Because the mechanical strength of hydrogels can be tuned by the water content and molecular weight, hydrogels have been utilized as scaffolds for soft tissues, such as nerve tissue, to harder tissues, such as bone tissue [2]. In addition, hydrogels show high permeability to liquid, which allows the diffusion of glucose, oxygen and other nutrients to cells adhered in the bulk hydrogel [3]. 3-dimensional tissues fabricated on hydrogel tissue scaffolds have applications in tissue engineering including the regeneration of spinal cords, but also can be useful in organ-on-a-chip, drug screening, and diagnostics [4–8]. Existing methods for processing bulk hydrogel for the fabrication of tissue scaffolds include, multiphoton polymerization, soft lithography, molding using sacrificial materials, and laser processing [9]. Femtosecond lasers are powerful tools for the fabrication of a 3-dimensional channel in bulk materials including hydrogels. Liu et al. fabricated a channel in collagen and showed the adhesion of human HT1080 fibroblasts aligned along the channels [10]. Sarig-Nadir et al. fabricated channels, with 4–17 μm width and 10–80 μm height, in a bulk poly(ethylene glycol) diacrylate (PEGDA) based hydrogel, one of the most typical hydrogels, by irradiating femtosecond or nanosecond laser pulses [11]. Dorsal root ganglion cells adhered in the channels and aligned along the direction of the channel. Because PEGDA does not induce high cell adhesion on its own, PEGDA mixed with fibrinogen cysteines was used for the tissue scaffold.

Gold nanoparticles have a high biocompatibility and allow the adsorption of proteins which enhance cell adhesion [12]. In recent years, it has been reported that the patterning of gold nanoparticles on the surface of hydrogels accelerates cell adhesion in monolayer cultures. Ren et al. patterned 10 µm wide arrays consisting of 20 nm gold nanoparticles on the surface of PEGDA hydrogels by microcontact printing. Murine fibroblast L-929 cells were reported to have only adhered to the region with patterned gold nanoparticles [13].

It has been reported that by focusing femtosecond laser pulses into the bulk of a material that contains metal ions, metal particles generated by photoreduction dispersed around the focal point. Tosa et al. irradiated a femtosecond laser pulse at a repetition rate of 82 MHz into a film consisting of gold chloride and poly (4-styrene sulfonic acid) [14]. In the fabricated structure, no gold nanoparticles were observed at the center of the focus but were precipitated around the focal point, which were considered to be attributable to the gradient of the electric field to thrust the nanoparticles to the edge of the laser beam. Bellec et al., used a femtosecond laser pulse at a repetition rate of 9.44 MHz to fabricate a silver structure by irradiating the pulse into silver doped phosphate glass [15]. The resulting structure consisted of clusters of silver nanoparticles at the edge of the laser beam. The photodissociation of silver nanoparticles in the focal point were discussed to explain the distribution. To the best of our knowledge, the simultaneous fabrication of hollow channels and gold nanoparticles around the channels by laser irradiation has yet to be reported.

In this paper, we demonstrate the simultaneous fabrication of hollow channels surrounded by gold nanoparticles in hydrogels by femtosecond laser irradiation. The bright-field microscope images and absorption spectra show the reduction of gold ions around the channels, induced by femtosecond laser irradiation. In addition, the digital microscope images and Raman spectroscopy results indicate that the center of the channels were void of PEGDA matrix.

2. Materials and Methods

2.1. Hydrogel Preparation

Hydrogels were prepared by dissolving 0.1 g of PEGDA (molecular weight 4000) (Polysciences, Warrington, PA, USA) in 1 mL of pure water, containing 1% Irgacure 2959 initiator (Sigma-Aldrich Co. LLC, St. Louis, MI, USA). The PEGDA solution was stirred with a magnetic stirrer for 30 min until completely dissolved. The solution was added to a silicon mold with dimensions of 8 mm × 11 mm × 2.5 mm. The PEGDA solution was irradiated with an ultraviolet (UV) lamp (LUV-16, AXEL, Osaka, Japan) with a central wavelength of 365 nm for 1 h to induce photo-crosslinking. The crosslinked hydrogels were immersed in a gold chloride solution (Fujifilm Wako Chemicals, Osaka, Japan) for 10 min to induce the uptake of gold ions into the hydrogel matrix. The concentrations of gold chloride ($AuCl_4$) solution used in this paper was 1, 2, 4, or 8 mg/mL. After immersing the fabricated hydrogels in water for 1 day, black India ink was injected into the channels with a syringe (Terumo, Tokyo, Japan) to assess whether the channels are hollow. The channels with black India ink, were observed with a bright-field microscope.

2.2. Laser Parameters

A Ti:sapphire chirped pulse amplification (CPA) laser system (Libra, Coherent. Inc., Santa Clara CA, USA) generated linearly polarized femtosecond laser pulses with a pulse duration of 100 fs, at a central wavelength of 800 nm, at a repetition rate of 1 kHz. PEGDA hydrogels were placed onto a glass slide on a XY stage. Femtosecond laser pulses were focused into PEGDA hydrogels with an objective lens (60×, NA 0.7). The setup of the experiment is shown in Figure 1. In hydrogels that were immersed in gold chloride solution, all structures were fabricated with a scanning speed of 200 µm/s and 2 scans. In order to prevent the shrinkage of the hydrogel while fabricating channels, the same concentration of gold chloride solution that the hydrogel was immersed in, was dropped on the hydrogel. Pulse energies of 2–10 µJ and gold chloride concentrations of 1–8 mg/mL were used to fabricate structures in PEGDA hydrogels. The channels were observed with a bright-field microscope (Eclipse Ti-E, Nikon, Tokyo, Japan). The hydrogels were cut by an iron gel cutter to observe the cross-sections of the hydrogels. The widths of the channels were measured from the bright-field microscope images of the top view of the channels by using ImageJ. For the fabrication of a channel with a larger width, a structure consisting of 3 parallel channels at an interval of 50 µm was fabricated at a pulse energy of 10 µJ and gold chloride solution concentration of 4 mg/mL.

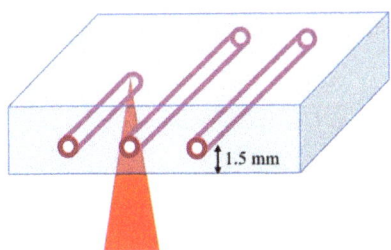

Figure 1. Schematic diagram of the simultaneous fabrication of hollow channels and surrounding gold nanoparticles.

2.3. Absorption Spectrum Measurements

The absorption spectra of the channels were measured to evaluate the relative concentration of gold nanoparticles formed around the channels. A white light was irradiated onto gratings consisting of channels inside a hydrogel. The transmission light was coupled to an optic fiber connected to an ultraviolet-visible-near infrared spectrophotometer (USB4000, Ocean Optics, Largo, FL, USA). A glass slide was used for the background correction.

2.4. Digital Microscope

A digital microscope (HRX-01, HIROX, Tokyo, Japan) was used to acquire the bright-field images of the cross-sectional view of the fabricated structures. Both the 2-dimensional and 2.5-dimensional image of the cross-sectional view of the structures were obtained.

2.5. Raman Spectroscopy

A laser-excited Raman spectrometer (InVia Raman Microscope, Renishaw, Wotton-under-Edge, UK) was used to analyze the chemistry of the fabricated structures. The excitation wavelength was 532 nm and the structures were analyzed for a wavenumber range of 100 to 3000 cm^{-1}. A 20× objective lens (NA 0.40) was used to focus the laser at the center of the cross-section of the structures and the unirradiated region of the hydrogel. From the wavelength and NA, the spatial resolution is estimated to be approximately 1.6 µm.

3. Results and Discussion

3.1. Fabrication of Hollow Channels Surrounded by Gold Nanoparticles

Figure 2a–c shows the bright-field microscope images of the structures fabricated at pulse energies of 2–10 µJ, a scan speed of 200 µm/s, and 2 scans. The concentrations of the gold chloride solution was 4 mg/mL for all structures. The top view and cross-sectional view of the fabricated structures are shown in Figure 2a–c, respectively. By immersing the hydrogel in gold chloride solution, a spontaneous color change from transparent to reddish color was observed in the hydrogel with fabricated channels, where the laser pulse was not irradiated, similar to the case of other studies using gold chloride solution [16]. In the cross-sectional view of the structures (Figure 2b), the change in color at the center of the structures was small, whereas the change in color around the structures was significant. The color of the structures changed from a light red to a darker red with the increase of pulse energy, suggesting that the density of the gold nanoparticles increased. As indicated in Figure 2c, the injected ink flowed into the fabricated channels. These results suggest that a hollow channel with gold nanoparticles distributed around the channel was fabricated. Figure 2d shows the relationship between the pulse energy and the widths of the fabricated channels. The widths of the channels increased with the increase in pulse energy. Even with the lowest pulse energy in the present study, 2 µJ, the intensity of the laser pulse was calculated to be 1.71×10^{15} W/cm^2, which is significantly higher than the reported threshold of the optical breakdown in water, which is in the range of 10^{13} W/cm^2 [17,18]. Therefore, it is highly probable that the laser induced plasma contributed to the fabrication of the structures.

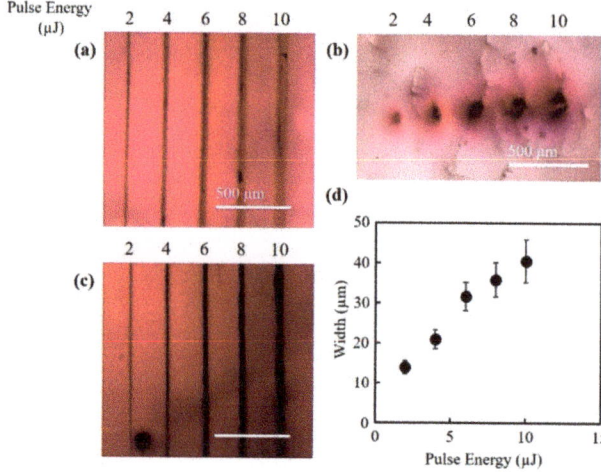

Figure 2. Fabrication of channels at different pulse energies. The bright-field microscope images of the (**a**) top view and (**b**) cross-sectional view of the fabricated channels. (**c**) The bright-field microscope images of the channels with ink. (**d**) The widths of the channels fabricated at different pulse energies.

Figure 3a,b show the top view and cross-sectional view of the structures fabricated at different concentrations of gold chloride solutions, respectively. All structures were fabricated at a pulse energy of 10 µJ, a scan speed of 200 µm/s and 2 scans. In hydrogels that were immersed in gold chloride solution concentrations of 1, 2, 4 mg/mL, the regions where there was no laser irradiations exhibited a reddish color. The color became darker with the increase of the gold chloride solution concentration. However, in hydrogels that were immersed in a gold chloride solution with a concentration of 8 mg/mL, the color of the unirradiated hydrogel was a bluish purple. The discussions for the result will be described in the next section. In hydrogels that were immersed in gold chloride solution concentration of 1 mg/mL, ink did not penetrate the channels. From Figure 3c, it can be observed that the ink penetrated the fabricated structures at gold chloride solution concentrations of 2, 4, and 8 mg/mL. The results suggest that hollow channels were fabricated, similar to Figure 2c. The widths of the channels increased with the increase in gold chloride solution concentration (Figure 3d). It is thought that in hydrogels that were immersed in higher concentrations of gold chloride solution, more gold nanoparticles were generated spontaneously or by photoreduction with primary laser pulses. The widths of the channels probably increased due to the change of absorption coefficient of the hydrogel by the nanoparticle generation. When a laser pulse with an intensity above the threshold of the optical breakdown of water is irradiated into a hydrogel, electrons in the water are excited in a short time, which results in the generation of plasma at the focus of the laser pulse. Because the water in the focus flows outward, the pressure at the focus becomes exceedingly low, and consequently the water undergoes a phase change from water to vapor, which results in the generation of a cavitation bubble [19]. As the cavitation bubble expands, the polymer bonds of the hydrogel would be physically dissociated, leading to hydrogel degradation. The dissociation of the bonds and the generation of shockwave are probably the mechanisms behind the fabrication of hollow channels in hydrogels [20]. The gold nanoparticles surrounding the channels were formed by the reduction of gold ions. It has been reported that, by inducing multiphoton reduction with the irradiation of a femtosecond laser pulse, metal ions have been reduced into metal nanoparticles in a hydrogel [16,21,22]. The reduction of metal ions is attributable to the electron donation induced through multiphoton absorption (multiphoton reduction) and/or thermal reduction from the heat accumulation caused by high repetition rates laser pulse. The photoreduction of metal ions can also be induced by radicals from photoinitiator. Izquierdo-Lorenzo et al. used biphenyl-(2,4,6-trimethylbenzoyl)-phosphine oxide (TPO) as a photoinitiator to fabricate gold nanoparticles by two-photon photoreduction [23]. In addition, the reduction of metal ions can be induced by a plasma produced by a focused laser pulse. Tasche et al. discussed that the electrons at the boundary of the plasma and silver nitrate solution reduced the silver ions to create silver nanoparticles [24]. The gold nanoparticles in the structures formed in the present study were distributed to regions larger than the focal spot size of the laser pulse. As described above, the laser intensities in the present study were higher than the threshold of the optical breakdown in water, which could result in the generation of plasma. Possible explanations are that either the gold nanoparticles generated by multiphoton reduction were pushed to the periphery by the expansion of the plasma and subsequent shockwave, or that the gold ions were reduced by plasma ion radiation at the interface of plasma and hydrogel.

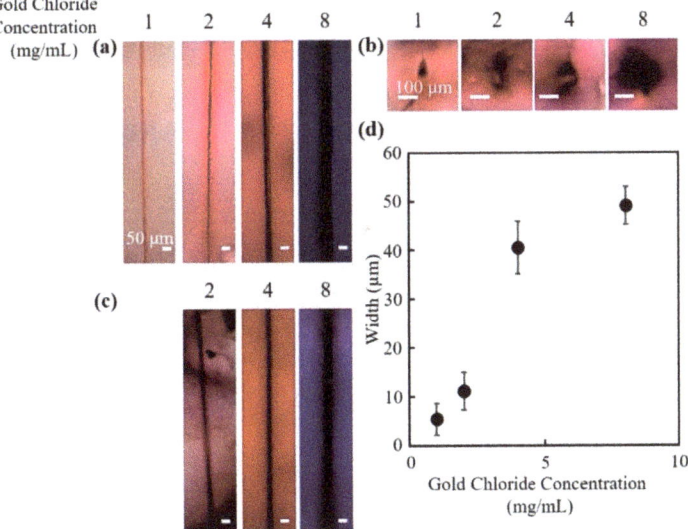

Figure 3. Fabrication of channels at different gold chloride concentrations. The bright-field microscope images of the (**a**) top view and (**b**) cross-sectional view of the fabricated channels. (**c**) The bright-field microscope images of the channels with ink. (**d**) The widths of the channels fabricated at different gold chloride concentrations.

3.2. Absorption Spectra of the Fabricated Channels

In order to measure the relative density of the gold nanoparticles, the absorption spectrum of the structures was obtained. The absorption spectrum of the structures that were fabricated with a pulse energy of 2–10 µJ, a scan speed of 200 µm/s and 2 scans are shown in Figure 4. The increase in pulse energy led to an increase in the peak height of the absorption spectrum. The results suggest that the number of formed gold nanoparticles increased with the increase of pulse energy. In addition, the absorbance peak was approximately 540 nm, which corresponds to the resonance wavelength of gold nanoparticles. The diameter of the gold nanoparticles was calculated from the equation below, where λ_{SPR} is the absorbance peak [25].

$$d = \frac{\ln\left(\frac{\lambda_{SPR}-512}{0.0216}\right)}{6.53} \tag{1}$$

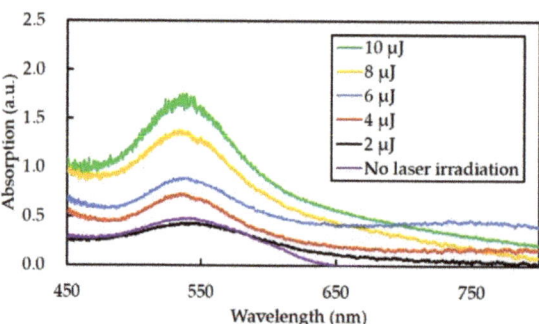

Figure 4. The absorption spectrum of channels fabricated at different pulse energies.

From Equation (1), the diameter of the gold nanoparticles was calculated to be approximately 67 nm. In previous studies, we have reported that the diameter of gold nanoparticles fabricated by multiphoton photoreduction was approximately 10 nm [16]. From the unaltered color transmission electron microscopy (TEM) images in our previous studies, it is thought that the calculated diameter in the present study is probably the value for the gold nanoparticle in the form of aggregation [16].

Figure 5 shows the absorption spectrum of the channels fabricated in hydrogels with different gold chloride solution concentrations. The peak height increased with the increase of gold chloride solution concentration, which indicates that the number of gold nanoparticles increased. In gold chloride solution concentrations of less than and including 4 mg/mL, the absorbance peak of the fabricated channels was approximately 540 nm. However, in gold chloride solution concentrations of 8 mg/mL, the absorbance peak of the channels was approximately 560 nm. The diameter of the gold cluster calculated from the absorbance wavelength was estimated to be 92 nm. The result suggests that the size of the gold nanoparticles, probably in the aggregated form, increased at a high concentration of gold chloride solution. The red shift of the absorption peak corresponds to the results shown in Figure 3, in which the color of the hydrogel changed to bluish purple. In multiphoton photoreduction, subsequent laser pulses irradiated to the gold nanoparticles would induce growth and aggregation of nanoparticles due to the plasmonic-enhanced photoreduction and photothermal effect. The results suggest that when the concentration of gold chloride solution was high, the gold ion in the vicinity of the laser focus was not depleted, which led to the formation of nanoparticles with a larger diameter.

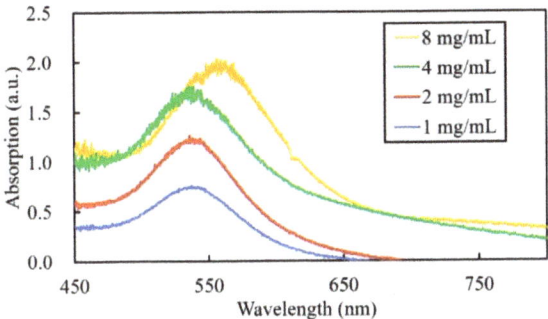

Figure 5. The absorption spectrum of channels fabricated at different gold chloride concentrations.

3.3. Fabrication of Structures with Different Number of Pulses

To investigate the interactions between subsequent laser pulses and gold nanoparticles, gold nanoparticles were formed with a different number of laser pulses. The laser pulse was irradiated in a setup similar to the setup shown in Figure 1, but without the laser scanning. The laser pulses at a pulse energy of 10 µJ were irradiated into the hydrogel with a gold chloride concentration of 4 mg/mL. Figure 6 shows that when 1 and 10 pulses were irradiated, modification of the hydrogel in the focal spot can be observed while no identifiable color change was observed. In contrast, when 50 or more laser pulses were irradiated, the color of the irradiated region changed to red. The region where color change was observed expanded with the increase of irradiated laser pulses. Since the repetition rate of the laser pulses was 1 kHz in the present study, heat accumulation by succeeding laser pulses may be negligible. The expansion of the region may be attributed to the absorption of subsequent laser pulses to the gold nanoparticles formed by primary laser pulses.

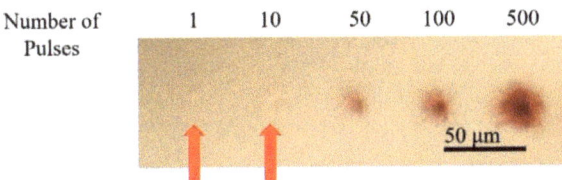

Figure 6. The bright-field microscope images of structures fabricated by irradiating a different number of pulses. The arrows indicate the modification of hydrogel, by laser irradiation, without significant color change.

3.4. Anaylsis of Structures Consisting of Parallel Channels

Figure 7a shows the bright-field microscope image of structures consisting of 3 parallel channels. The color of the irradiated regions is dark red, which indicates that a significant number of gold ions were reduced by laser irradiation. Black ink penetrated the structure as shown in Figure 7b. In the case of structures consisting of parallel channels, the ink penetrated the structure easily without the use of a syringe. Figure 7c,d show the 2-dimensional and 2.5-dimensionnal images of the cross-sectional view of the structures taken by a digital microscope. The hydrogel was cut perpendicularly to the structures for the observation. The focal point of the microscope did not coincide for the center of the structure and the periphery, which supports the result that a hollow channel was fabricated by the material removal.

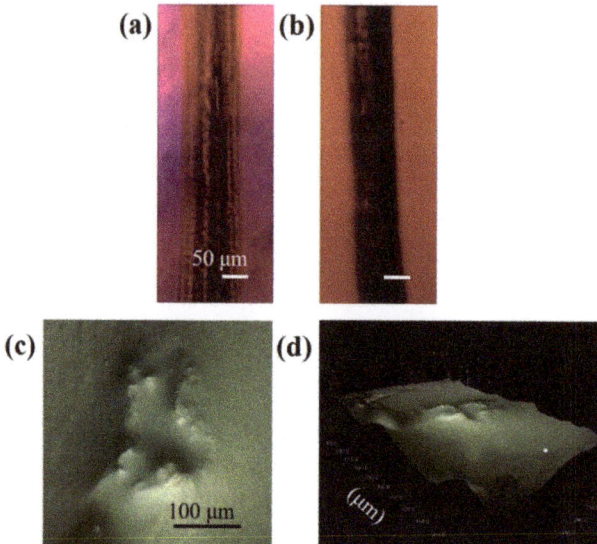

Figure 7. Fabrication of wide channels with applications in tissue engineering, microfluidics, and drug delivery. The bright-field microscope images of the fabricated channels (**a**) without ink and (**b**) with ink. The (**c**) 2-dimensional and (**d**) 2.5-dimensional image of the cross-sectional view of the channels captured by a digital microscope.

Figure 8 shows the result of Raman spectroscopy analyses for the structures consisting of 3 parallel channels. The hydrogel was cut perpendicularly, then the unirradiated region and the center of the structure were analyzed. In unirradiated regions of the hydrogel, the spectrum peaks corresponding to C–O–C (850 cm^{-1}), C=O (1720 cm^{-1}), and C–H (2950 cm^{-1}), respectively were observed, indicating the presence of PEGDA. No peaks corresponding to bonds present in PEGDA were observed at the center of the structures, and only a single peak at 1650 cm^{-1}, corresponding to H–O–H bonds was observed. These results also indicate that the center of the structures was void, i.e., without PEGDA matrix.

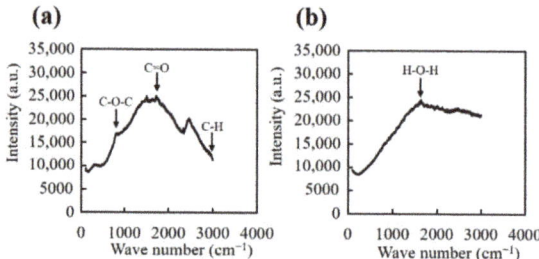

Figure 8. Raman spectra of (**a**) the bulk hydrogel and (**b**) the center of the structures.

4. Conclusions

In conclusion, the structures consisted of hollow channels which were surrounded by gold nanoparticles. The hollow channels and gold nanoparticles were formed in a single step of irradiating a femtosecond laser pulse in PEGDA hydrogels which contained gold ions. To the best of our knowledge, this is the first example of fabricating a hollow channel surrounded by gold nanoparticles in a hydrogel, by femtosecond laser irradiation. Given that hydrogels and gold nanoparticles have a high biocompatibility, this research has many applications in tissue engineering, microfluidics, and drug delivery.

Author Contributions: Conceptualization, M.T.; methodology, I.T., A.K., and M.T.; investigation, I.T. and A.K.; writing, I.T. and M.T.; project administration, M.T. All authors have read and agreed to the published version of the manuscript.

Funding: This work was supported in part by Ministry of Education, Culture, Sports, Science and Technology (MEXT) KAKENHI (18H03551).

Acknowledgments: The authors appreciate HIROX, Co. Ltd. for digital microscopic observation.

Conflicts of Interest: The authors declare no conflict of interest.

References

1. Butcher, D.T.; Alliston, T.; Weaver, V.M. A tense situation: Forcing tumour progression. *Nat. Rev. Cancer* **2009**, *9*, 108–122. [CrossRef]
2. Engler, A.J.; Sen, S.; Sweeney, H.L.; Discher, D.E. Matrix Elasticity Directs Stem Cell Lineage Specification. *Cell* **2006**, *126*, 677–689. [CrossRef] [PubMed]
3. Cruise, G.M.; Scharp, D.S.; Hubbell, J.A. Characterization of permeability and network structure of interfacially photopolymerized poly(ethylene glycol) diacrylate hydrogels. *Biomaterials* **1998**, *19*, 1287–1294. [CrossRef]
4. Chen, M.B.; Srigunapalan, S.; Wheeler, A.R.; Simmons, C.A. A 3D microfluidic platform incorporating methacrylated gelatin hydrogels to study physiological cardiovascular cell-cell interactions. *Lab Chip* **2013**, *13*, 2591–2598. [CrossRef] [PubMed]
5. Cuchiara, M.P.; Allen, A.C.B.; Chen, T.M.; Miller, J.S.; West, J.L. Multilayer microfluidic PEGDA hydrogels. *Biomaterials* **2010**, *31*, 5491–5497. [CrossRef]
6. Moore, M.J.; Friedman, J.A.; Lewellyn, E.B.; Mantila, S.M.; Krych, A.J.; Ameenuddin, S.; Knight, A.M.; Lu, L.; Currier, B.L.; Spinner, R.J.; et al. Multiple-channel scaffolds to promote spinal cord axon regeneration. *Biomaterials* **2006**, *27*, 419–429. [CrossRef]
7. Chen, X.; Zhao, Y.; Li, X.; Xiao, Z.; Yao, Y.; Chu, Y.; Farkas, B.; Romano, I.; Brandi, F.; Dai, J. Functional Multichannel Poly(Propylene Fumarate)-Collagen Scaffold with Collagen-Binding Neurotrophic Factor 3 Promotes Neural Regeneration After Transected Spinal Cord Injury. *Adv. Healthc. Mater.* **2018**, *7*, 1800315. [CrossRef]
8. Pawelec, K.M.; Hix, J.; Shapiro, E.M.; Sakamoto, J. The mechanics of scaling-up multichannel scaffold technology for clinical nerve repair. *J. Mech. Behav. Biomed. Mater.* **2019**, *91*, 247–254. [CrossRef]
9. Papadimitriou, L.; Manganas, P.; Ranella, A.; Stratakis, E. Biofabrication for neural tissue engineering applications. *Mater. Today Bio* **2020**, *6*, 100043. [CrossRef]
10. Liu, Y.; Sun, S.; Singha, S.; Cho, M.R.; Gordon, R.J. 3D femtosecond laser patterning of collagen for directed cell attachment. *Biomaterials* **2004**, *26*, 4597–4605. [CrossRef]
11. Sarig-Nadir, O.; Livnat, N.; Zajdman, R.; Shoham, S.; Seliktar, D. Laser Photoablation of Guidance Microchannels into Hydrogels Directs Cell Growth in Three Dimensions. *Biophys. J.* **2009**, *96*, 4743–4752. [CrossRef] [PubMed]
12. De Paoli Lacerda, S.H.; Park, J.J.; Meuse, C.; Pristinski, D.; Becker, M.L.; Karim, A.; Douglas, J.F. Interaction of gold nanoparticles with common human blood proteins. *ACS Nano* **2010**, *4*, 365–379. [CrossRef] [PubMed]
13. Ren, F.; Yesildag, C.; Zhang, Z.; Lensen, M. Surface Patterning of Gold Nanoparticles on PEG-Based Hydrogels to Control Cell Adhesion. *Polymers* **2017**, *9*, 154. [CrossRef] [PubMed]
14. Tosa, N.; Bosson, J.; Vitrant, G.; Baldeck, P.; Stephan, O. Fabrication of metallic nanowires by two-photon absorption. *Sci. Study Res.* **2006**, *7*, 899–904.
15. Bellec, M.; Royon, A.; Bousquet, B.; Bourhis, K.; Treguer, M.; Cardinal, T.; Richardson, M.; Canioni, L. Beat the diffraction limit in 3D direct laser writing in photosensitive glass. *Opt. Express* **2009**, *17*, 10304–10318. [CrossRef]
16. Machida, M.; Niidome, T.; Onoe, H.; Heisterkamp, A.; Terakawa, M. Spatially-targeted laser fabrication of multi-metal microstructures inside a hydrogel. *Opt. Express* **2019**, *27*, 14657–14666. [CrossRef] [PubMed]

17. Fan, C.H.; Sun, J.; Longtin, J.P. Breakdown threshold and localized electron density in water induced by ultrashort laser pulses. *J. Appl. Phys.* **2002**, *91*, 2530–2536. [CrossRef]
18. Schaffer, C.B.; Nishimura, N.; Glezer, E.N.; Kim, A.M.-T.; Mazur, E. Dynamics of femtosecond laser-induced breakdown in water from femtoseconds to microseconds. *Opt. Express* **2002**, *10*, 169–203. [CrossRef]
19. Noack, J.; Vogel, A. Laser-induced plasma formation in water at nanosecond to femtosecond time scales: Calculation of thresholds, absorption coefficients, and energy density. *IEEE J. Quantum Electron.* **1999**, *35*, 1156–1167. [CrossRef]
20. Pradhan, S.; Keller, K.A.; Sperduto, J.L.; Slater, J.H. Fundamentals of Laser-Based Hydrogel Degradation and Applications in Cell and Tissue Engineering. *Adv. Healthc. Mater.* **2017**, *6*, 1700681–1700709. [CrossRef]
21. Kang, S.; Vora, K.; Mazur, E. One-step direct-laser metal writing of sub100nm 3D silver nanostructures in a gelatin matrix. *Nanotechnology* **2015**, *26*, 121001–121007. [CrossRef] [PubMed]
22. Terakawa, M.; Torres-Mapa, M.L.; Takami, A.; Heinemann, D.; Nedyalkov, N.N.; Nakajima, Y.; Hördt, A.; Ripken, T.; Heisterkamp, A. Femtosecond laser direct writing of metal microstructure in a stretchable poly(ethylene glycol) diacrylate (PEGDA) hydrogel. *Opt. Lett.* **2016**, *41*, 1392–1395. [CrossRef] [PubMed]
23. Izquierdo-Lorenzo, I.; Jradi, S.; Adam, P.M. Direct laser writing of random Au nanoparticle three-dimensional structures for highly reproducible micro-SERS measurements. *Rsc Adv.* **2014**, *4*, 4128–4133. [CrossRef]
24. Tasche, D.; Weber, M.; Mrotzek, J.; Gerhard, C.; Wieneke, S.; Möbius, W.; Höfft, O.; Viöl, W. In situ investigation of the formation kinematics of plasma-generated silver nanoparticles. *Nanomaterials* **2020**, *10*, 555. [CrossRef]
25. Haiss, W.; Thanh, N.T.K.; Aveyard, J.; Fernig, D.G. Determination of size and concentration of gold nanoparticles from UV-Vis spectra. *Anal. Chem.* **2007**, *79*, 4215–4221. [CrossRef]

Publisher's Note: MDPI stays neutral with regard to jurisdictional claims in published maps and institutional affiliations.

© 2020 by the authors. Licensee MDPI, Basel, Switzerland. This article is an open access article distributed under the terms and conditions of the Creative Commons Attribution (CC BY) license (http://creativecommons.org/licenses/by/4.0/).

Article

Specific Features of Reactive Pulsed Laser Deposition of Solid Lubricating Nanocomposite Mo–S–C–H Thin-Film Coatings

Vyacheslav Fominski [1],*, Dmitry Fominski [1], Roman Romanov [1], Mariya Gritskevich [1], Maxim Demin [2], Petr Shvets [2], Ksenia Maksimova [2] and Alexander Goikhman [2]

[1] National Research Nuclear University MEPhI (Moscow Engineering Physics Institute), Kashirskoe sh., 31, 115409 Moscow, Russia; dmitryfominski@gmail.com (D.F.); limpo2003@mail.ru (R.R.); mgritskevich@yandex.ru (M.G.)
[2] Immanuel Kant Baltic Federal University, A. Nevskogo St 14, 236016 Kaliningrad, Russia; sterlad@mail.ru (M.D.); pshvets@kantiana.ru (P.S.); xmaksimova@gmail.com (K.M.); aygoikhman@gmail.com (A.G.)
* Correspondence: vyfominskij@mephi.ru

Received: 20 November 2020; Accepted: 6 December 2020; Published: 8 December 2020

Abstract: This work investigates the structure and chemical states of thin-film coatings obtained by pulsed laser codeposition of Mo and C in a reactive gas (H_2S). The coatings were analysed for their prospective use as solid lubricating coatings for friction units operating in extreme conditions. Pulsed laser ablation of molybdenum and graphite targets was accompanied by the effective interaction of the deposited Mo and C layers with the reactive gas and the chemical states of Mo- and C-containing nanophases were interdependent. This had a negative effect on the tribological properties of Mo–S–C–H nanocomposite coatings obtained at H_2S pressures of 9 and 18 Pa, which were optimal for obtaining MoS_2 and MoS_3 coatings, respectively. The best tribological properties were found for the Mo–S–C–H_5.5 coating formed at an H_2S pressure of 5.5 Pa. At this pressure, the $x = S/Mo$ ratio in the MoS_x nanophase was slightly less than 2, and the a-C(S,H) nanophase contained ~8 at.% S and ~16 at.% H. The a-C(S,H) nanophase with this composition provided a low coefficient of friction (~0.03) at low ambient humidity and 22 °C. The nanophase composition in Mo–S–C–H_5.5 coating demonstrated fairly good antifriction properties and increased wear resistance even at −100 °C. For wet friction conditions, Mo–S–C–H nanocomposite coatings did not have significant advantages in reducing friction compared to the MoS_2 and MoS_3 coatings formed by reactive pulsed laser deposition.

Keywords: reactive pulsed laser deposition; solid lubricants; nanocomposite; molybdenum sulfides; coefficient of friction; wear; diamond-like carbon

---

1. Introduction

Researchers and practitioners turned their attention to solid lubricating coatings based on transitional metal dichalcogenides (TMDs), such as MoS_2, WS_2, $MoSe_2$, and WSe_2, in the 1980s [1,2]. This was due to the need to improve qualitatively the tribological properties (low coefficient of friction, durability, wear resistance) of friction units operating in a vacuum or the inert environment of a spacecraft. By that time, sufficiently effective technologies for the deposition of such coatings (mainly ion sputtering) had already been developed. This made it possible to regulate flexibly the modes of deposition and the composition of the coatings [3,4]. At the initial stage of research into these coatings, the emphasis was on finding the optimal conditions for the deposition of monophase (pure) TMD coatings and for obtaining their required composition and structure. Soon the main problem of such coatings became apparent: it was low wear resistance, especially at high contact loads. Consequently, the focus of research shifted towards the search for nanocomposite and multilayer (nanolayer) coatings. The wear resistance of nanocomposite coatings can be significantly increased by combining the plastic TMD phase with a harder/stronger or corrosion-resistant component (metals, hard carbon, metal carbides or nitrides) [5–8].

Despite the breakthroughs in the formation of nanocomposite TMD-based coatings with improved tribological characteristics, the problem of obtaining solid lubricating coatings to reduce friction under various operating conditions remains topical. This is due to both the growing demand for such coatings in traditional and new hi-tech industries (space technology, vacuum/cryogenic technology, micromechanics, etc.), and the interest in new processes that can bring about a radical change in the patterns of friction and wear. Modern studies show that it is necessary to regulate the architecture of the coating, i.e., morphology and phase composition, at the nanoscale to improve the tribological properties of nanocomposite TMD-based coatings [9–14]. In this case, triboinduced processes in the contact layer can change significantly and new nanophases can form in the tribofilm.

Nanocomposite coatings containing the TMD nanophase and hard carbon/diamond-like carbon/graphene have always stirred the intense interest of researchers since the C-based nanophase causes both the strengthening of the coating and contributes to the manifestation of the "chameleon" effect upon changing friction conditions [5,9,15]. In high air humidity, nanocomposite coatings can be used for a wider array of applications. During triboactivated interaction of nanodiamonds with the ultrafine (2D) MoS_2 phase, onion-like inclusions can form in the tribolayer. This leads to a decrease in the friction coefficient in vacuum to very low values (~0.005) [16]. Certain composition of coatings containing nanosized MoS_2 phases and diamond-like carbon demonstrate superlubricity: reduced wear can be achieved in air due to the formation of a tribofilm containing nanoscrolls of graphene-like material [17]. Many researchers note a significant effect of such graphene-like carbon nanoscrolls on friction and wear [18,19].

The majority of works on the production and study of nanocomposite TMD-based films analyse the method of deposition by ion (magnetron) sputtering of multisector targets. The method of ion sputtering has been used in many experiments and is still being perfected [20,21]. Pulsed laser deposition (PLD) is also used to form sufficiently high-quality solid lubricating TMD-based coatings [22–25]. This method differs from the more traditional magnetron sputtering since PLD gives precise control of the growth rate of the coating (up to one monolayer of the material) [26]. It allows researchers to obtain numerous combinations of different materials under controlled vacuum conditions; it is also possible to supplement deposition by the implantation of high-energy ions for ion mixing [27,28]. Yet, the PLD method has a deficiency due to the specificity of pulsed laser ablation of TMD targets. It is difficult to achieve the deposition of pure vapour (plasma) during the ablation of MoS_2 targets prepared by pressing MoS_2 powder. The ablation of the MoS_2 target results as a rule in the formation of microparticles/microdroplets, the deposition of which contributes to the formation of the porous structure of the coating [29,30]. During pulsed laser ablation of $MoSe_2$ and WSe_2 targets, explosive boiling of the material and the formation of liquid metal droplets can occur. However, for these materials, most of the particles on the surface of the coating are round in shape and ~10–100 nm in size [31–33]. Such nanoparticles do not have any noticeable negative effect on the formation of high-quality monophase and nanocomposite coatings with solid lubricating components $MoSe_x$ and WSe_x [24,27,33]. Still, these nanoparticles may complicate the formation of multilayer coatings with layer thickness controlled at the nanoscale [34].

The aim of this work is to study the composition, structure, and tribological properties of Mo–S–C–H thin-film coatings formed by reactive PLD (RPLD) on steel substrates at room temperature. Reactive PLD makes it possible to form smoother and more uniform layers of a molybdenum sulphide MoS_x and alter the ratio of elements x = Mo/S over a wide range (1 ≤ ≤ 4) [35]. Varying the composition of the coatings has proven to be an important factor for the development of their specific applications. For instance, the MoS_3 coatings of clustered type demonstrated improved antifriction properties when tested at low temperatures (−100 °C) and low humidity. The deficiency of these coatings is their reduced wear resistance compared to MoS$_2$ coatings. As shown above, the wear resistance of such coatings can be improved through the formation of a nanocomposite material containing both the MoS$_x$ nanophase and the nanophase of hard (diamond-like) carbon. During the RPLD of a Mo–S–C–H nanocomposite coating from a target containing sectors of pure molybdenum and graphite, the deposition of a pulsed laser plume occurs in a reactive medium—hydrogen sulphide (H$_2$S). Fominski et al. [35] revealed the dependence of the composition of solid lubricate MoS$_x$ thin-film coatings on the pressure of H$_2$S. The influence of H$_2$S on the composition and properties of carbon thin films obtained by RPLD remains unexplored, which creates difficulties in choosing optimal conditions for obtaining high-quality nanocomposite coatings Mo–S–C–H.

It was found in the work that during the ablation of graphite target in H$_2$S, the deposited a-C(S,H) films effectively captured S and H atoms. The tribological properties of nanocomposite Mo–S–C–H coatings under various sliding conditions were largely determined by the properties of amorphous a-C(S,H) nanophases, which strongly degraded with an increase in S content. The Mo–S–C–H coating formed at a relatively low H$_2$S pressure (~5.5 Pa) turned out to be the most promising. Overall, these coatings outperformed MoS$_2$ and MoS$_3$ monophase coatings in their tribological characteristics when tested in dry friction conditions at room and low (-100 °C) temperatures. In humid air at room temperature, the Mo–S–C–H coatings did not show a noticeable improvement of low friction properties in comparison with the MoS$_2$ and MoS$_3$ monophase coatings previously obtained by RPLD and studied in [35].

At first glance, the RPLD technique is not quite suitable for large area deposition, especially onto shaped work pieces used in practice. Moreover, the use of H$_2$S gas requires special safety measures since it is explosive and toxic. However, several important issues should be highlighted in the research of this technique. Within the framework of a fundamental problem, if new nanomaterials with unique/interesting properties would be formed by RPLD, these results could initiate improving more conventional deposition techniques (e.g., ion-sputter deposition). From a practical point of view, the solvent of environmental safety problem arising due to H$_2$S is not very complicated. In the case of the unique results of the RPLD application, the area of the covered surface can be increased by moving/rotating the processed parts under a laser-induced plume. Also, the use of pulsed electric fields applied to a shaped work piece would improve the uniformity of RPLD treatment. Obviously, the RPLD technique can be applied (and even difficult to replace) when processing small-sized parts or the inner surface of pipes/rings.

2. Materials and Methods

Reactive pulsed laser deposition technique for MoS$_x$ thin-film coating formation was analysed in detail in [35]. The Mo target was ablated by nanosecond laser pulses with a radiation wavelength of 1064 nm. The pulse energy did not exceed 85 mJ at a pulse repetition rate of 25 Hz. A laser fluence of ~20 J/cm^2 ensured efficient evaporation of the Mo target without any noticeable formation of a droplet fraction. When obtaining Mo–S–C–H films, a graphite target was placed next to the Mo target. The ablation time of the Mo target was 8 s, and the C target—4 s. Before the ablation, the film deposition chamber was evacuated with a vacuum pump to a pressure no higher than 10^{-3} Pa. Then, hydrogen sulphide was introduced into the chamber to a predetermined pressure. The pressures were chosen taking into account the results of [35]. Fominski et al. [35] established that for obtaining films MoS$_{1.5}$, MoS$_2$ and MoS$_3$, the pressure of hydrogen sulphide had to be maintained at 5.5, 9, and 18 Pa, respectively. The prepared nanocomposite coatings were designated taking into account the H$_2$S pressure used: Mo–S–C–H_5.5, Mo–S–C–H_9 and Mo–S–C–H_18.

To reveal the effect of H$_2$S on the carbon nanophase, additional experiments on the deposition of carbon films from a graphite target at the same H$_2$S pressures were carried out. To better understand the effect of hydrogen sulphide on the deposition rate of Mo and C, Mo-C films were obtained under vacuum conditions (at a residual gas pressure of 10^{-3} Pa).

We used polished discs made of 95Cr18 stainless steel (C content of 0.95% and Cr content of 18%) and polished silicon wafers as substrates for the deposition of thin-film coatings. The total deposition time of the Mo–S–C–H coatings on the steel substrates was 40 min and on silicon substrates, 20 min. The total thickness of Mo–S–C–H thin-film coatings on steel substrates was ~300-400 nm.

The substrates were kept at room temperature during the formation of the coatings. Before the deposition of the Mo–C–S–H coatings on the steel substrates, a SiC sublayer was deposited on the surface of the substrates. The sublayer was obtained by PLD from the SiC target under vacuum conditions. It was considered that the silicon carbide sublayer can increase the adhesion of C-based coatings to the steel substrate [36,37].

To determine the atomic composition of the coatings, Rutherford back-scattering spectroscopy (RBS) and elastic recoil detection analysis (ERDA) techniques were used. The energy of helium ions in the analysing beam was 1.5 MeV, and the detector resolution was 20 keV. The RBS spectra were recorded in the configuration $\alpha = 0°$, $\beta = 20°$, $\theta = 160°$. The ERDA spectra were recorded in the configuration $\alpha = 80°$, $\beta = 80°$, $\theta = 20°$. The measured spectra data were processed using the Simnra software (Max-Planck-Institut für Plasmaphysok, Garching bei München, Germany). The surface morphologies of the Mo–S–C–H coatings were studied using scanning electron microscopy (SEM, Tescan LYRA 3, Brno, Czech Republic) before and after friction testing. The crystal structure of the coatings was examined by grazing incidence X-ray diffraction (XRD) using an angle of 5° and Cu Kα radiation in an Ultima IV (Rigaku, Tokyo, Japan) diffractometer. The chemical states of the coatings were studied by X-ray photoelectron spectroscopy (XPS). The XPS spectra were obtained by a Theta Probe Thermo Fisher Scientific spectrometer (Madison, WI 53711, USA) with a monochromatic Al Kα X-ray source (1486.7 eV) and an X-ray spot size of 400 μm. The photoelectron take-off angle was 50° with respect to the surface plane. The spectrometer energy scale was calibrated using Au4f7/2 core level lines located at E = 84.0 eV.

The structure of the coatings before and after the friction tests was studied by micro-Raman spectroscopy (MRS). Raman spectra of the samples were collected using a Horiba Jobin Yvon micro-Raman spectrometer LabRam HR800 (Horiba, Kyoto, Japan) with a 100× magnification lens. Measurements were conducted at room temperature in air. A He-Ne laser with a 632.8 nm wavelength was used to excite Raman scattering. The irradiation power density on the sample was chosen to avoid any structural changes or phase degradation in the films. The typical measurement conditions involved a laser power of ~1 mW and a laser spot with a diameter of ~30 μm.

To study the structure of the nanocomposite coatings at the nanoscale level, thin Mo–S–C–H films were deposited on NaCl substrates. The conditions for obtaining thin Mo–S–C–H films reproduced the conditions for obtaining coatings on steel discs. Thin films were studied using transmission electron microscopy (TEM, including high-resolution HRTEM) and selected area diffraction (SAED) in a JEM-2100 microscope (JEOL Ltd., Tokyo, Japan). The films deposited on NaCl crystals were first planted in water using a metal mesh and then transferred to the microscope to obtain a planar image.

The friction testing of thin-film coatings was carried out with the help of an Anton Paar TRB3 tribometer (Anton Paar GmbH, Graz, Austria) in the reciprocating motion mode, using a steel ball (100Cr6) with a diameter of 6 mm as a counterbody. The load on the ball was 1 N, and the Hertzian contact stress was ~660 MPa. The average speed of the ball over a substrate with a Mo–S–C–H coating was 1cm/s. The length of the wear track was 5 mm. A detailed description of the technique and setup for friction testing can be found in [35]. Three conditions were selected for testing, differing in ambient humidity and substrate temperature. The first tests were carried out at 22 °C in air at a related humidity (RH) ~58% (wet friction conditions). The second tests were carried out at a reduced atmospheric humidity (RH ~8%, dry friction condition), which was achieved by pumping argon through the testing chamber. The sample temperature was 22 °C. The third series of tests was carried out at low humidity (RH ~8%) and the sample was cooled to −100 °C (low temperature/dry friction conditions). The wear tracks were studied by MRS, SEM, optical microscopy, and optical profilometry.

3. Results

3.1. Composition of Mo–S–C–H Films Obtained by RPLD

Figure 1 shows the experimental and simulated RBS and ERDA spectra for Mo–C and Mo–S–C–H films obtained by pulsed laser codeposition of molybdenum and carbon under vacuum conditions and in H_2S gas with different pressures. For the film obtained under vacuum, mathematical processing of the spectra showed that the composition of this film was described by the formula $C_{0.81}Mo_{0.16}H_{0.03}$. There was almost no sulphur in the bulk of o–C–H film. A small amount of sulphur was found at the boundary of the Mo–C film with the Si substrate and on the surface of the Mo–C–H film. This was possibly due to the fact that the walls of the deposition chamber were not subjected to any special treatment prior to the formation of the Mo–C–H films. The walls of the chamber were covered with a thin S-containing film, formed during previous experiments on RPLD of MoS_x films. Sulphur atoms desorbing from the walls of the chamber could have deposited on the surface of the Si plate during vacuum pumping of the chamber and on the surface of the Mo–C–H film—after its deposition and storage (for some time) in the chamber. The presence of a small amount of hydrogen in the Mo–C–H films was possibly due to the interaction of the growing film with residual water vapour in the deposition chamber.

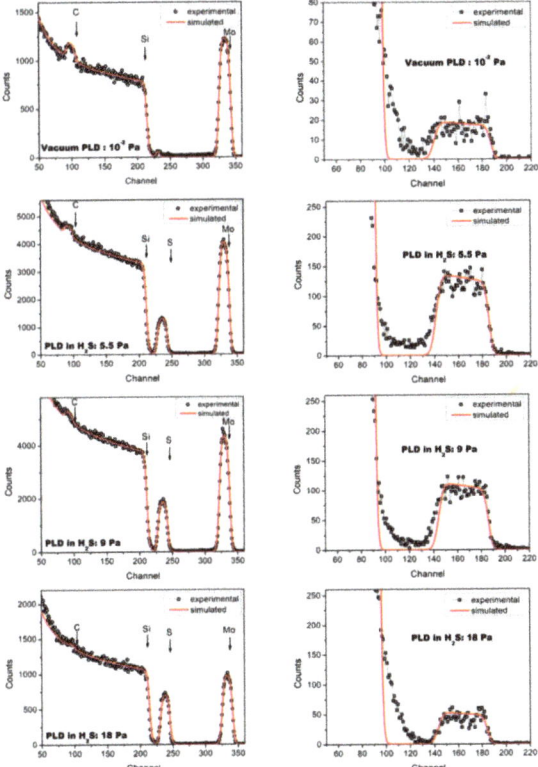

Figure 1. Experimental and simulated RBS (**left**) and ESDA (**right**) spectra of the films prepared on Si substrates by pulsed laser co-deposition of carbon and molybdenum under vacuum conditions (residual gas pressure was ~10^{-3} Pa) and in H_2S gas with different pressures. The RBS spectrum of the film deposited in vacuum contains a peak at the channel number 230. This peak is due to scattering of ions by sulfur atoms that have been adsorbed on the surface of the Si substrate before the film deposition.

According to the RBS data, the thickness of the Mo–C–H film obtained during 20 min of deposition was 8.2×10^{17} atom/cm^2. In reactive H$_2$S gas, S atoms penetrated the deposited Mo–C–S–H film. Despite this fact, the overall deposition rate of atoms decreased. For hydrogen sulphide pressures of 5.5, 9, and 18 Pa, the composition of the films was described by the formulas $C_{0.49}Mo_{0.13}S_{0.28}H_{0.1}$, $C_{0.405}Mo_{0.13}S_{0.37}H_{0.095}$, and $C_{0.24}Mo_{0.115}S_{0.55}H_{0.095}$ respectively. The thickness of these films was 6.2×10^{17}, 5.8×10^{17}, and 5.2×10^{17} atom/cm^2. For ablation of graphite and molybdenum targets, 1.5×10^4 laser pulses were used, which were divided into series of 100 pulses for the C target and 200 pulses for the Mo target. The ablation of the C target resulted in the deposition of ~2×10^{15} atom/cm^2 in hydrogen sulphide. This was sufficient for the formation of approximately one monolayer of amorphous carbon, given the atomic density in amorphous carbon of ~$(1 \div 1.7) \times 10^{23}$ atom/cm^3. The same estimates for Mo showed that a $0.6 \div 1$ MoS$_x$ monolayer could be formed after 200-pulse ablation of the Mo target in hydrogen sulphide.

Figure 2 shows a change in the rate of deposition of various elements in the Mo–C–H and Mo–S–C–H films following an increase in the hydrogen sulphide pressure. As it is shown, H$_2$S gas did not only ensure the saturation of the films with S atoms, but also had a significant effect on the rate of carbon deposition. At a pressure of 18 Pa, the rate of carbon deposition dropped almost fivefold compared to the rate of deposition in a vacuum. Under the same conditions, the rate of deposition of Mo atoms decreased only twofold. This could be explained by the difference of collisions when light (C) and heavy (Mo) atoms moved through H$_2$S gas. The relatively heavy Mo atoms changed their trajectory only slightly and slowly lost their kinetic energy in collisions with the relatively light H$_2$S molecules. When colliding with the H$_2$S molecules, the lighter C atoms are scattered at large angles and leave the area where the deposition on the substrate took place. The possibility of reactive collisions of carbon ions with H$_2$S molecules cannot be ruled out. These collisions may have resulted in the formation of volatile hydrocarbon molecules. This is a possible explanation as to why an increase in the H$_2$S pressure did not result in a discernible change in the rate of the saturation of the films with hydrogen.

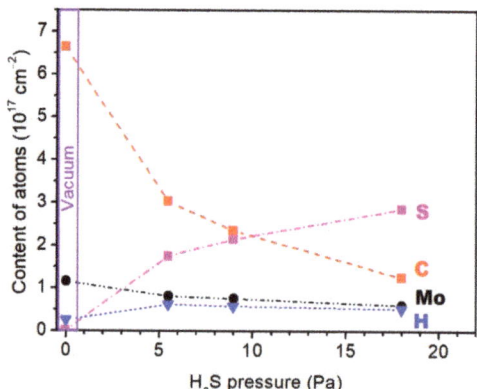

Figure 2. Influence of hydrogen sulfide pressure on the composition of Mo-S-C-H films which were obtained by pulsed laser codeposition of carbon and molybdenum in 20 min.

The analysis of the RBS and ERDA data for a-C(S,H) films formed during pulsed laser ablation of graphite in H$_2$S confirmed the assumption about the effective interaction of the ablated flux of carbon atoms and H$_2$S molecules (see Figure S1, Supplementary Materials). An increase in the pressure of hydrogen sulphide resulted in a noticeable increase in the concentration of sulphur in the a-C (S, H) films. The compositions of the films at hydrogen sulphide pressures of 5.5, 9, and 18 Pa were described by the formulas $C_{0.71}S_{0.15}H_{0.14}$, $C_{0.61}S_{0.26}H_{0.13}$, and $C_{0.42}S_{0.4}H_{0.18}$. The thicknesses of the films were 2.2×10^{18}, 1.7×10^{18} and 1.0×10^{18} atom/cm^2 respectively. During the PLD of carbon films in vacuum (residual gas pressure ~10^{-3} Pa), their thickness was 2.88×10^{18} atom/cm^2, and the H atoms concentration did not exceed 1 at.%.

3.2. Morphology, Structure, and Chemical State of Mo–S–C–H Films Obtained by RPLD

The use of RPLD for obtaining Mo–S–C–H thin films made it possible to produce sufficiently smooth and dense coatings, the morphology of which, according to the results of SEM studies (Figure 3), did not depend on the H$_2$S pressure. There were individual round-shaped particles of a submicron size on the surface of the Mo–S–C–H coatings. These particles could have formed because of the deposition of droplets formed during the ablation of the Mo and graphite targets. Such particles were found in a-C(S,H) films produced by RPLD from a graphite target in H$_2$S gas (Figure S2, Supplementary Materials).

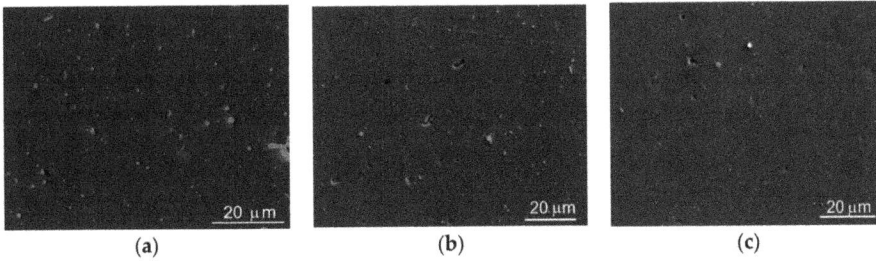

Figure 3. SEM images of the surface of Mo–S–C–H thin-film coatings obtained by reactive pulsed laser deposition (PLD) on steel substrates at the following H$_2$S pressures: (**a**) 5.5; (**b**) 9; (**c**) 18 Pa.

The results of the XRD of Mo–S–C–H thin-film coatings (Figure 4) show the deposition of Mo particles upon ablation of the Mo target. The X-ray diffraction pattern of the Mo–S–C–H_5.5 coating had a weak intensity peak, which corresponded to the (110) reflection for a body-centred cubic Mo lattice. The peak was practically invisible in the X-ray diffraction patterns of the Mo–S–C–H_9 and Mo–S–C–H_18 coatings. This could be attributed to the fact that with an increase in the pressure of H$_2$S gas, the surface of the Mo target interacted with the reactive gas. Following this interaction, molybdenum sulphides could form on the target surface; this changed to a certain extent the mechanism of pulsed laser ablation of the Mo target.

XRD studies showed that at all selected pressures of H$_2$S, the Mo–S–C–H thin-film coatings had an amorphous structure with a broadened diffraction peak in the angle range from 35° to 50°. With greater hydrogen sulphide pressure, the intensity of this peak noticeably weakened. This indicated an increased disordering of the structure following a rise in the sulphur concentration. This type of XRD pattern has been extensively described in the literature; in most experiments, the Mo–S–C coatings have been obtained by ion sputtering/codeposition from MoS$_2$ and graphite targets (for example, [15,38–40]). For TMD coatings having an amorphous structure, this broad peak is usually explained by the formation of nanosize inclusions with a hexagonal lattice of the 2H-MoS$_2$ type [41]. In the cases when there was no peak at angles 2θ~13°, but there was a peak in the 2θ range from 35° to 50°, the turbostratic stacking of (10L) planes into Type I texture was supposed. With this texture, the basal planes (002) are oriented perpendicular to the surface of the substrate [42]. The absence of a peak at 2θ~13° shows that the reactive PLD of Mo–S–C–H_5.5 films in H$_2$S may not have caused the formation of a self-assembled multilayer structure MoS$_x$/a-C (doped with Mo/S/H) with a periodicity in the nanometer scale, as it was the case during the magnetron sputtering of graphite and MoS$_2$ targets in Ar/N$_2$ gases [14,39]. In XRD patterns for the Mo–S–C–H_18 coatings, a weak-intensity and a very broad band appears at 2θ~15°. This shows that a MoS$_x$ nanophase with a high sulphur concentration (

≥ 3) formed in the structure of these coatings. Such coatings are characterized by an XRD pattern with two strongly broadened bands at 2θ~15° and 2θ~40° [35,43].

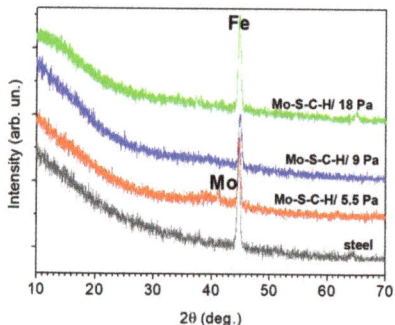

Figure 4. In plane grazing incidence X-ray diffraction patterns of Mo-S-C-H thin-film coatings obtained on steel substrates by the reactive PLD at various pressures of H_2S gas. For comparison, X-ray diffraction pattern for the bare steel substrate is shown.

HRTEM studies of the Mo–S–C–H thin films confirmed their amorphous structure. Only in the Mo–S–C–H_5.5 films obtained at the lowest H_2S, pressure, MoS_2 nanocrystallites with laminar packing of atomic planes were found in some local regions. The size of these crystallites did not exceed 10 nm, and they were surrounded by an amorphous matrix. The concentration of MoS_2 nanocrystallites in the Mo–S–C–H_5.5 film was not high, and their structure probably had a turbostratic character with a high degree of disordered local atomic packing. This was confirmed by the SAED pattern, consisting only of diffusely broadened rings (Figure 5), as well as by the results of the Raman studies of Mo–S–C–H films.

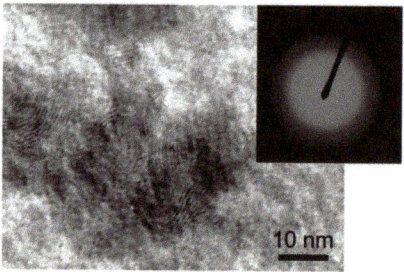

Figure 5. High-resolution TEM image of the Mo-S-C-H thin film obtained by reactive PLD at an H_2S pressure of 5.5 Pa.

The Raman spectrum for the Mo–S–C–H_9 coating in the frequency range of 100–600 cm^{-1} was in many respects similar to the spectrum of the Mo–S–C–H_5.5 coating (Figure 6). Broad peaks at 350 and 402 cm^{-1} indicated the formation of the MoS_2 nanophase with a disordered atomic packing. The appearance in the spectrum of the Mo–S–C–H_9 coating of weak-intensity and broad peaks at ~200 and ~500 cm^{-1} suggested that, along with the MoS_x nanophase, Mo_3–S clusters could form. When such clusters are combined into a polymer-like network, MoS_x compounds are formed, in which ≥ 3 (Mo_3S_{12}/Mo_3S_{13}-type). The composition of Mo_3-S clusters includes three Mo atoms connected in the Mo_3–S triangle through monomers and/or dimers of S atoms (S^{2-}/S_2^{2-}). With a sufficiently ordered packing of atoms in such clusters, narrow peaks are observed in the indicated frequency range; they correspond to various sulphur ligands [35,43–45]. In the Raman spectrum for the Mo–S–C–H_18 coating, the Mo_3–S clusters corresponded to peaks at the following vibration modes: ν(Mo-Mo) at

~210 cm^{-1}, ν(Mo-S)$_{coupled}$ at ~330 cm^{-1}, ν(Mo-S$_{apical}$) at ~450 cm^{-1}, ν(S-S)$_{terminal}$ at ~520 cm^{-1}, and ν(S-S)$_{bridging}$ at 550 cm^{-1}. In addition to these peaks, the spectrum of this coating exhibited peaks at 360, 380, and 401 cm^{-1}, which could be due to atomic vibrations in the defective MoS$_2$ nanophase.

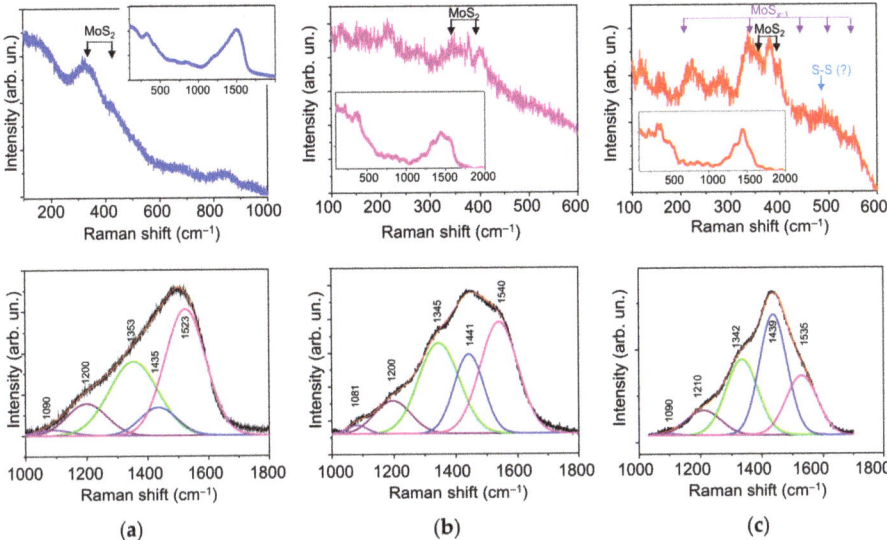

Figure 6. Raman spectra for the Mo–S–C–H thin-film coatings obtained by reactive PLD in the H$_2$S gas at pressures of (**a**) 5.5, (**b**) 9, and (**c**) 18 Pa. The regions of Raman shifts corresponding to resonance light scattering by MoS$_x$- and a-C(S,H)-based nanophases are shown at the top and bottom, respectively. The model of spectrum decomposition into the indicated peaks is discussed in the text. Inserts show the Raman spectra in the region from 100 to 2000 cm^{-1} that allows a correct comparison of the peak intensities for different nanophases.

When choosing a model for the decomposition of the Raman spectra for C-based nanophase in Mo–S–C–H films, we took into account the changes in the Raman spectra for a-C(S,H) films with increasing hydrogen sulphide pressure. The Raman spectra for a-C(S,H) films are shown in Figure S3 (Supplementary Materials). Figure S3 shows that, as the H$_2$S pressure grows, i.e., with an increase in the concentration of sulphur and hydrogen in a-C(S,H) films, the contribution to the Raman spectra of the two peaks at frequencies of ~1220 cm^{-1} and ~1440 cm^{-1} rises as well. The intensity of these peaks in films having the highest concentration of S exceeded the intensity of the D (at ~1340 cm^{-1}) and G (at ~1530 cm^{-1}) peaks characteristic of pure a-C films. In this case, with an increase in the sulphur concentration, the I_D/I_G ratio grew, which indicated an increase in the disordering (defectiveness) of atomic packing in graphite clusters.

Comparative analysis of the Raman spectra for the C-based nanophase in Mo–S–C–H and a-C (S,H) films showed (Figure 6) that the addition of Mo atoms to the depositing flux did not cause significant changes in the Raman spectra for the C-based nanophase. This was confirmed by the fact that the spectra of Mo–S–C–H coatings and a-C (S, H) films in the frequency range of 1000–1800 cm^{-1} were similar in many respects. An increase in the H$_2$S pressure during RPLD of the Mo–S–C–H coatings led to an increase in the contribution to the spectrum of lines at ~1220 and ~1440 cm^{-1}. Our analysis of the published data on Raman studies of Mo–S–C films formed by codeposition under magnetron sputtering (including the reactive one in an Ar/CH$_4$ mixture) showed that the spectra of these films did not have properties characteristic of the spectra of the Mo–S–C–H films produced by RPLD. In the spectra of Mo–S–C films for the C-based nanophase, the positions of the D and G peaks, as well as the ratio of their intensity, tended to change, but new peaks did not appear [14,15,38,39,46,47]. Changes in the Raman spectra were caused by the influence of the MoS$_2$ nanophase on the sp^2/sp^3 ratio (graphitization) and the level of mechanical stresses in the a-C(H) nanophase. More significant changes in the Raman spectrum of the carbon component in the a-C(S,H) films were found when using chemical vapour deposition in H$_2$S, as well as during magnetron sputtering and pulsed laser ablation of composite targets made of a mixture of powders (MoS$_2$, sulphur, graphite) [48–50]. Unfortunately, these works do not contain a sufficiently detailed analysis of the Raman spectra. Therefore, to investigate the C-based nanophase in the a-C(S,H) and Mo–S–C–H films obtained by RPLD, we used the approach proposed by Takeuchi et al. [51] for organic carbon sulphur materials.

Takeuchi et al. [51] have identified a class of organic carbon sulphur materials, the Raman spectrum of which has peaks at ~1250, 1350, 1440, and 1590 cm^{-1}. The position of each peak has a tolerance of ±50 cm^{-1}. The structure of such materials depends on the ratio of the intensities of these peaks. If the peak at 1400 cm^{-1} is the most intense, there is a large amount of the sp^3 component of the G-band, and the majority of the carbon component form an undeveloped graphene (C-C) skeleton. Other peaks correspond to the sp^3 component of the D band (~1250 cm^{-1}), the sp^2 component of the D band (~1350 cm^{-1}), and the sp^2 component of the G band (~1590 cm^{-1}). The S-S bond stretching vibration should peak at ~480 cm^{-1}. This peak is present in the Raman spectrum of the Mo–S–C–H_18 coating (Figure 6c). The low intensity of this peak shows that RPLD was not effective for the formation of sulphur clusters. The process of dispersing sulphur in the carbon nanophase turned out to be more productive and caused a change in the local packing of carbon atoms and in the structure of the carbon skeleton. With a rise in the H$_2$S pressure, i.e., with an increase in the concentration of sulphur, the contribution from the sp^3 states caused by the introduction of sulphur in the structure of the carbon skeleton grew. At the same time, an increase in the intensity of the I_D peak at 1350 cm^{-1} (compared to I_G at 1540 cm^{-1}) was indicative of a growing number of defects in the atomic packing of pure graphite clusters. Low intensity peak at 1080 cm^{-1} should be introduced for better fitting of the Raman spectrum.

The RBS technique made it possible to determine the concentration of sulphur in the nanocomposite Mo–S–C–H coatings. This technique nevertheless does not distinguish between the sulphur content in MoS$_x$ and a-C(S,H) nanophases. Therefore, XPS measurements were carried out. Figure 7 shows the XPS spectra, revealing chemical bonds in the surface layer of the Mo-S-C-H coatings formed by RPLD at various H$_2$S pressures. Decomposition of the Mo 3d spectrum showed that the chemical state of Mo atoms did not undergo significant changes with the increasing pressure of H$_2$S. The Mo 3d spectra contained Mo3d$_{5/2}$-Mo3d$_{/2}$ doublets, corresponding to Mo^{2+}, Mo^{4+}, Mo^{5+}, and Mo^{6+}. The electron binding energies for the Mo3d$_{5/2}$ peaks at such valences of molybdenum were 228.6, 229.2, 230.3, and 232.6 eV, respectively. The dominance of the Mo^{4+} doublet indicated the effective formation of MoS$_2$ and/or MoS$_x$ compounds (with packing Mo$_3$-C), in which ≥ 3 [35,43,52,53]. The Mo^{5+} doublet may have been a result of the binding in the MoS$_3$ compound with a linear packing of atoms into Mo-S$_3$ clusters [35,53,54]. The presence of a Mo^{6+} doublet with a low relative peak intensity indicated weak surface oxidation and the formation of Mo-O compounds [35,55]. With increasing H$_2$S pressure, the intensity of the Mo^{6+} doublet weakened even more due to the chemical properties of hydrogen sulphide, which is a strong reducing agent. An increase in the H$_2$S pressure caused a decrease in the contribution of the Mo^{2+} doublet, which corresponded to the Mo–C (Mo$_2$C) bonds [55,56]. This was due to both an increase in the total sulphur concentration in the Mo–S–C–H films with a rise in the H$_2$S pressure and, probably, due to an increased chemical activity of radicals formed upon activation of H$_2$S by a laser plasma, compared with carbon atoms in a laser plasma from a graphite target. Considering the small contribution of the Mo-C states, the effect of the carbide nanophase on the properties of M–S–C–H coatings was not considered in this work.

In the decomposition of the C 1s spectra for Mo–S–C–H coatings, it was assumed that C atoms could form chemical bonds with each other (C=C binding energy 284.6 eV and C-C binding energy 285.5 eV), with S atoms (C-S energy bonds 286.5 eV), and Mo atoms (C-Mo binding energy 283.6 ÷ 284.2 eV) [36,37,56,57]. The peak with the highest binding energy (~289 eV) is usually attributed to C-O bonds [57]. The analysis of the C 1s spectra showed that, at all H_2S pressures, the peak corresponding to the sp^2 bonds of C atoms dominated. As the H_2S pressure grew, the contribution from the peak corresponding to C-S bonds increased too. In this case, the contribution of the peak at 285.5 eV, corresponding to sp^3 bonds of carbon atoms, slightly decreased. A thin film of organic contaminants containing CH_x molecules may form on the surface of Mo–S–C–H coatings after being blown out of the PLD chamber. The presence of this film could have a definite effect on the results of studying the chemical state of carbon in Mo–S–C–H coatings, first of all, it could have increase the intensity of the XPS peak binding energy of ~284.5 eV.

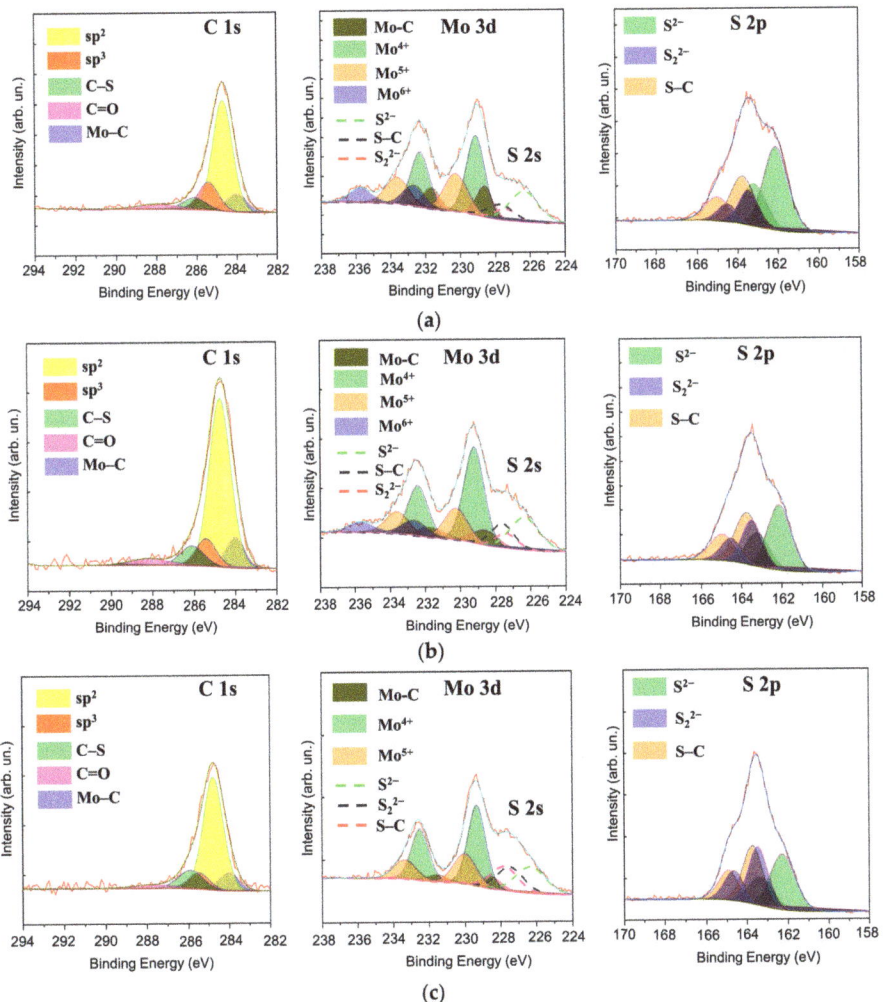

Figure 7. XPS spectra of C 1s, Mo 3d, and S 2p which were measured on the surface of the Mo–S–C–H thin-film coatings obtained by reactive PLD at H_2S pressures of (**a**) 5.5, (**b**) 9, and (**c**) 18 Pa. For Mo 3d spectra, the S 2s spectra of different S species are indicated.

The deconvolution of the S 2p spectra for Mo–S–C–H coatings allowed us to assume that S atoms can form chemical bonds with Mo atoms, which are characteristic of MoS_x compounds with different values of x, as well as of chemical bonds with C atoms. The most common approach to analysing the chemical state of S atoms in molybdenum sulphides is the separation of two doublets S $2p_{3/2}$-S $2p_{1/2}$ having "low" and "high" binding energies. A doublet with a low binding energy (the binding energy of the S $2p_{3/2}$ peak does not exceed 162.3 eV, and the S $2p_{1/2}$—163.5 eV) usually corresponds to the S^{2-} species characteristic of the MoS$_x$ compound [35,58]. A doublet with a high binding energy (the binding energy of the S $2p_{3/2}$ peak is ~162.8 ÷ 163.4.4 eV) corresponds to the S_2^{2-}-species, which are characteristic of MoS$_x$ compounds, where clusters of Mo-S$_3$ and/or Mo$_3$-S are formed due to a high concentration of sulphur (> 2), [35,43,52,53]. S atoms in the chemical bond with C atoms (-S-C-S-) correspond to the doublet S S $2p_{3/2}$-S $2p_{1/2}$, in which the binding energy of the S $2p_{3/2}$ peak is 163.6 eV, and the S $2p_{1/2}$ peak equals 165.2 eV [57]. The choice of a model for describing the chemical state of S in a carbon matrix seems to be quite problematic since the binding energies of S atoms can strongly depend on the configuration of the nearest atoms. Thus, for a certain configuration of chemical bonds, the spin-orbit splitting of the S 2p state does not occur. Our analysis of the published results of XPS studies of sulphur-doped carbon materials has shown that even in the absence of spin-orbit splitting, a band at ~163.5 eV can dominate in the XPS spectra of S 2p (for example, a configuration of the C-$S_{1\div 2}$-C type), together with which a band at 165.0 eV (for instance, a configuration of the -C=S-type) appears [59,60].

The application of the chosen model of the decomposition of the S 2p spectra showed that an increase in the H$_2$S pressure resulted in a decrease in the concentration of S^{2-} states, and the contribution to the S_2^{2-} states increased. The contribution of the species corresponding to the C-S bonds increased as well. The calculation of the ratio x = S/Mo, taking into account S species ($S^{2-} + S_2^{2-}$) associated with Mo, and Mo species (Mo^{4+} + Mo^{5+}) associated with S, showed that it was approximately 1.8, 2.5, and 4.0 for the Mo–S–C–H coatings obtained at pressures of 5.5, 9, and 18 Pa respectively. The composition of the C component in these coatings was described by the approximate formulas $C_{0.78}S_{0.08}H_{0.14}$, $C_{0.73}S_{0.11}H_{0.16}$, and $C_{0.62}S_{0.18}H_{0.2}$. We assumed that H atoms are concentrated mainly in the a-C(S,H) nanophase. The calculated composition of the a-C(S,H) nanophase for Mo–S–C–H composite films differed from the composition of a-C(S,H) coatings obtained by RPLD at similar H$_2$S pressures. The concentration of S atoms in the nanophase was lower than in the monophase thin-film coating. This could be attributed to the fact that S atoms deposited on the surface of the growing layer from the gas phase during ablation of a graphite target can be captured during the formation of the MoS nanophase in the course of the subsequent ablation of the Mo target. Our calculations have shown that, as a result of the codeposition of Mo and C in reactive gas, the ratio = S/Mo exceeds the ratio obtained earlier for MoS$_x$ films produced by RPLD of molybdenum in H$_2$S at the same gas pressures.

3.3. Tribological Properties of Mo–S–C–H Films Obtained by RPLD

Figure 8 shows the results of measuring the average coefficient of friction as a function of the sliding cycle number of the steel counterbody over the Mo–C–S–H coatings in humid air. The Mo–C–S–H_5.5 coating obtained at the lowest H$_2$S pressure turned out to be the most wear-resistant. The endurance of this coating exceeded 10^3 cycles. For other coatings, the wear resistance did not exceed 200 cycles. The analysis of the wear tracks and the wear scar showed (Figure 9) that the low wear resistance of the Mo–C–S–H_9 and Mo–C–S–H_18 coatings is mainly due to the weak adhesion of the coatings to the substrate. The sliding of the counterbody caused the cracking of these coatings accompanied by the separation of microplates. Microplates accumulated around the track and adhered to the counterbody as well (Figure 9b,c).

The minimum value of the friction coefficient for the Mo–S–C–H_5.5 coating was 0.08; it was achieved after 10 sliding cycles. After 100 cycles, the coefficient of friction rapidly increased to 0.22 (±0.05), and this value remained constant throughout the entire testing period. A profilometric study of the wear crater showed that the wear rate of Mo–S–C–H_5.5, when tested in a humid atmosphere, was ~9 × 10^{-7} mm^3/N m.

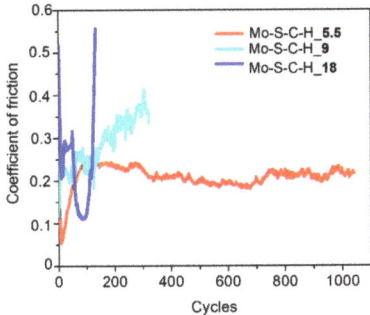

Figure 8. Characteristic evolution of the friction coefficient as a function of the cycle number for the Mo–S–C–H thin-film coatings obtained by reactive PLD at the pressures of H_2S gas of 5.5, 9 and 18 Pa. Pin-on-disk tribometer testing was conducted in wet friction conditions (RH ~58%) at 22 °C.

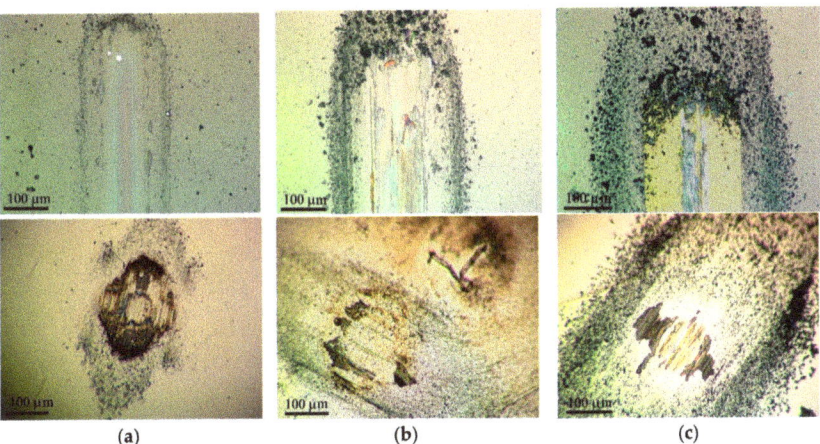

Figure 9. Optical images of wear tracks and wear scars formed on the steel substrates and steel balls for the Mo–S–C–H thin-film coatings obtained by reactive PLD at the different pressures of H_2S gas: (**a**) 5.5, (**b**) 9, and (**c**) 18 Pa. Pin-on-disk tribometer testing was conducted in wet friction conditions (RH ~58%) at 22 °C. The test durations are indicated in Figure 8.

The Mo–S–C–H_5.5 coating showed better tribological properties compared to Mo–S–C–H_9 and Mo–S–C–H_18 coatings when tested in a dry atmosphere at room and low temperatures. Figure 10 demonstrates that, in dry friction conditions at 22 °C, the average coefficient of friction after the running-in period gradually increased from 0.03 (±0.05) to 0.05 (±0.05) with an increase in the test duration from 10 to 4×10^3 cycles. The coating showed good adhesion to the substrate (Figure 11a). The wear rate of this coating was ~3×10^{-7} mm^3/N m. The Mo–C–S–H_9 coating also had fairly good antifriction properties and durability despite its poor adhesion to the substrate. For this coating, the average coefficient of friction did not exceed 0.08 during the entire testing period (i.e., 4×10^3 cycles). Weak adhesion of the coating to the substrate manifested itself in the formation of coating delamination areas in the track area. Coating separation caused the formation of micro-scales, which accumulated at the edges of the track (Figure 11b). The Mo–S–C–H_18 coating had poor tribological properties: it began to deteriorate immediately after the testing had started because of its weak adhesion to the substrate. Microscopic analysis showed that the loose fragments of the coating effectively adhered to the counterbody (Figure 11c).

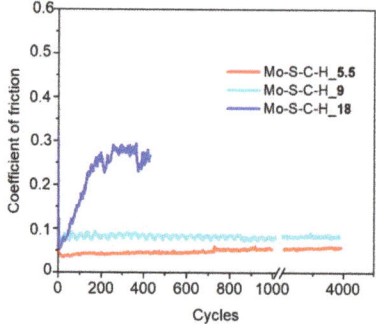

Figure 10. Characteristic evolution of the friction coefficient as a function of the cycle number for the Mo-S-C-H thin-film coatings obtained by reactive PLD at the pressures of H_2S gas of 5.5, 9, and 18 Pa. Pin-on-disk tribometer testing was conducted in dry friction conditions (air + Ar mixture, RH ~8%) at 22 °C.

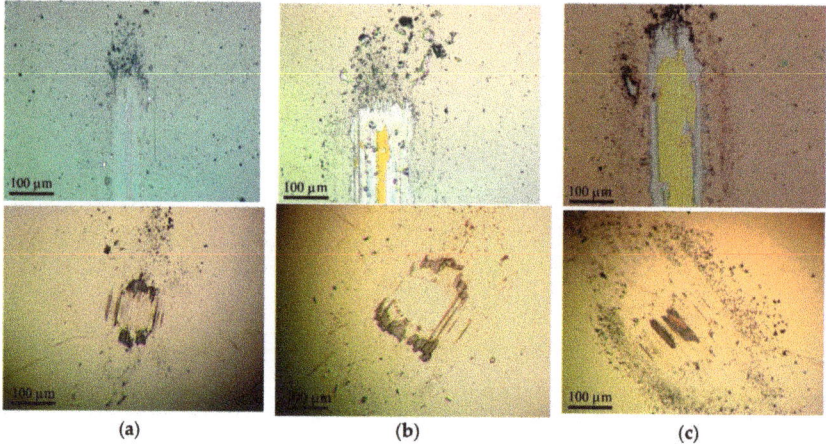

Figure 11. Optical images of wear tracks and wear scars formed on the steel substrates and steel balls for the Mo-S-C-H thin-film coatings obtained by reactive PLD at the different pressures of H_2S gas: (**a**) 5.5, (**b**) 9, and (**c**) 18 Pa. Pin-on-disk tribometer testing was conducted in dry friction conditions (air + Ar mixture, RH ~8%) at 22 °C. The test durations are indicated in Figure 10.

During testing in dry friction conditions at −100 °C, the average coefficient of friction for the Mo–S–C–H_5.5 coating did not exceed 0.08 (±0.1) over 10^3 sliding cycles (Figure 12). A shallow track formed on the coating surface, and the wear rate of the coating did not exceed 1.6×10^{-7} mm^3/N m (Figure 13a). The sliding of the counterbody over the Mo–C–S–H_9 and Mo-C-S-H_18 coatings was accompanied by noticeable changes in the average coefficient of friction in the range from 0.05 to 0.25 (Figure 12). In this case, the coatings retained their continuity, but they could deform and crack (Figure 13b,c). Sliding of the ball on the Mo–C–S–H_9 and Mo–C–S–H_18 coatings caused more intensive wear of the counterbody than sliding on the Mo–S–C–H_5.5 coating.

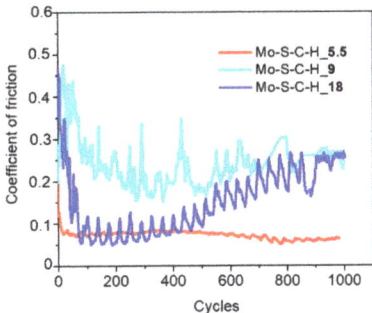

Figure 12. Characteristic evolution of the friction coefficient as a function of the cycle number for the Mo-S-C-H thin-film coatings obtained by reactive PLD at the pressures of H_2S gas of 5.5, 9, and 18 Pa. Pin-on-disk tribometer testing was conducted in dry friction conditions at −100 °C.

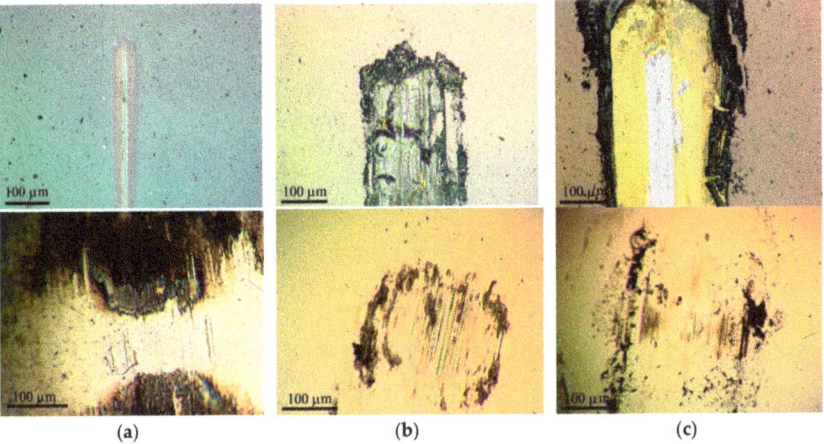

Figure 13. Optical images of wear tracks and wear scars formed on the steel substrates and steel balls for the Mo-S-C-H thin-film coatings obtained by reactive PLD at the different pressures of H_2S gas: (a) 5.5, (b) 9, and (c) 18 Pa. Pin-on-disk tribometer testing was conducted in dry friction conditions at −100 °C. The test durations are indicated in Figure 12.

4. Discussion

Our comparison of the tribological properties of the Mo–S–C–H coatings obtained at different pressures of hydrogen sulphide showed that an increase in pressure negatively affected both the average coefficient of friction and the wear resistance of the coatings under various tribological testing conditions. To explain this result, it is necessary to assess the possible effect of the nanophases of these coatings on the tribological properties. A change in the conditions of RPLD caused significant changes in both the MoS_x and C-based nanophase. An increase in the H_2S pressure caused a rise in the x = S/Mo ratio from ~1.8 to ~4.0. Fominski et al. [35] found that increasing x to 4 can significantly worsen the tribological properties of MoS_x coatings. For this reason, the generally unsatisfactory properties of the Mo–S–C–H_18 coatings could be caused by inclusions of the MoS_4 nanophase. At $2 \leq \leq 3$, the properties of MoS_x coatings depend on the conditions of tribological tests. In a humid atmosphere, after 400 cycles of sliding, MoS_2 and MoS_3 monophase coatings had the friction coefficient of ~0.2 and ~0.12, respectively. Additional studies of MoS_x thin film coatings formed by the RPLD at H_2S gas pressure of 5.5 Pa revealed unsatisfactory tribological properties. Under various tribotest conditions, these coatings began to break down almost immediately after the start of the counterbody sliding (Supplementary Materials, Figure S4). The Mo–S–C–H_9 coating, containing MoS$_{\sim2.5}$ nanophase, did not show good antifriction properties in a humid atmosphere, which indicates the possibility of a negative effect of another nanophase, which is part of this coating—the a-C(S,H) component formed at H$_2$S pressure of 9 Pa. The antifriction properties of the Mo–S–C–H_5.5 coating containing the MoS$_{1.8}$ nanophase correlated rather well with the properties of a single-layer MoS$_2$ coating.

For comparing the tribological properties of the Mo–S–C–H coatings produced by RPLD with those of Mo–S–C coatings obtained by more traditional deposition methods (mainly, magnetron sputtering), it is necessary to take into account a number of important factors, such as the composition of the coatings, air humidity and the load applied on the counterbody [15,61]. In the friction tests of the Mo–S–C–H_9 and Mo–S–C–H_18 coatings in a humid atmosphere with an increased load on the ball (5 N), the average friction coefficient was ~0.066, and the endurance did not change with an increase in the load and was 100 cycles. The friction coefficient for the MoS$_2$ and Mo–S–C coatings at an increased carbon concentration (≥30 at.%) in a humid atmosphere (RH ≥ 50%) varied from ~0.1 to 0.3, and the wear rate varied from ~5 × 10^{-7} mm^3/N m to 20 × 10^{-7} mm^3/N m. The best performance is achieved only by alloying these coatings with metals (Ti, Pb, and others) [62–64]. Thus, it can be assumed that the Mo–S–C–H_5.5 nanocomposite coating, when tested in a humid atmosphere (RH ≥ 50%), is inferior in its tribological properties only to the best samples of the MoS$_2$ and Mo–S–C coatings doped with metals.

During friction testing in dry friction conditions at room temperature, the coefficient of friction for the MoS$_2$ and MoS$_3$ coatings was 0.08 and 0.1 respectively after 400 cycles of sliding [35]. To compare, the Mo–S–C–H_9 coating containing the MoS$_{\sim2.5}$ nanophase also provided a fairly stable and low coefficient of friction (~0.08). However, the Mo–S–C–H_5.5 coating containing the MoS$_{1.8}$ nanophase provided a more effective decrease in the friction coefficient (down to 0.03), which was probably due to the positive effect of the a-C(S,H) component formed at an H$_2$S pressure of 5.5 Pa. The friction coefficient value of ~0.05 has been noted in many studies of pure TMD and nanocomposite TMD+C coatings when tested under dry friction conditions (~5% ≤ RH ≤ ~30%) at moderate loads on the counterbody (for example, [15,47,61,65–67]). The wear rate can be reduced to ~2 × 10^{-7} mm^3/N m, which is achieved by alloying with metals. Fundamentally lower values of the friction coefficient (~0.005 ÷ 0.01) during sliding in dry friction conditions are achieved by creating nanoscale layers of MoS$_2$ and a-C(H) [17,68]. The friction coefficient for the Mo–S–C–H_5.5 coating equal to 0.03 at a humidity of RH ~8% turned out to be slightly lower than the values for the MoS$_x$ and Mo–S–C coatings formed by magnetron sputtering. Yet this coating was clearly inferior to the MoS$_2$/a-C() nanolayer coatings exhibiting superlubricity properties. This could be caused by both the suboptimal composition of the coating nanocomponents and by the fact that the a-C(S,H) nanophase formed during RPLD did not provide ultralow friction in combination with the MoS$_{1.8}$ nanophase.

For the MoS$_2$ and MoS$_3$ monophase coatings, the average coefficient of friction under extreme test conditions (−100 °C) was 0.18 and 0.08 respectively after 400 cycles of sliding [35]. The Mo–S–C–H_9 coating containing the MoS$_{\sim2.5}$ nanophase provided sliding with a higher coefficient of friction (0.2–0.3). A lower and stable coefficient of friction (~0.08) was determined for the Mo–S–C–H_5.5 coating containing the MoS$_{1.8}$ nanophase. At this stage, it is difficult to do a comparative analysis of the tribological properties of Mo–S–C–H coatings obtained by RPLD and coatings of the same type obtained by other techniques since there is no information on tribotests of various coatings at −100 °C in the literature. A comparison of the tribological properties of the monophase MoS$_x$ and nanocomposite Mo–S–C–H coatings indicates a significant effect of the a-C(S,H) phase on the tribological properties of the Mo–S–C–H nanocomposite coatings under friction at low temperatures.

A comparative analysis of the tribological properties of the Mo–S–C–H nanocomposite coatings obtained by RPLD with the properties of MoS$_x$ coatings (also obtained by RPLD) and TMD+C coatings prepared by more traditional techniques showed the importance of collecting additional information on the tribological properties of a-C (S,H) coatings formed by RPLD. A review of the literature shows that sulphur can significantly change the tribological and mechanical properties of a-C(S,H) coatings, and the effect of the introduction of sulphur depends on its concentration and the concentration of hydrogen to a considerable degree [69–71]. The tribological properties of a-C(S,H) coatings formed by RPLD require further research. Our results of tribological studies of a-C(S,H) thin-film coatings obtained by RPLD on steel substrates are presented in Supplementary Materials, Figures S6–S11.

It is important to note that pure a-C coatings prepared by the RPLD in a vacuum delaminated from the substrate one to two days after the sample had been taken out into the air from the deposition chamber. The SiC sublayer failed to provide sufficient adhesion of the a-C coating to the substrate, which can be explained by a high level of mechanical stress in the carbon film. Tribotests of a-C(S,H) thin-film coatings under various friction conditions showed that an increase in the concentration of sulphur due to an increase in H_2S pressure had a negative effect on both antifriction properties and fracture resistance. Even if the $MoS_{2.5}$ nanophase in the Mo-S-C_9 coating could provide good antifriction properties under certain conditions, the a-C(S,H)_9 phase, having an approximate composition of $C_{0.73}S_{0.11}H_{0.16}$, would not let it happen. The degradation of the a-C(S,H)_9 and a-C(S,H)_18 coatings occurred by cracking and pilling off from the substrate. It can clearly be seen in the shape of the wear debris formed after friction test. These were mainly microplates accumulating at the edges of the track and adhering to the counterbody.

When under dry friction conditions, only the a-C(S,H)_5.5 coating demonstrated good antifriction properties. These coatings were produced from graphite by reactive PLD at an H_2S pressure of 5.5 Pa. The friction coefficient did not exceed 0.03 during 10^3 cycles of the sliding of the ball. In that case, the surface of the a-C(S,H)_5.5 coating underwent a slight wear, and the wear scar on the surface of the steel counterbody was just incipient. Obviously, the qualitative tribological properties of the Mo–S–C–H_5.5 nanocomposite coating under dry friction conditions were due to the influence of the a-C(S,H) phase. Under other friction conditions, the tribological properties of the Mo–S–C–H_5.5 coating depended on the synergistic effect of the formation of a composition of the MoS and a-C(S,H) nanophases. Under friction in a humid atmosphere, the $MoS_{\sim1.8}$ + a-C(S,H) combination provided the coefficient of friction characteristic of both phases, but changed the wear mechanism of the a-C(S,H) phase, preventing its cracking and delamination from the substrate. At low temperatures ($-100\,°C$), the synergy effect of the $MoS_{\sim1.8}$ and a-C(S,H) phases caused a rather low coefficient of friction and high wear resistance. The friction coefficient was found to be lower than the values typical for the $MoS_{x\sim2}$ and a-C(S,H)_5.5 thin-film coatings.

The Raman studies of the wear track on the nanocomposite Mo–S–C–H_5.5 coating showed (Figures 14–16) that sliding friction of steel counterbody caused subtle changes in the Raman spectra measured in a middle region of the track. This indicates that the triboinduced changes occurred in a very thin near-surface layer of the coating. These changes manifested in an increase in the contribution of the peak at 1433 cm^{-1}. This could be due to the accumulation (an increase in the concentration) of S atoms in the surface layer of the coating during friction. No changes in the structure of the MoS_x nanophase were found. Comparison of Raman peaks for the Mo–S–C–H_5.5 coating before (Figure 6a) and after the friction testing showed no noticeable changes in the spectra in the frequency range of 800–1000 cm^{-1}. This indicated a high resistance of Mo–S–C–H_5.5 coatings to oxidation in a humid atmosphere.

RS analysis of the wear debris accumulated near the counterbody reversal points showed (Figures 14 and 15) that the chemical state of the wear debris, and hence the chemical state of the tribofilm, depended on the tribotest conditions. Under friction in a humid atmosphere, the wear debris contained the crystalline phase of MoS_2. This was indicated by the appearance of rather narrow peaks at ~370 and 405 cm^{-1} caused by first-order reflection for the 2H-MoS_2 phase. Formed during triboinduced crystallization, this phase can cause peaks at ~520 and 650 cm^{-1}, which appear due to the second-order vibration modes of MoS_2 [64]. Another peak at ~950 cm^{-1} may have occurred due to the formation of a Fe-Mo-O compound (for example, $FeMoO_4$) as a result of the tribochemical reaction of the surface of the steel counterbody with the surface of the Mo–S–C–H_5.5 coating in the presence of adsorbed water molecules. The Raman spectrum of the wear debris contained peaks that could be attributed to the Fe-S phase. The Raman spectrum of the Fe-S nanoparticulate phase has the most intense peaks at ~215, 323, and 463 cm^{-1} [72]. Further research is needed to confirm the formation of the Fe-S phase. Figure 14 demonstrates the peaks for the Fe-S nanophase, but they are indicated by a question mark (?).

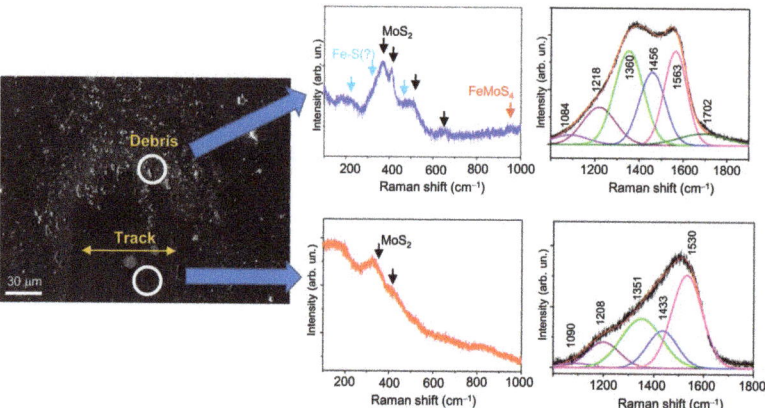

Figure 14. SEM image and Raman spectra for the Mo-S-C-H_5.5 thin-film coating subjected to pin-on-disk test at 22 °C in wet friction conditions (RH ~58%).

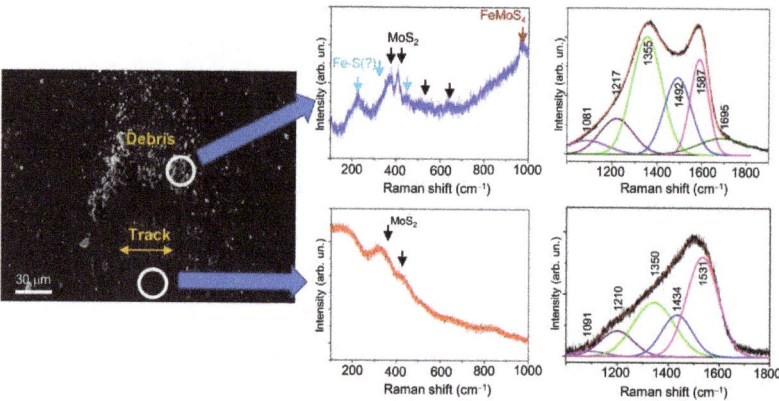

Figure 15. SEM image and Raman spectra for the Mo-S-C-H_5.5 thin-film coating subjected to pin-on-disk test at 22 °C in dry friction conditions (RH ~8%).

In addition to the MoS$_2$, FeMoO$_4$ and, possibly, Fe-S phases, the wear debris contained a graphite-like phase, which corresponded to a doublet containing broadened G (at ~1563 ÷ 1587 cm^{-1}) and G (at 1355 ÷ 1360 cm^{-1}) peaks. The formation of such a C-based phase is typical of triboinduced changes in TMD+C coatings (see, for example, [15,24,46,73]). An important factor influencing friction is the local environment and atom packing in that phase. The Raman spectra for the wear particles formed on the Mo-SC-H_5.5 coating after friction in a humid and dry atmosphere, differ in both the position of the G peak and the width of the G and D peaks. Under friction in a dry atmosphere, these peaks turn out to be narrower, and the G peak shifts to higher frequencies up to 1587 cm^{-1}. It shows that during friction in a humid atmosphere, graphitization of the surface layer of the coating was insufficient, and the tribofilm structure was highly disordered. Probably, this was due to the fact that, at increased air humidity, the contact of the coating with the counterbody was modified because of the adsorption of water molecules and slowed graphitization. This resulted a relatively high coefficient of friction (~0.2) for the Mo–S–C–H_5.5 coating in a humid atmosphere.

The analysis of published data on the triboinduced graphitization of the interface layer for TMD+C and a-C(H) coatings and their comparison with the results of this work showed that S atoms incorporated into the a-C(H) phase during the RPLD of Mo–S–C–H_5.5 films did not have a significant effect on the formation of tribofilms during friction in a dry atmosphere. The relatively narrow G and D peaks located at 1355 and 1580 cm^{-1} respectively, correspond to the graphite/graphene-like packing of atoms with a laminar structure [46,71,73–75]. The incorporation of S atoms into the graphene structure may not cause noticeable changes in the Raman spectrum of graphene [76]. Still, under certain conditions when graphene interacts with sulphur, a band may appear at ~1440 cm^{-1} in the graphene Raman spectrum [77], the nature of which was considered in the above analysis of the Raman spectra of Mo–S–C–H and a-C(S,H) films with a high sulphur concentration. During sliding friction against the Mo–S–C–H_5.5 coating in a dry atmosphere, along with the graphitization of the a-C(S,H) phase, crystallization of the MoS$_2$ nanophase occurred; therefore, this phase could not compromise the positive antifriction properties of the a-C(S,H) phase.

During the tests of the Mo–S–C–H_5.5 coating at −100 °C, the formation of wear debris and their accumulation on the sample surface did not occur in an intensive way. This may be attributed to the fact that these particles were very small and effectively adhered to the counterbody surface (Figure 13a). When measuring the Raman spectrum on a single small particle, light was registered; it was resonantly scattered by both the particle and the coating (Figure 16). The main contribution to the spectrum from the particle was the appearance of weak peaks at ~1340 and 1579 cm^{-1}, that indicated weak graphitization of the material in the wear debris. No signs of crystallization of the MoS$_x$ nanophase were found. It shows that that the mechanism of friction and wear of the Mo–S–C–H_5.5 coating at −100 °C differed from that for the MoS$_2$ coatings, which underwent effective crystallization under similar test conditions [35]. This could be due to the fact that inclusions of the MoS$_x$ nanophase into the amorphous a-C(S,H) matrix could reduce the level of mechanical stresses in the coating. The nanocomposite Mo–S–C–H_5.5 coatings wear out by the mechanism of layer-by-layer removal of the surface layer while the a-C(S,H) coatings showed a tendency to cracking. The presence of the MoS$_x$ phase in the nanocomposite coating could have an effect on the adsorption of water molecules on the coating surface at low temperatures, and possibly on the formation of water microcrystals. The influence of these factors on the tribological properties of Mo–S–C–H coatings at low temperatures require a further in-depth study.

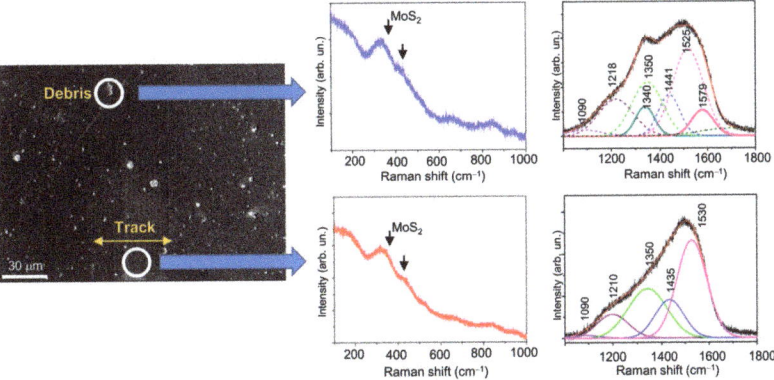

Figure 16. SEM image and Raman spectra for the Mo-S-C-H_5.5 thin-film coating subjected to pin-on-disk test in dry friction conditions at −100 °C.

5. Conclusions

Pulsed laser deposition of Mo–S–C–H thin-film nanocomposite coatings from Mo and graphite targets in H_2S reactive gas ensures effective saturation of the formed layers with S atoms. The penetration of S atoms causes the formation of the MoS_x nanophase, but also significantly changes the chemical state of the a-C(S,H) phase. In this work, we chose specific RPLD conditions under which the ablation time of the Mo target was twice as long as the ablation time of the C target. With an increase in the H_2S pressure from 5.5 to 18 Pa, the concentration of S in the MoS_x nanophase increased from $x\sim1.8$ to $x\sim4$. Significant changes were observed in the chemical composition of the a-C(S,H) phase. At 5.5 Pa, the composition was described by the formula $C_{0.78}S_{0.08}H_{0.14}$, and at 18 Pa, an increase in the concentration of sulphur caused the formation of $C_{0.62}S_{0.18}H_{0.2}$. For the coatings obtained at 5.5 and 18 Pa, the ratio of the number of atoms in the MoS_x and a-C(S,H) phases was ~40/60 and 60/40, respectively. At the lowest S concentration, the local packing of atoms in the MoS nanophase was close to the laminar packing characteristic of the turbostratic MoS_2 structure. In this case, the a-C(S,H) nanophase was amorphous with a predominance of sp^2 bonds between C atoms. An increase in the concentration of S atoms caused the formation of MoS_x clusters, in which the Mo_3-S packing began to dominate. In this case, the configuration of the local packing of atoms in the a-C(S,H) phase was significantly modified due to the efficient formation of C–S bonds.

The best tribological properties were found for the Mo–S–C–H_5.5 nanocomposite coatings obtained at an H_2S pressure of 5.5 Pa. At higher H_2S pressures, an increase in the concentration of S atoms both in the MoS_x nanophase and in the a-C(S,H) nanophase caused a noticeable deterioration in the tribological properties of the Mo–S–C–H_9 and Mo–S–C–H_18 nanocomposite coatings. The tribological properties of the Mo–S–C–H_5.5 thin-film coatings were superior to those of the MoS_2 coatings under various friction test conditions. However, in a humid atmosphere, the antifriction properties of the Mo–S–C–H_5.5 coating turned out to be worse than the properties of the MoS_3 coating. MoS_2 and MoS_3 coatings were also obtained by RPLD [35]. The friction and wear of Mo–S–C–H_5.5 coatings in a humid atmosphere at 22 °C and in a dry atmosphere at −100 °C were due to the synergy effect of the MoS_x and a-C(S,H) nanophases. Under dry friction conditions, the sufficiently high-quality tribological properties of Mo–S–C–H_5.5 resulted from the dominant influence of the a-C(S,H) phase, which is the most suitable for these conditions and has a relatively optimal chemical composition.

Supplementary Materials: The following are available online at http://www.mdpi.com/2079-4991/10/12/2456/s1, Figure S1: Experimental and simulated RBS (left) and ESDA (right) spectra of the films prepared on Si substrates by PLD of carbon under vacuum conditions and in H_2S gas with different pressures. The RBS spectrum of the carbon film deposited in vacuum contains (residual gas pressure was $\sim10^{-3}$ Pa) a peak at the channel number 200. This peak is due to scattering of ions by sulfur atoms that have been adsorbed on the surface of the Si substrate before the carbon film deposition, Figure S2: Typical SEM images (two magnifications) of a-C (S, H) film obtained on a steel substrate by reactive PLD in H_2S, Figure S3: Raman spectra for (a) a-C(H) and (b-d) a-C(S,H) films obtained by PLD on Si substrate (a) under vacuum conditions and in reactive H_2S gas at pressures of (b) 5.5, (c) 9, and (d) 18 Pa. The model of spectrum decomposition into the indicated peaks is discussed in the text of the article, Figure S4: The friction force (in relation with the normal force) evolution during reciprocate sliding of the counterbody over the MoS_x coating under different environmental conditions. The MoS_x coating was obtained on the steel substrate by reactive PLD at H_2S pressure of 5.5 Pa, Figure S5: Optical images of wear tracks formed on the steel substrates for MoS_x thin-film coating obtained by reactive PLD at H_2S pressure of 5.5 Pa. Pin-on-disk tribometer testing was conducted in (a) wet friction conditions (RH ~50%) at 22 °C, (b) dry friction conditions (RH ~8%) at 22 °C, and (c) dry friction conditions at −100 °C. The test durations are indicated in Figure S4, Figure S6: Characteristic evolution of the friction coefficient as a function of the cycle number for a-C(S,H) thin-film coatings obtained by reactive PLD at the pressures of H_2S gas of 5.5, 9 and 18 Pa. Pin-on-disk tribometer testing was conducted in wet air (RH ~58%) at 22 °C, Figure S7: Optical images of wear tracks and wear scars formed on the steel substrates and steel balls for a-C(S,H) thin-film coatings obtained by reactive PLD at the different pressures of H_2S gas: (a) 5.5, (b) 9, and (c) 18 Pa. Pin-on-disk tribometer testing was conducted in wet air (RH ~58%) at 22 °C. The test durations are indicated in Figure S6, Figure S8: Characteristic evolution of the friction coefficient as a function of the cycle number for a-C(S,H) thin film coatings obtained by reactive PLD at the pressures of H_2S gas of 5.5, 9, and 18 Pa. Pin-on-disk tribometer testing was conducted in dry friction conditions (air+Ar mixture, RH ~8%) at 22 °C, Figure S9: Optical images of wear tracks and wear scars formed on the steel substrates and steel balls for a-C(S,H) thin film coatings obtained by reactive PLD at the different pressures of H_2S gas: (a) 5.5, (b) 9, and (c) 18 Pa. Pin-on-disk tribometer testing was conducted in dry friction conditions at 22 °C. The test durations are indicated in Figure S8, Figure S10: Characteristic evolution of the friction coefficient as a function of the cycle number for a-C(S,H) thin-film coatings obtained by reactive PLD at the pressures of H_2S gas of 5.5, 9 and 18 Pa. Pin-on-disk tribometer testing was conducted in dry friction conditions at −100 °C, Figure S11: Optical images of wear tracks and wear scars formed on the steel substrates and steel balls for a-C(S,H) thin-film coatings obtained by reactive PLD at the different pressures of H_2S gas: (a) 5.5, (b) 9, and (c) 18 Pa. Pin-on-disk tribometer testing was conducted in dry friction conditions at −100 °C. The test durations are indicated in Figure S10.

Author Contributions: Conceptualization and writing—original draft preparation, V.F.; methodology, M.D. and A.G.; validation, R.R.; investigation, D.F., M.G., K.M., and P.S. All authors have read and agreed to the published version of the manuscript.

Funding: This research was funded by the Russian Science Foundation, grant number 19-19-00081.

Acknowledgments: Sample characterization by Rutherford backscattering spectroscopy of ions and Raman spectroscopy has been done in REC "Functional Nanomaterials" with support from Ministry of Science and Higher Education of the Russian Federation (project FZWN-2020-0008).

Conflicts of Interest: The authors declare no conflict of interest.

References

1. Spalvins, T. A review of recent advances in solid film lubrication. *J. Vac. Sci. Technol. A* **1987**, *5*, 212–219. [CrossRef]
2. Fleischauer, P.D. Fundamental aspects of the electronic structure, materials properties and lubrication performance of sputtered MoS_2 films. *Thin Solid Films* **1987**, *154*, 309–322. [CrossRef]
3. Müller, C.; Menoud, C.; Mailat, M.; Hintermann, H.E. Thick compact MoS_2 coatings. *Surf. Coat. Technol.* **1988**, *36*, 351–359. [CrossRef]
4. Pope, L.E.; Panitz, J.K.G. The effect of hertzian stress and test atmosphere on the friction coefficients of MoS_2 coatings. *Surf. Coat. Technol.* **1988**, *36*, 341–350. [CrossRef]
5. Voevodin, A.A.; O'Neill, J.P.; Zabinski, J.S. WC/DLC/WS_2 nanocomposite coatings for aerospace tribology. *Tribol. Lett.* **1999**, *6*, 75–78. [CrossRef]
6. Hudec, T.; Mikula, M.; Satrapinskyy, L.; Roch, T.; Truchlý, M.; Jr, P.S.; Huminiuc, T.; Polcar, T. Structure, mechanical and tribological properties of Mo-S-N solid lubricant coatings. *Appl. Surf. Sci.* **2019**, *486*, 1–14. [CrossRef]
7. Gustavsson, F.; Bugnet, M.; Polcar, T.; Cavaleiro, A.; Jacobson, S. A high-resolution TEM/EELS study of the effect of doping elements on the sliding mechanisms of sputtered WS_2 coatings. *Tribol. Trans.* **2015**, *58*, 113–118. [CrossRef]
8. Zeng, C.; Pu, J.; Wang, H.; Zheng, S.; Wang, L.; Xue, Q. Study on atmospheric tribology performance of MoS_2–W films with self-adaption to temperature. *Ceram. Int.* **2019**, *45*, 15834–15842. [CrossRef]
9. Voevodin, A.A.; Zabinski, J.S. Supertough wear-resistant coatings with 'chameleon' surface adaptation. *Thin Solid Films* **2000**, *370*, 223–231. [CrossRef]
10. Mutyala, K.C.; Wu, Y.A.; Erdemir, A.; Sumant, A.V. Graphene-MoS_2 ensembles to reduce friction and wear in DLC-Steel contacts. *Carbon* **2019**, *146*, 524–527. [CrossRef]
11. Cho, D.-H.; Jung, J.; Kim, C.; Lee, J.; Oh, S.-D.; Kim, K.S.; Lee, C. Comparison of frictional properties of CVD-Grown MoS_2 and graphene films under dry sliding conditions. *Nanomaterials* **2019**, *9*, 293. [CrossRef] [PubMed]
12. Guo, J.; Peng, R.; Du, H.; Shen, Y.; Li, Y.; Li, J.; Dong, G. The application of nano-MoS_2 quantum dots as liquid lubricant additive for tribological behavior improvement. *Nanomaterials* **2020**, *10*, 200. [CrossRef] [PubMed]
13. Yaqub, T.B.; Vuchkov, T.; Sanguino, P.; Polcar, T.; Cavaleiro, A. Comparative study of DC and RF sputtered $MoSe_2$ coatings containing carbon—An approach to optimize stoichiometry, microstructure, crystallinity and hardness. *Coatings* **2020**, *10*, 133. [CrossRef]
14. Xu, J.; He, T.; Chai, L.; Qiao, L.; Wang, P.; Liu, W. Growth and characteristics of self-assembled MoS_2/Mo-S-C nanoperiod multilayers for enhanced tribological performance. *Sci. Rep.* **2016**, *6*, 25378. [CrossRef] [PubMed]
15. Polcar, T.; Cavaleiro, A. Review on self-lubricant transition metal dichalcogenide nanocomposite coatings alloyed with carbon. *Surf. Coat. Technol.* **2011**, *206*, 686–695. [CrossRef]
16. Berman, D.; Narayanan, B.; Cherukara, M.J.; Sankaranarayanan, S.K.R.S.; Erdemir, A.; Zinovev, A.; Sumant, A.V. *Operando* tribochemical formation of onion-like-carbon leads to macroscale superlubricity. *Nat. Commun.* **2018**, *9*, 1164. [CrossRef]

17. Yu, G.; Gong, Z.; Jiang, B.; Wang, D.; Bai, C. Superlubricity for hydrogenated diamond like carbon induced by thin MoS$_2$ and DLC layer in moist air. *Diam. Relat. Mater.* **2020**, *102*, 107668. [CrossRef]
18. Gong, Z.; Shi, J.; Zhang, B.; Zhang, J. Graphene nano scrolls responding to superlow friction of amorphous carbon. *Carbon* **2017**, *116*, 310–317. [CrossRef]
19. Berman, D.; Deshmukh, S.A.; Sankaranarayanan, S.K.R.S.; Erdemir, A.; Sumant, A.V. Macroscale superlubricity enabled by graphene nanoscroll formation. *Science* **2015**, *348*, 6238. [CrossRef]
20. Muratore, C.; Voevodin, A.A. Control of molybdenum disulfide plane orientation during coating growth in pulsed magnetron sputtering discharges. *Thin Solid Films* **2009**, *517*, 5605–5610. [CrossRef]
21. Tillmann, W.; Wittig, A.; Stangier, D.; Moldenhauer, H.; Thomann, C.-A.; Debus, J.; Aurich, D.; Bruemmer, A. Influence of the bias-voltage, the argon pressure and the heating power on the structure and the tribological properties of HiPIMS sputtered MoS$_x$ films. *Surf. Coat. Technol.* **2020**, *385*, 125358. [CrossRef]
22. Donley, M.S.; Murray, P.T.; Barber, S.A.; Haas, T.W. Deposition and properties of MoS$_2$ thin films grown by pulsed laser evaporation. *Surf. Coat. Technol.* **1988**, *36*, 329–340. [CrossRef]
23. Hu, J.J.; Bultman, J.E.; Muratore, C.; Phillips, B.S.; Zabinski, J.S.; Voevodin, A.A. Tribological properties of pulsed laser deposited Mo–S–Te composite films at moderate high temperatures. *Surf. Coat. Technol.* **2009**, *203*, 2322–2327. [CrossRef]
24. Grigoriev, S.N.; Fominski, V.Y.; Romanov, R.I.; Gnedovets, A.G. Tribological properties of gradient Mo-Se-Ni-C thin films obtained by pulsed laser deposition in standard and shadow mask configurations. *Thin Solid Films* **2014**, *556*, 35–43. [CrossRef]
25. Theiler, G.; Gradt, T.; Österle, W.; Brückner, A.; Weihnacht, V. Friction and endurance of MoS$_2$/ta-C coatings produced by Laser Arc deposition. *Wear* **2013**, *297*, 791–801. [CrossRef]
26. Fominski, V.Y.; Romanov, R.I.; Nevolin, V.N.; Fominski, D.V.; Komleva, O.V.; Popov, V.V. Formation of ultrathin MoS$_2$ films using laser-based methods. *J. Phys. Conf. Ser.* **2019**, *1238*, 1–5. [CrossRef]
27. Fominski, V.Y.; Romanov, R.I.; Gusarov, A.V.; Celis, J.-P. Pulsed laser deposition of antifriction thin-film MoSe$_x$ coatings at the different vacuum conditions. *Surf. Coat. Technol.* **2007**, *201*, 7813–7821. [CrossRef]
28. Fominski, V.Y.; Grigoriev, S.N.; Gnedovets, A.G.; Romanov, R.I. Specific features of ion-initiated processes during pulsed laser deposition of MoSe$_2$ coatings in pulsed electric fields. *Tech. Phys. Lett.* **2012**, *38*, 683–686. [CrossRef]
29. Walck, S.D.; Zabinski, J.S.; Donley, M.S.; Bultman, J.E. Evolution of surface topography in pulsed-laser-deposited thin films of MoS$_2$. *Surf. Coat. Technol.* **1993**, *62*, 412–416. [CrossRef]
30. Fominski, V.; Demin, M.; Fominski, D.; Romanov, R.; Goikhman, A.; Maksimova, K. Comparative study of the structure, composition, and electrocatalytic performance of hydrogen evolution in MoS$_{x\sim 2+\delta}$/Mo and MoS$_{x\sim 3+\delta}$ films obtained by pulsed laser deposition. *Nanomaterials* **2020**, *10*, 201. [CrossRef]
31. Fominski, V.Y.; Grigoriev, S.N.; Gnedovets, A.G.; Romanov, R.I. On the mechanism of encapsulated particle formation during pulsed laser deposition of WSe$_x$ thin-film coatings. *Tech. Phys. Lett.* **2013**, *39*, 312–315. [CrossRef]
32. Grigoriev, S.N.; Fominski, V.Y.; Romanov, R.I.; Volosova, M.A.; Shelyakov, A.V. Pulsed laser deposition of nanocomposite MoSe$_x$/Mo thin-film catalyst for hydrogen evolution reaction. *Thin Solid Films* **2015**, *592*, 175–181. [CrossRef]
33. Fominski, V.Y.; Nevolin, V.N.; Romanov, R.I.; Titov, V.I.; Scharff, W. Tribological properties of pulsed laser deposited WSe$_x$(Ni)/DLC coatings. *Tribol. Lett.* **2004**, *17*, 289–294. [CrossRef]
34. Muratore, C.; Hu, J.J.; Voevodin, A.A. Tribological coatings for lubrication over multiple thermal cycles. *Surf. Coat. Technol.* **2009**, *203*, 957–962. [CrossRef]
35. Fominski, V.; Demin, M.; Nevolin, V.; Fominski, D.; Romanov, R.; Gritskevich, M.; Smirnov, N. Reactive pulsed laser deposition of clustered-type MoS$_x$ ($x \sim 2$, 3, and 4) films and their solid lubricant properties at low temperature. *Nanomaterials* **2020**, *10*, 653. [CrossRef]
36. Cemin, F.; Boeira, C.D.; Figuero, C.A. On the understanding of the silicon-containing adhesion interlayer in DLC deposited on steel. *Tribol. Int.* **2016**, *94*, 464–469. [CrossRef]
37. Grigoriev, S.N.; Fominski, V.Y.; Romanov, R.I.; Shelyakov, A.V.; Volosova, M.A. Effect of energy fluence and Ti/W co-deposition on the structural, mechanical and tribological characteristics of diamond-like carbon coatings obtained by pulsed Nd:YAG laser deposition on a steel substrate. *Surf. Coat. Technol.* **2014**, *259*, 415–425. [CrossRef]

38. Niakan, H.; Zhang, C.; Yang, L.; Yang, Q.; Szpunar, J.A. Structure and properties of DLC–MoS$_2$ thin films synthesized by BTIBD method. *J. Phys. Chem. Solids* **2014**, *75*, 1289–1294. [CrossRef]
39. Zhang, X.; Xu, J.; Chai, L.; He, T.; Yu, F.; Wang, P. Carbon and nitrogen co-doping self-assembled MoS$_2$ multilayer films. *Appl. Surf. Sci.* **2017**, *406*, 30–38. [CrossRef]
40. Gu, L.; Ke, P.; Zou, Y.; Li, X.; Wang, A. Amorphous self-lubricant MoS$_2$-C sputtered coating with high hardness. *Appl. Surf. Sci.* **2015**, *331*, 66–71. [CrossRef]
41. Weise, G.; Mattern, N.; Hermann, H.; Teresiak, A.; Bächer, I.; Brückner, W.; Bauer, H.-D.; Vinzelberg, H.; Reiss, G.; Kreissig, U.; et al. Preparation, structure and properties of MoS$_x$ films. *Thin Solid Films* **1997**, *298*, 98–106. [CrossRef]
42. Moser, J.; Lévy, F. Random stacking in MoS$_{2-x}$ sputtered thin films. *Thin Solid Films* **1994**, *240*, 56–59. [CrossRef]
43. Lee, C.-H.; Lee, S.; Lee, Y.-K.; Jung, Y.C.; Ko, Y.-I.; Lee, D.C.; Joh, H.-I. Understanding on the origin of formation and active sites for thiomolybdate [Mo$_3$S$_{13}$]$^{2-}$ clusters as hydrogen-evolution catalyst through the selective control of sulfur atoms. *ACS Catal.* **2018**, *8*, 5221–5227. [CrossRef]
44. Wu, Q.; Abraham, A.; Wang, L.; Tong, X.; Takeuchi, E.S.; Takeuchi, K.J.; Marschilok, A.C. Electrodeposition of MoS$_x$: Tunable fabrication of sulfur equivalent electrodes for high capacity or high power. *J. Electrochem. Soc.* **2020**, *167*, 050513. [CrossRef]
45. Escalera-López, D.; Lou, Z.; Rees, N.V. Benchmarking the activity, stability, and inherent electrochemistry of amorphous molybdenum sulfide for hydrogen production. *Adv. Energy Mater.* **2019**, *9*, 1802614. [CrossRef]
46. Xu, J.; Chai, L.; Qiao, L.; He, T.; Wang, P. Influence of C dopant on the structure, mechanical and tribologicalproperties of r.f.-sputtered MoS$_2$/a-C composite films. *Appl. Surf. Sci.* **2016**, *364*, 249–256. [CrossRef]
47. Wu, Y.; Li, H.; Ji, L.; Liu, L.; Ye, Y.; Chen, J.; Zhou, H. Structure, mechanical, and tribological properties of MoS$_2$/a-C:H composite films. *Tribol. Lett.* **2013**, *52*, 371–380. [CrossRef]
48. Filik, J.; Lane, I.M.; May, P.W.; Pearce, S.R.J.; Hallam, K.R. Incorporation of sulfur into hydrogenated amorphous carbon films. *Diam. Relat. Mater.* **2004**, *13*, 1377–1384. [CrossRef]
49. Honglertkongsakul, K.; May, P.W.; Paosawatyanyong, B. Effect of temperature on sulfur-doped diamond-like carbon films deposited by pulsed laser ablation. *Diam. Relat. Mater.* **2011**, *20*, 1218–1221. [CrossRef]
50. Saeheng, A.; Tonanon, N.; Bhanthumnavin, W.; Paosawatyanyong, B. Sulphur doped DLC films deposited by DC magnetron sputtering. *Can. J. Chem. Eng.* **2012**, *90*, 909–914. [CrossRef]
51. Takeuchi, T.; Kojima, T.; Kageyama, H. Carbon Sulfur Materials and Methods for Producing Same. U.S. Patent 10,710,960 B2, 14 July 2020.
52. Deng, Y.; Ting, L.R.L.; Neo, P.H.L.; Zhang, Y.-J.; Peterson, A.A.; Yeo, B.S. Operando Raman spectroscopy of amorphous molybdenum sulfide (MoS$_x$) during the electrochemical hydrogen evolution reaction: Identification of sulfur atoms as catalytically active sites for H$^+$ reduction. *ACS Catal.* **2016**, *6*, 7790–7798. [CrossRef]
53. Lassalle-Kaiser, B.; Merki, D.; Vrubel, H.; Gul, S.; Yachandra, V.K.; Hu, X.; Yano, J. Evidence from in situ X-ray absorption spectroscopy for the involvement of terminal disulfide in the reduction of protons by an amorphous molybdenum sulfide electrocatalyst. *J. Am. Chem. Soc.* **2015**, *137*, 314–321. [CrossRef] [PubMed]
54. Lince, J.R.; Pluntze, A.M.; Jackson, S.A.; Radhakrishnan, G.; Adams, P.M. Tribochemistry of MoS$_3$ nanoparticle coatings. *Tribol. Lett.* **2014**, *53*, 543–554. [CrossRef]
55. Lu, X.F.; Yu, L.; Zhang, J.; Lou, X.W. Ultrafine dual-phased carbide nanocrystals confined in porous nitrogen-doped carbon dodecahedrons for efficient hydrogen evolution reaction. *Adv. Mater.* **2019**, *31*, 1900699. [CrossRef] [PubMed]
56. Chen, M.; Ma, Y.; Zhou, Y.; Liu, C.; Qin, Y.; Fang, Y.; Guan, G.; Li, X.; Zhang, Z.; Wang, T. Influence of transition metal on the hydrogen evolution reaction over nano-molybdenum-carbide catalyst. *Catalysts* **2018**, *8*, 294. [CrossRef]
57. Wang, C.; Guo, Z.; Shen, W.; Zhang, A.; Xu, Q.; Liu, H.; Wang, Y. Application of sulfur-doped carbon coating on the surface of Li$_3$V$_2$(PO$_4$)$_3$ composite facilitate li-ion storage as cathode materials. *J. Mater. Chem. A* **2015**, *3*, 6064–6072. [CrossRef]
58. Fominski, V.Y.; Markeev, A.M.; Nevolin, V.N.; Prokopenko, V.B.; Vrublevski, A.R. Pulsed laser deposition of MoS$_x$ films in a buffer gas atmosphere. *Thin Solid Films* **1994**, *248*, 240–246. [CrossRef]

59. Su, Y.; Zhang, Y.; Zhuang, X.; Li, S.; Wu, D.; Zhang, F.; Feng, X. Low-temperature synthesis of nitrogen/sulfur co-doped three-dimensional graphene frameworks as efficient metal-free electrocatalyst for oxygen reduction reaction. *Carbon* **2013**, *62*, 296–301. [CrossRef]
60. Paraknowitsch, J.P.; Wienert, B.; Zhang, Y.; Thomas, A. Intrinsically Sulfur- and Nitrogen-Co-doped Carbons from Thiazolium Salts. *Chem. Eur. J.* **2012**, *18*, 15416–15423. [CrossRef]
61. Polcar, T.; Nossa, A.; Evaristo, M.; Cavaleiro, A. Nanocomposite Coatings of Carbon-based and Transition Metal Dichalcogenides Phases: A Review. *Rev. Adv. Mater. Sci.* **2007**, *15*, 118–126. Available online: http://www.ipme.ru/e-journals/RAMS/no_21507/cavaleiro.pdf (accessed on 7 December 2020).
62. Wang, X.; Wang, T.; Ye, M.; Wang, L.; Zhang, G. Microstructure and tribological properties of GLC/MoS$_2$ composite films deposited by magnetron sputtering. *Diam. Relat. Mater.* **2019**, *98*, 107471. [CrossRef]
63. Li, L.; Lu, Z.; Pu, J.; Wang, H.; Li, Q.; Chen, S.; Zhang, Z.; Wang, L. The superlattice structure and self-adaptive performance of C–Ti/MoS$_2$ composite coatings. *Ceram. Int.* **2020**, *46*, 5733–5744. [CrossRef]
64. Zhao, X.; Lu, Z.; Zhang, G.; Wang, L.; Xue, Q. Self-adaptive MoS$_2$-Pb-Ti film for vacuum and humid air. *Surf. Coat. Technol.* **2018**, *345*, 152–166. [CrossRef]
65. Cao, H.; Wen, F.; Kumar, S.; Rudolf, P.; de Hosson, J.T.M.; Pei, Y. On the S/W stoichiometry and triboperformance of WS$_x$C(H) coatings deposited by magnetron sputtering. *Surf. Coat. Technol.* **2019**, *365*, 41–51. [CrossRef]
66. Wu, Y.; Li, H.; Ji, L.; Ye, Y.; Chen, J.; Zhou, H. A long-lifetime MoS$_2$/a-C:H nanoscale multilayer film with extremely low internal stress. *Surf. Coat. Technol.* **2013**, *236*, 439–443. [CrossRef]
67. Fominski, V.Y.; Grigor'ev, S.N.; Romanov, R.I.; Nevolin, V.N. Effect of the pulsed laser deposition conditions on the tribological properties of thin-film nanostructured coatings based on molybdenum diselenide and carbon. *Tech. Phys.* **2012**, *57*, 516–523. [CrossRef]
68. Gao, K.; Lai, Z.; Jia, Z.; Zhang, B.; Wei, X.; Zhang, J. Bilayer a-C:H/MoS$_2$ film to realize superlubricity in open atmosphere. *Diam. Relat. Mater.* **2020**, *108*, 107973. [CrossRef]
69. Freyman, C.A.; Chen, Y.; Chung, Y.-W. Synthesis of carbon films with ultra-low friction in dry and humid air. *Surf. Coat. Technol.* **2006**, *201*, 164–167. [CrossRef]
70. Moolsradoo, N.; Watanabe, S. Deposition and tribological properties of sulfur-doped DLC Films deposited by PBII method. *Adv. Mater. Sci. Eng.* **2010**, *168*, 1–7. [CrossRef]
71. Huang, P.; Qi, W.; Yin, X.; Choi, J.; Chen, X.; Tian, J.; Xu, J.; Wu, H.; Luo, J. Ultra-low friction of a-C:H films enabled by lubrication of nanodiamond and graphene in ambient air. *Carbon* **2019**, *154*, 203–210. [CrossRef]
72. Matamoros-Veloza, A.; Cespedes, O.; Johnson, B.R.G.; Stawski, T.M.; Terranova, U.; de Leeuw, N.H.; Benning, L.G. A highly reactive precursor in the iron sulfide system. *Nat. Commun.* **2018**, *9*, 1–7. [CrossRef]
73. Cai, S.; Guo, P.; Liu, J.; Zhang, D.; Ke, P.; Wang, A.; Zhu, Y. Friction and wear mechanism of MoS$_2$/C composite coatings under atmospheric environment. *Tribol. Lett.* **2017**, *65*, 79. [CrossRef]
74. Wang, K.; Yang, B.; Zhang, B.; Bai, C.; Mou, Z.; Gao, K.; Yushkov, G.; Oks, E. Modification of a-C:H films via nitrogen and silicon doping: The way to the superlubricity in moisture atmosphere. *Diam. Relat. Mater.* **2020**, *107*, 107873. [CrossRef]
75. Yin, X.; Wu, F.; Chen, X.; Xu, J.; Wu, P.; Li, J.; Zhang, C.; Luo, J. Graphene-induced reconstruction of the sliding interface assisting the improved lubricity of various tribo-couples. *Mater. Des.* **2020**, *191*, 108661. [CrossRef]
76. Liang, J.; Jiao, Y.; Jaroniec, M.; Shi Qiao, S.Z. Sulfur and nitrogen dual-doped mesoporous graphene electrocatalyst for oxygen reduction with synergistically enhanced performance. *Angew. Chem. Int. Ed.* **2012**, *51*, 11496–11500. [CrossRef] [PubMed]
77. Bautista-Flores, C.; Arellano-Peraza, J.S.; Sato-Berrú, R.Y.; Camps, E.; Mendoza, D. Sulfur and few-layer graphene interaction under thermal treatments. *Chem. Phys. Lett.* **2016**, *665*, 121–126. [CrossRef]

Publisher's Note: MDPI stays neutral with regard to jurisdictional claims in published maps and institutional affiliations.

© 2020 by the authors. Licensee MDPI, Basel, Switzerland. This article is an open access article distributed under the terms and conditions of the Creative Commons Attribution (CC BY) license (http://creativecommons.org/licenses/by/4.0/).

Article

Laser Printing of Plasmonic Nanosponges

Sergey Syubaev [1,2], Stanislav Gurbatov [1,2], Evgeny Modin [3], Denver P. Linklater [4,5], Saulius Juodkazis [4,6], Evgeny L. Gurevich [7] and Aleksandr Kuchmizhak [1,2,*]

[1] Institute of Automation and Control Processes, Far Eastern Branch, Russian Academy of Sciences, 690041 Vladivostok, Russia; trilar@bk.ru (S.S.); gurbatov_slava@mail.ru (S.G.)
[2] Far Eastern Federal University, 690041 Vladivostok, Russia
[3] CIC NanoGUNE BRTA, Avda Tolosa 76, 20018 Donostia-San Sebastian, Spain; e.modin@nanogune.eu
[4] Optical Sciences Center and ARC Training Centre in Surface Engineering for Advanced Materials (SEAM), School of Science, Swinburne University of Technology, John st., Hawthorn, VIC 3122, Australia; dlinklater@swin.edu.au (D.P.L.); saulius.juodkazis@gmail.com (S.J.)
[5] School of Science, RMIT University, Melbourne, VIC 3000, Australia
[6] World Research Hub Initiative (WRHI), School of Materials and Chemical Technology, Tokyo Institute of Technology, 2-12-1, Ookayama, Meguro-ku, Tokyo 152-8550, Japan
[7] Laser Center (LFM), University of Applied Sciences Munster, Stegerwaldstraße 39, 48565 Steinfurt, Germany; gurevich@fh-muenster.de
* Correspondence: alex.iacp.dvo@mail.ru

Received: 15 November 2020; Accepted: 1 December 2020; Published: 4 December 2020

Abstract: Three-dimensional porous nanostructures made of noble metals represent novel class of nanomaterials promising for nonlinear nanooptics and sensors. Such nanostructures are typically fabricated using either reproducible yet time-consuming and costly multi-step lithography protocols or less reproducible chemical synthesis that involve liquid processing with toxic compounds. Here, we combined scalable nanosecond-laser ablation with advanced engineering of the chemical composition of thin substrate-supported Au films to produce nanobumps containing multiple nanopores inside. Most of the nanopores hidden beneath the nanobump surface can be further uncapped using gentle etching of the nanobumps by an Ar-ion beam to form functional 3D plasmonic nanosponges. The nanopores 10–150 nm in diameter were found to appear via laser-induced explosive evaporation/boiling and coalescence of the randomly arranged nucleation sites formed by nitrogen-rich areas of the Au films. Density of the nanopores can be controlled by the amount of the nitrogen in the Au films regulated in the process of their magnetron sputtering assisted with nitrogen-containing discharge gas.

Keywords: laser ablation; noble-metal films; magnetron sputtering; nanosecond laser pulses; porous nanostructures; plasmonics; nanosponges

1. Introduction

Three-dimensional (3D) percolated porous nanostructures made of noble metals and having large surface-to-volume ratio have drawn significant attention due to their remarkable physicochemical properties allowing to use them for various important applications ranging from the photo- or electro-catalysis, water splitting, hydrogen storage to bio- and chemosensing via surface-enhanced effects [1–5]. For most of the suggested applications, the surface-to-volume ratio defined by the distribution, density and size of the pores within the 3D nanostructure is of crucial importance.

Recently, optical properties of 3D porous nanostructures have become a hot topic [6–10]. Specifically, the porous Au nanoparticles (also referred to as nanosponges) were shown to demonstrate polarization-dependent scattering as well as to support long-lived electron emission associated with localized and propagating surface plasmon modes having remarkably high quality factors. These optical properties make such structures appealing for various optical and nonlinear optical applications including random lasing, enhanced photo-emission, harmonic and supercontinuum light generation, as well as single-molecule biosensing based on metal-enhanced fluorescence, surface-enhanced Raman scattering (SERS), infrared absorption (SEIRA), etc. Multiple sensing applications are benefited from the random (but highly dense) arrangement of the plasmon-mediated electromagnetic (EM) hot spots within the structure allowing to obtain the spectrally broadband signal enhancement over the entire visible and near-IR spectral range [11–14]. Indeed, incorporation of the nanopores into the bulk plasmonic nanostructures with a sub-wavelength overall size provides more intense SERS signal due to multiple hot spots and enlarged surface area increasing probability for analyte molecules to reach these hot spots [15–18]. Nevertheless, upon excitation with EM radiation, both the arrangement of the nanosized pores and the overall nanostructure geometry govern the resulting response [19,20]. From this point of view, the plasmonic properties of the resulting 3D porous nanostructures, their general geometric shape as well as porosity are to be adjusted simultaneously that still remains challenging.

State-of-the-art methods for porous nanostructure fabrication generally require complicated multi-step fabrication protocols as dealloying and soft- or hard-template synthesis [21–27], where accurate management of the reaction conditions (temperature, composition, precursors, etc.) is crucial. Furthermore, the minimization of the surface free energy typically leads to generally spherical-shaped nanostructures. Seed-mediated growth provides simple and versatile method allowing to produce arbitrary-shaped porous nanostructures [28]. However, it's often problematically to obtain only one desired geometry because of internal structural variations of the seeds as well as local variations of the reaction environment. Alternatively, liquid-free lithography-based approaches as electron- or ion-beam milling [29] are suitable for high-precision formation of geometrically-diverse nanostructures. However, the need for upscaling of the fabrication procedure creates an economically justified barrier for lithography-based techniques being applied for porous nanostructure fabrication and large-area replication. In addition, post-processing is also required to impart porosity into the nanostructures.

Herein, we applied scalable easy-to-implement nanosecond (ns) laser ablation of nitrogen-rich Au films to fabricate parabola-shaped nanobumps containing multiple nanopores inside. The nanopores were found to originate from laser-induced explosive boiling and coalescence of the nucleation sites formed by nitrogen-rich areas of the Au film, while the nanopore density can be controlled by amount of nitrogen used as discharge gas for magnetron sputtering of the Au films. Most of the nanopores hidden beneath the nanobump surface can be further uncapped using gentle etching of the nanobumps by an Ar-ion beam to form 3D plasmonic nanosponges promising for various nonlinear optical and sensing applications.

2. Materials and Methods

2.1. Deposition of Au Films Assisted with Various Discharge Gases

Au films with a thickness of 150 ± 5 nm were deposited onto silica glass substrates without any adhesion sub-layer using a custom-built magnetron sputtering system and three discharge gases: argon, nitrogen and purified air. Deposition was performed at 10^{-2} mbar and fixed applied voltage of 2.5 kV. At the same time, the current was maintained at a constant value of 25 mA by dynamically adjusting the discharge gas pressure that allowed to fix the sputtering rate at ≈ 1 nm·s^{-1} for all gases.

2.2. Characterization of Au Films

The actual thickness and average roughness of the films were controlled by an atomic-force microscopy (AFM, Nano-DST, Pacific Nanotechnology, Santa Clara, CA, USA) . Optical spectroscopic measurements performed with an integrating sphere spectrometer confirmed the identical reflectance for all Au films evaporated with different discharge gases (Cary 5000, Agilent Technologies, Santa Clara, CA, USA). The surface chemical composition of the Au films was carefully studied with X-ray Photoelectron Spectroscopy (XPS). XPS spectra were collected using a Kratos Axis Nova instrument (Kratos Analytical Inc., Manchester, UK) with a monochromatic Al Kα source (source energy 1486.69 eV) at a power of 150 W. Elemental identification was carried out using survey spectra collected at a pass energy of 160 eV with 1 eV steps. A Shirley algorithm was used to measure the background core-level spectra, and chemically distinct species in the high-resolution regions of the spectra were fitted with synthetic Gaussian Lorentzian components after removing the background (using the CasaXPS software, v. 2.3.15). High-resolution XPS scans were performed in the N1s and Au4f regions.

2.3. Fabrication of Porous Nanostructures and Nanosponges

Precise pointed ablation of the Au films was performed with second-harmonic (wavelength of 532 nm), ns (pulse duration of 7 ns) laser pulses generated by an Nd:YAG laser system (Brio GRM Gaussian, Quantel, France). The laser radiation was focused into a sub-micrometer spot on the sample surface using a dry objective (Nikon, 50x Plan Fluor, Tokyo, Japan) with a numerical aperture NA of 0.8 (optical spot diameter of $1.22\lambda(NA)^{-1} \approx 0.8$ μm on the sample surface). The sample was mounted onto a PC-driven nanopositioning platform (ANT series, Aerotech GmbH., Nurnberg, Germany) allowing spot-by-spot laser printing of the computer-generated patterns with the movement repeatability better than 100 nm. The laser fluence was monitored by a pyroelectric photodetector (Ophir Optronics, Jerusalem, Israel) and adjusted by a PC-driven attenuator (Standa, Vilnius, Lithuania). All nanostructures were produced under identical ambient conditions upon single-pulse laser irradiation.

Additionally, to reveal the nanopores hidden beneath the surface, the laser-printed nanobumps were also post-processed via etching with an accelerated Ar-ion beam (IM4000, Hitachi, Tokyo, Japan) at acceleration voltage of 3 kV, gas flow of 0.15 cm^3/min and discharge current of 105 μA. Such parameters were previously calibrated to provide relatively slow removal rate of $\approx$1 nm/s [15,30], allowing to avoid excessive heating, melting or deformation of the Au film and laser-printed nanostructures.

2.4. Characterization of Laser-Printed Nanostructures

Scanning electron microscopy (SEM) was performed with a Helios Nanolab 450 FIB-SEM (Thermo Fisher Scientific, Waltham, MA, USA). High-resolution surface characterization was conducted at an accelerating voltage of 5 kV and electron beam current 100 pA. Signal channels with secondary (SE) and back-scattered electrons (BSE) were simultaneously collected and analysed. Despite the lack of topographical information, the escape depth for BSE is greater than that for SE improving material/density sensitivity. This allows visualisation of the nanopores under the surface with high contrast and spatial resolution.

To shed light onto the internal structure of the nanobumps (nanosponges), we involved a focused ion beam technique to prepare single cross-sectional cuts and serial cuts that were subsequently combined into a 3D reconstruction. After defining the area of interest, deposition (starting from electron beam-induced deposition to prevent surface damage and followed ion beam-induced deposition) of the Pt protective layers was performed. The thickness of the layers was chosen taking into account the morphology and smoothness of the certain laser-printed structure and varied between 150 nm (for nanobumps) and 1000 nm (for through holes). Slicing was carried out using an Ga-ion beam at an accelerating voltage of 30 kV and beam current of 30 pA. After producing subsequent FIB cut, high-resolution SEM image was automatically acquired at 5 kV and 50 pA. The resulting image had a field of view of 2.5×1.6 μm at a 1536×1024 pixel resolution, which corresponds to the pixel size of 2.4×1.6 nm. A series of 111 slices were acquired and measured slice thickness was 12 nm with standard deviation of 6.54 nm. Further data processing (including alignment, filtration, and visualization) was performed using Aviso 8.1 software (Thermo Fisher Scientific, Waltham, MA, USA).

Three-dimensional finite-difference time-domain (FDTD) calculations were undertaken to reveal local structure of the EM fields excited near the isolated plasmonic nanosponge by linearly-polarized laser pump at 532, 632 and 1030 nm. The pump wavelengths within rather broad spectral range were chosen to highlight potential applications, where the enhanced and localized plasmon-mediated fields are highly demanding including SERS substrates and enhancement of nonlinear optical effects. Representative 3D model of the nanosponge was reconstructed using high resolution top- and side-view SEM images. The linearly polarized laser radiation was modeled to pump the nanosponge from the top. Elementary cell size was $1 \times 1 \times 1$ nm^3, while the computational volume was limited by the perfectly matched layers. Dielectric permittivity of Au was modeled according to the data from [31].

3. Results and Discussion

In this paper we considered three types of glass-supported Au films produced via magnetron sputtering in various discharges gases—argon, nitrogen and purified air (≈80% of nitrogen). Produced films had the same thickness (150 ± 5 nm) and showed identical diffuse reluctance as well as AFM verified surface roughness of about 2 ± 0.7 nm. This guaranteed identical coupling of the incident ns-laser pulse energy to all types of metal films under study (Figure 1a). Such laser pulse induces thermalisation of charge carriers in the metal film that results in its local melting accompanied by detachment from the substrate via relaxation of the thermal-generated stress or evaporation at the interface between the film and the substrate [32,33]. At a pulse energy that is smaller than the ablation threshold ($F_{th} \approx 0.17$ J/cm^2 [30,33–35]), detached metal shell resolidifies before its rupture forming parabola-shaped surface protrusion (also referred to as nanobump; Figure 1b,c). Typical laser-printed nanobump on the surface of 150-nm thick Au film sputtered with nitrogen discharge gas is illustrated by corresponding top- and side-view SEM images. The former image combines the signals from the SE and BSE detectors giving useful information regarding either a surface morphology or difference in chemical composition/density. The latter allowed to reveal nanoscale pores under the surface of the nanobump that can be also visualized by producing its FIB cross-sectional cuts (Figure 1d).

At elevated laser fluence ($F > 0.23$ J/cm^2), rupture of the nanobump led to formation of the through hole in the metal film. In this case, the multiple nanopores can be identified within the resolidified rim surrounding the microhole as revealed by SEM visualization of the FIB cuts (see Figure 2a). Generally, for the fixed composition of the Au film, higher laser fluences produced the nanopores of the larger size. Similarly, the larger nanopores typically appeared closer to the nanobump center (see Figure 2d–e). Taking into account the Gaussian-shaped intensity profile of the irradiating laser beam, the nanopore formation appears to be driven by the local temperature (that will be discussed later in the text) that is higher in the metal film section coinciding with the beam center. To enrich information regarding the density and geometry of the nanopores, multiple FIB cuts were merged to build exact 3D model of through hole and the surrounding rim (Figure 2b,c). This 3D model clearly indicates broad size distribution of the nanopores ranging from 10 to 150 nm. Also, average size of the nanopores in the rim increases towards the center of the microhole. The larger nanopores can have irregular shape and reach the size ≈150 nm (Figure 2c), while the smaller nanopores far from the rim walls preserve spherical-like geometry.

In part, broad size distribution of the nanopores could be explained by merging (coalescence) of the closest nanopores growing from neighbouring randomly distributed nucleation centers. Origin of such nucleation centers will be discussed somewhat later in the paper in the context of chemical composition of the metal film fabricated with different discharge gases. Earlier coalescence was suggested as a leading mechanism of the water bubbles growing on dispersed gold nanoparticles heated by incident laser radiation [36,37]. As two bubbles with radii r merge, the gain in the surface energy $E_+ = 4\pi r^2 \sigma(2 - 2^{2/3})$ compensates the work of viscous forces needed to move the melt over a distance $\sim r$ at a velocity, which can be estimated as $v \sim r/t$. Here, t is the time needed for this melt relocation and $\sigma = 1.1 \, N/m$ is the surface tension [38]. Using Newton's law of viscosity to calculate the energy dissipation force $F = \eta v A/\delta_h$, where $A = 4\pi r^2$—the surface area of the bubble, $\eta = 4 \times 10^{-3}$ Pa s—viscosity of liquid gold [38], and $\delta_h \sim r$—the characteristic length scale of the flow, we estimate the energy loss as $E_- = 4\pi \eta r^3/t$. Equalizing E_+ and E_- we estimate the time for two bubbles to merge $t \sim \eta r/\sigma \approx$ 10–100 ps (estimated for r = 3–30 nm), which is less than the time gold film remains liquid.

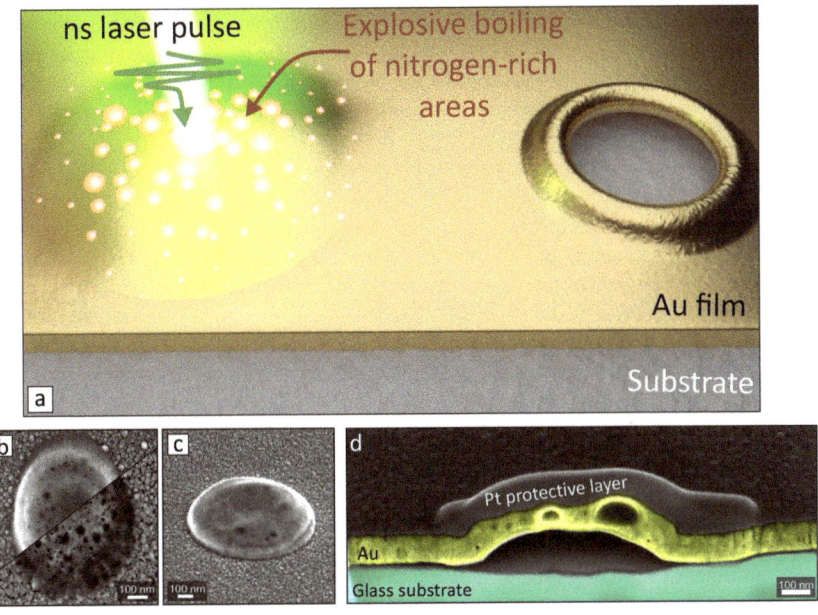

Figure 1. (a) Sketch illustrating direct laser printing of porous nanostructures on a glass-supported Au film. Single-pulse laser ablation at a near-threshold fluence produces a parabola-shaped nanobump (nanosponge), while the explosive boiling of randomly distributed nitrogen-rich sites create a nanoscale pores inside irradiated area. High-intense laser pulse drills a through hole where the nanopores can be found in the surrounding resolidified rim. (b,c) Representative top- and side-view SEM images of the isolated laser-printed nanosponge produced at F = 0.15 J/cm^2 in the Au film evaporated with nitrogen discharge gas. The top-view image is divided into two parts (recorded at different e-beam acceleration voltage) to illustrate multiple nanopores hidden beneath the nanobump surface. (d) False-color SEM images of the cross-sectional FIB cuts made through the center of such nanobump.

This time can be estimated assuming that the heat accumulated by the illuminated spot is removed mostly by the thermal conductivity. The heated volume can be estimated as a cylinder of the radius $r \approx 0.5$ µm (the typical lateral size of the nanobump) and the height $h = 0.15$ µm (the film thickness). The areas of the side $S_{side} = 2\pi r h$ and the bottom $S_{bottom} = \pi r^2$ surfaces are comparable, but the thermal conductivity of gold $\lambda_{Au} = 300$ W/(Km) is larger than that of glass $\lambda_{glass} = 0.5$ W/(Km), so that only the heat flow through the side wall decreases the temperature of the molten surface. This heat flow can be estimated as $\frac{dE}{dt} = \lambda_{Au} S_{side} \nabla T \approx 2\pi h \lambda (T_m - T_0)$, where $T_m \approx 1.4 \times 10^3$ K is the melting point of Au and $T_0 = 300$ K. This heat flow should remove the energy accumulated by the surface in one pulse $E_p = F\pi r^2$, hence the characteristic cooling time scale is $t \sim \frac{Fr^2}{2h\lambda T_m} \approx 5 \cdot 10^{-9}$ s, i.e., is also comparable to the laser pulse duration. This estimation also agrees with previously reported studies [39]. Hence, coalescence of the bubbles provides a plausible explanation of the observed distribution of the pore sizes.

Noteworthy, different growth times of the pores can be also considered as alternative way to explain the large size variation of the nanopores. In particular, the pores started growing short after the onset of the melting had more time to develop than that ones started just before the resolidification. This assumption cannot be accepted because of the low growth velocity of the bubbles. Wang et al. [37] reported that during the initial bubble nucleation phase the bubble size R grows with time as $R \propto t^{1/6}$ (we notice that at later stages in degassed water this dependency drops to $R \propto t^{0.07}$, whereas in air-equilibrated water $R \propto t^{1/3}$). Hence, to get one order of magnitude difference in the nanopore radii, the times should differ by six orders magnitude. If the earliest possible nucleation starts after ten picoseconds after the start of the laser pulse (time comparable with the electron-phonon coupling times in metals), then the smaller pores formation must start several microseconds later, which is impossible because the resolidification happens on a time scale of several nanoseconds.

In a similar way, the broad size distribution can not be explained by local temperature fluctuations (pores at hot spots grow quicker) as the $R(t) \propto T^{1/3}$ [37]. Hence, three-orders-of-magnitude fluctuation of the local temperature in gold separated by a distances much less than one micrometer is obviously not possible. Spontaneous merging (coalescence) of several nanopores into the larger one can also stimulate rupture of the nanobumps and can explain formation of the through nanoholes previously reported for AuPd films processed with ns laser pulses [40]. The average pore size was observed to grow with the laser intensity, as it was shown in [15]. This can be explained by cooperative action of two mechanisms: (1) each pore grows more rapidly, since the temperature is higher and the pore radius $R(t) \propto T^{1/3}$. (2) The metal film remains molten for a longer time and the coalescence-based growth is stopped by the resolidification after a longer time interval $t \propto T$, so that more pores can merge together. Assuming that the surface temperature is proportional to the intensity, we expect the average observed pore size to grow with the intensity.

Remarkably, the density and the average size of the nanopores started to grow from randomly distributed nucleation centers were found to depend on the type of discharge gas used for the film fabrication. More specifically, negligible amount of nanopores was found for the nanobumps produced on the surface of Au films sputtered with Ar (see Figure 2d). The maximal density of the nanopores was observed for the Au film evaporated with nitrogen discharge gas, decreasing for the Au film produced in purified air (Figure 2e,f). Taking into account same morphology and light absorbing characteristics of all mentioned Au films produced with different discharge gases, the amount of nitrogen in the film could be considered as a driving parameter that allows to control density of the nanopores. The molecular nitrogen can be bounded to the metal film surface as well as form chemically stable AuN phase upon magnetron sputtering as it was reported by several previous studies [41–43].

XPS measurements were performed to understand the effect of the discharge gas used for magnetron sputtering on the resulting chemical composition of the Au films (see Methods for details). Analyses of the deconvolved high resolution spectra taken in the N1s region revealed near-zero signal for films produced with Ar discharge gas. However, a pronounced signal for Au films fabricated with purified air and nitrogen (Figure 3b–d) were also observed. Deconvolution of the latter two spectra revealed several peaks with their binding energies at $\approx$401, 399 and 397.3 eV. The first high-energy peak can be assigned either to carbonitride [43] or to molecular nitrogen (that can be adsorbed on the metal film surface having multiple grains and cracks as well as be trapped beneath the surface [44,45]). The two remaining peaks can be attributed to chemically stable gold nitride phases AuN$_x$ as well as to oxynitrides [41–43]. All of the different Au films demonstrated similar Au4f XPS spectra with low-intense shoulders (marked by blue areas in the Figure 3a) shifted towards larger binding energies near both low- (Au4$f_{7/2}$) and high-energy (Au4$f_{5/2}$) peaks. These shoulders can be indicative of stable chemical compounds like surface bonded carbon (AuC) or nitrogen (AuN$_x$) as shown in previous studies [43]. However, the similarity of the signals obtain from different Au films does not clarify the exact chemical nature of these low-intense shoulders where signals from different compounds could overlap. To confirm such peak assignment, the Au film sputtered with nitrogen was further annealed at 200 °C for 2 h. Thermal annealing was expected to remove thermodynamically unstable compounds (such as oxynitrides) as well as most of the molecular nitrogen from the near surface layer probed by XPS. N1s core-level spectra of the annealed Au film demonstrates the only remaining peak with a binding energy of 398.8 eV that can be presumably attributed to AuN$_x$.

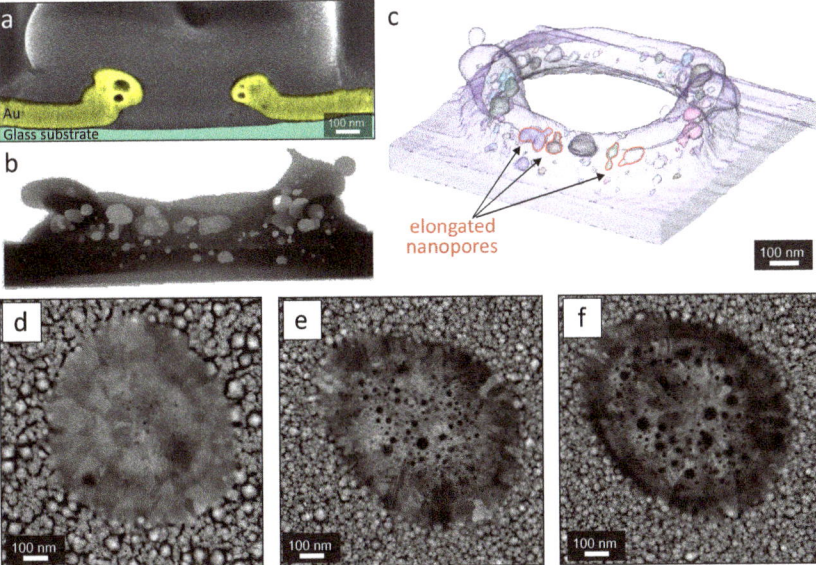

Figure 2. (a) False-color SEM images of cross-sectional FIB cuts made through the center of the microhole. The hole was produced at $F = 0.25$ J/cm^2 in the Au film evaporated with nitrogen discharge gas. (b,c) Distribution of the nanopores in the rim around a through hole visualized by tomographic 3D model reconstructed using serial FIB cuts. Several elongated irregular-shaped nanopores are highlighted in the figure. (d–f) Top-view SEM images of the nanobumps produced under single-pulse irradiation of the Au films ($F \approx 0.15$ J/cm^2) produced with argon (d), purified air (e) and nitrogen (f) discharge gases, respectively.

Considering the chemical composition of the Au film produced with different discharge gases we can suggest the following scenario regarding formation of the nanopores upon ns-pulse laser ablation. Such laser pulse with a near-threshold fluence rapidly heats up the Au film to temperatures above 10^3 K. Such temperature jump is expected to remove all molecular nitrogen from the exposed area as well as induce an explosive boiling of the N- or AuN$_x$-rich areas of the film. The data regarding melting/boiling temperatures of gold nitrides is weakly discussed in a literature. For example, annealing at 90 °C was shown to remove AuN$_x$ from the film [41], while present studies clearly showed signature of AuN$_x$ remaining even after annealing at 200 °C. Anyway, both melting/boiling temperatures of gold nitrides are expected to be much lower comparing to those for the pure gold. The light absorption should be initiated at the Au grain boundaries where also nitrogen adsorption should occur. The ionisation potential of Au is 9.23 eV while that of nitrogen is 14.53 eV (for O—13.62 eV). The laser pulse driven avalanche ionisation of gold is seeding the energy deposition which is evolving into run away ablation (melting, evaporation, ionisation).

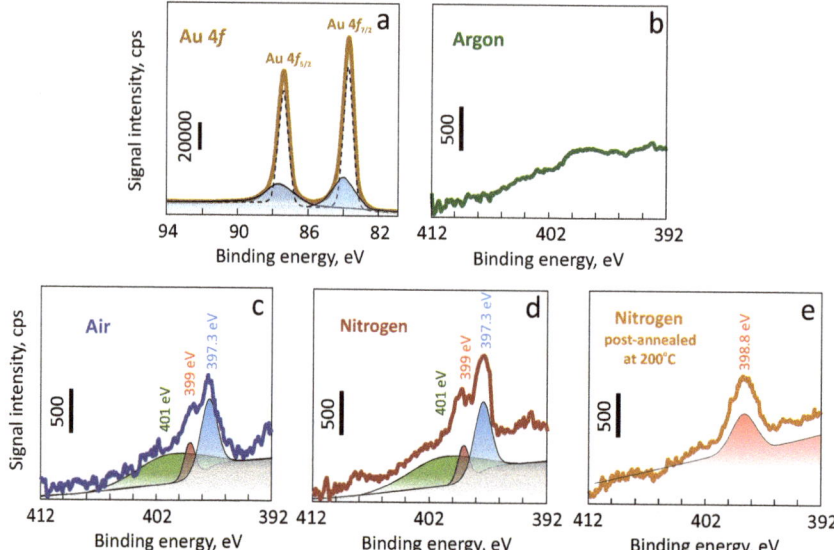

Figure 3. XPS characterization of the Au films produced using different discharge gases. (**a**) Representative Au4f photoemission spectra measured from the glass-supported Au film produced with nitrogen discharge gas. (**b**–**e**) N 1s core-level photoemission spectra of the Au films produced using magnetron sputtering assisted with various discharge gases: argon (**b**), purified air (**c**), nitrogen (**d**), nitrogen followed by thermal annealing at 200 °C for 2 h (**e**). Deconvolution of the obtained spectral signal allowed for identification of several characteristic peaks highlighted by the colored areas in (**a**,**c**–**e**).

The general geometry of the surface structure produced by direct laser ablation is defined by the laser irradiation parameters (fluence, pulse width and beam profile [46]) as well as by the thickness/composition of the metal film [33,47]. Our results clearly show that advanced chemical engineering of the metal film composition gives additional degree of freedom allowing to modify morphology of the laser-printed structures at the nanoscale, namely, incorporate the nanopores and control their density. However, a large amount of the nanopores is typically hidden beneath the top metal shell of the nanobump that limits the potential range of practically relevant applications. To make morphology of the laser-printed nanobumps more functional, we applied gentle etching of the laser-printed nanobumps with unfocused Ar-ion beam schematically illustrated on the Figure 4a. The processing parameters were calibrated to ensure gradual removal of the Au film without melting and deformation of the nanobumps geometry (see Materials and Methods). Two representative SEM images in the Figure 4b,c compare nanoscale surface morphology of the nanobump before and after its etching with Ar beam for 20 min. As can be seen, etching reveals the multiple hidden nanopores making the nanobump shell surface perforated with multiple through nanoholes. Besides the rather random arrangement of the nanopores (nanoholes), general geometry of the produced nanosponges reproduces well from pulse to pulse (see Figure 4d). From the simple geometrical consideration it is clear that the incident electromagnetic radiation can be efficiently absorbed by such nanosponges that produce enhanced electromagnetic fields via coupling to propagating and localized surface plasmons. Being combined with the mentioned broad size distribution of the nanopores, the 3D nanosponges with their lateral size of about visible-light wavelength are expected to be efficient for local EM field enhancement within rather broad spectral range spanning from visible to near-infrared. To illustrate this, we calculated squared electromagnetic field amplitude E^2/E_0^2 (E_0 is the amplitude of the incident EM field) near the isolated plasmonic nanosponge pumped at 532, 632 and 1030 nm (see Materials and Methods). The cross-sectional E^2/E_0^2 maps shown in the

Figure 4e indicate that the nanosponge supports densely arranged EM hot spots upon broadband EM excitation. Multi-fold enhancement of the EM field amplitude comes from coupling of the incident radiation to the localized plasmon resonances of the isolated nanoscale surface features on the top side of the Au shell (like nanopores, spikes and cracks clearly seen in the Figure 4c) as well as plasmons propagating within overall micro-scale sponge geometry. Propagating plasmons can couple to or re-excite the localized ones appearing as the localized EM hot spots on the opposite size of the opaque 150-nm thick Au shell. Owing to weak radiate decay, the certain localized plasmon modes in such plasmonic nanosponges were found to survive for a long time producing substantial field localization and enhancement [8]. Alternatively, most of the modes can dissipate via heating of the nanosponge. These features will potentially allow to use such nanosponges for various applications as thermo and nonlinear plasmonics (enhanced higher harmonic and white light generation), bio-imaging and visualization, photothermal conversion, sensing based on nonlinear optical effects as well as SEIRA- and SERS-based sensing [48–52].

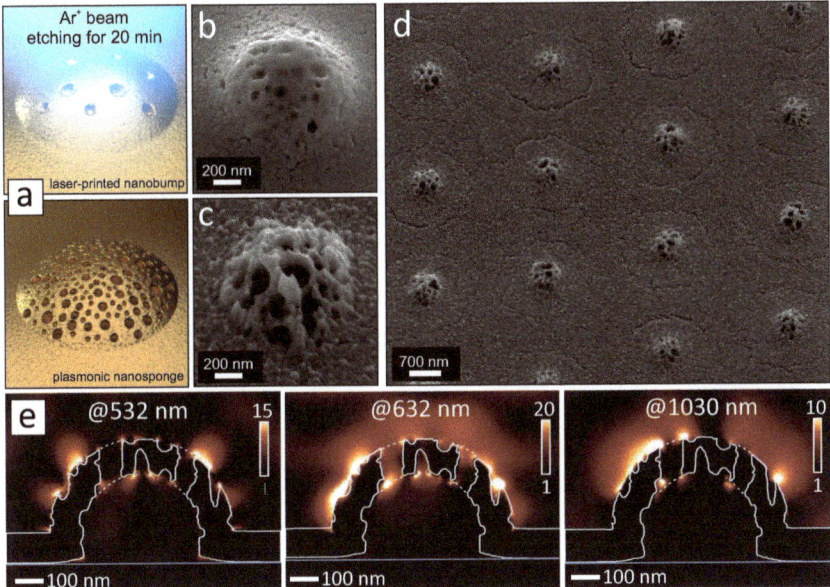

Figure 4. (**a**) Schematic illustration of the nanosponge fabrication. (**b**,**c**) Representative side-view SEM images comparing typical morphology of the nanobump before and after its etching with Ar-ion beam for 20 min. The nanobump was produced at $F = 0.165$ J/cm^2 on the surface of Au film evaporated with nitrogen discharge gas. (**d**) SEM image of the nanosponge array showing how the nanosponge morphology reproduces from pulse to pulse. (**e**) Calculated E^2/E_0^2 maps near the isolated Au nanosponge pumped at 532, 632 and 1030 nm.

4. Conclusions and Outlook

Here, we propose a simple approach allowing fabrication of porous plasmonic nanostructures using direct ns-laser ablation followed by Ar-ion beam etching. The nanopores were found to form through the explosive evaporation/boiling of the nitrogen-rich metal film areas exposed by a ns laser pulse. Detailed XPS and SEM analysis confirmed that the density of the nanopores correlates with the initial amount of nitrogen in the Au films that were fabricated by magnetron sputtering assisted with different discharge gases. Demonstrated unique porous nanostructures, parabola-shaped nanobumps perforated with multiple nanoholes, are expected to be useful for various applications where the plasmon-mediated EM fields are of mandatory importance as nonlinear plasmonic and chemo-/biosensing.

In a broader context, advanced chemical engineering of the metal film composition suggested in this paper being combined with an adjustment of the laser ablation process (as optimization of fluence, pulse width, laser beam shaping, etc.) is expected to provide facile way for fabrication of unique nanostructures. Also, elaborated strategy being applied to more complicated material combinations (like metal alloys or metal-dielectric materials) will further enrich the potential types of nanostructures as well as their application range [53–55]. For example, in a similar way nanosponges can be produced from co-sputtered noble-metal films forming nano-alloys with engineered permittivity as demonstrated for Au-Cu-Ag [56]. Along with plasmonic hot spot engineering, local potential defined by the nearest atomic composition was shown to modify physical and chemical adsorption of the analyte molecules affecting their characteristic SERS and SEIRA signals and sensor performance [57]. Insights into formation of metallic glasses and high entropy alloys with even large number of constituent materials [58] will benefit from a nanoscale control of re-melting and phase-explosions demonstrated in this work.

Author Contributions: Conceptualization, A.K. and S.S.; methodology, S.G., E.M. and D.P.L.; software, E.M.; validation, A.K. and S.J.; formal analysis, A.K., E.L.G. and S.S.; investigation, S.S., E.M. and D.P.L.; resources, A.K.; data curation, A.K.; writing—original draft preparation, S.S., A.K., E.M., E.L.G. and S.J.; writing—review and editing, A.K., E.L.G. and S.J.; visualization, S.G.; supervision, A.K.; project administration, A.K.; funding acquisition, A.K. All authors have read and agreed to the published version of the manuscript.

Funding: This work was supported by Russian Foundation for Basic Research (projects nos. 20-32-70056) and Ministry of Science and Higher Education (0262-2019-0001).

Conflicts of Interest: The authors declare no conflict of interest.

References

1. Ryu, Y.; Kang, G.; Lee, C.W.; Kim, K. Porous metallic nanocone arrays for high-density SERS hot spots via solvent-assisted nanoimprint lithography of block copolymer. *RSC Adv.* **2015**, *5*, 76085–76091. [CrossRef]
2. Liu, G.; Li, K.; Zhang, Y.; Du, J.; Ghafoor, S.; Lu, Y. A facile periodic porous Au nanoparticle array with high-density and built-in hotspots for SERS analysis. *Appl. Surf. Sci.* **2020**, *5*, 146807. [CrossRef]
3. Zhang, T.; Bai, Y.; Sun, Y.; Hang, L.; Li, X.; Liu, D.; Lyu, X.; Li, C.; Cai, W.; Li, Y. Laser-irradiation induced synthesis of spongy AuAgPt alloy nanospheres with high-index facets, rich grain boundaries and subtle lattice distortion for enhanced electrocatalytic activity. *J. Mat. Chem. A* **2018**, *6*, 13735–13742. [CrossRef]
4. Eid, K.; Wang, H.; Malgras, V.; Alothman, Z.; Yamauchi, Y.; Wang, L. Trimetallic PtPdRu Dendritic Nanocages with Three-Dimensional Electrocatalytic Surfaces. *J. Phys. Chem. C* **2015**, *119*, 19947–19953. [CrossRef]
5. Santos, G.; Ferrara, F.; Zhao, F.; Rodrigues, D.; Shih, W. Photothermal inactivation of heat-resistant bacteria on nanoporous gold disk arrays. *Opt. Mater. Express* **2016**, *6*, 1217–1229. [CrossRef]
6. Vidal, C.; Sivun, D.; Ziegler, J.; Wang, D.; Schaaf, P.; Hrelescu, C.; Klar, T.A. Plasmonic Horizon in Gold Nanosponges. *Nano Lett.* **2018**, *18*, 1269–1273. [CrossRef]
7. Vidal, C.; Wang, D.; Schaaf, P.; Hrelescu, C.; Klar, T.A. Optical Plasmons of Individual Gold Nanosponges. *ACS Photonics* **2015**, *2*, 1436–1442. [CrossRef]
8. Hergert, G.; Vogelsang, J.; Schwarz, F.; Wang, D.; Kollmann, H.; Groß, P.; Lienau, C.; Runge, E.; Schaaf, P. Long-lived electron emission reveals localized plasmon modes in disordered nanosponge antennas. *Light Sci. Appl.* **2017**, *6*, e17075. [CrossRef]
9. Zhang, X.; Zheng, Y.; Liu, X.; Lu, W.; Dai, J.; Lei, D.; MacFarlane, D. Hierarchical Porous Plasmonic Metamaterials for Reproducible Ultrasensitive Surface—Enhanced Raman Spectroscopy. *Adv. Mater.* **2015**, *6*, 1090–1096. [CrossRef]
10. Zhong, J.; Chimeh, A.; Korte, A.; Schwarz, F.; Yi, J.; Wang, D.; Zhan, J.; Schaaf, P.; Runge, E.; Lienau, C. Strong Spatial and Spectral Localization of Surface Plasmons in Individual Randomly Disordered Gold Nanosponges. *Nano Lett.* **2018**, *18*, 4957–4964. [CrossRef]

11. Genç, A.; Patarroyo, J.; Sancho-Parramon, J.; Arenal, R.; Duchamp, M.; Gonzalez, E.; Henrard, L.; Bastùs, N.; Dunin-Borkowski, R.; Puntes, V.; et al. Tuning the Plasmonic Response up: Hollow Cuboid Metal Nanostructures. *ACS Photonics* **2016**, *3*, 770–779. [CrossRef]
12. Garcia-Leis, A.; Torreggiani, A.; Garcia-Ramos, J.; Sanchez-Cortes, S. Hollow Au/Ag nanostars displaying broad plasmonic resonance and high surface-enhanced Raman sensitivity. *Nanoscale* **2015**, *7*, 13629–13637. [CrossRef]
13. Schubert, I.; Huck, C.; Kröber, F.P.; Neubrech, F.; Pucci, A.; Toimil-Molares, M.; Trautmann, C.; Vogt, J. Porous Gold Nanowires: Plasmonic Response and Surface—Enhanced Infrared Absorption. *Adv. Opt. Mater.* **2016**, *4*, 1838–1845. [CrossRef]
14. Garoli, D.; Calandrini, E.; Bozzola, A.; Ortolani, M.; Cattarin, S.; Barison, S.; Tomaa, A.; Angelis, F.D. Boosting infrared energy transfer in 3D nanoporous gold antennas. *Nanoscale* **2017**, *9*, 915–922. [CrossRef]
15. Syubaev, S.; Nepomnyashchiy, A.; Mitsai, E.; Pustovalov, E.; Vitrik, O.; Kudryashov, S.; Kuchmizhak, A. Fabrication of porous microrings via laser printing and ion-beam post-etching. *Appl. Phys. Lett.* **2017**, *111*, 083102. [CrossRef]
16. Li, K.; Liu, G.; Zhang, S.; Dai, Y.; Ghafoor, S.; Huang, W.; Zu, Z.; Lu, Y. A porous Au–Ag hybrid nanoparticle array with broadband absorption and high-density hotspots for stable SERS analysis. *Nanoscale* **2019**, *11*, 9587–9592. [CrossRef]
17. Calandrini, E.; Giovannini, G.; Garoli, D. 3D nanoporous antennas as a platform for high sensitivity IR plasmonic sensing. *Opt. Express* **2019**, *27*, 25912–25919. [CrossRef]
18. Ruffino, F.; Grimaldi, M.G. Nanoporous Gold-Based Sensing. *Coatings* **2020**, *10*, 899. [CrossRef]
19. Rao, W.; Wang, D.; Kups, T.; Baradács, E.; Parditka, B.; Erdélyi, Z.; Schaaf, P. Nanoporous Gold Nanoparticles and Au/Al$_2$O$_3$ Hybrid Nanoparticles with Large Tunability of Plasmonic Properties. *ACS Appl. Mater. Interfaces* **2017**, *9*, 6273–6281. [CrossRef]
20. Schwarz, F.; Runge, E. Towards Optimal Disorder in Gold Nanosponges for Long?Lived Localized Plasmonic Modes. *Ann. Phys.* **2017**, *529*, 1600234. [CrossRef]
21. Liu, K.; Bai, Y.; Zhang, L.; Yang, Z.; Fan, Q.; Zheng, H.; Yin, Y.; Gao, C. Porous Au–Ag nanospheres with high-density and highly accessible hotspots for SERS analysis. *Nano Lett.* **2016**, *16*, 3675–3681. [CrossRef] [PubMed]
22. Li, G.G.; Lin, Y.; Wang, H. Residual silver remarkably enhances electrocatalytic activity and durability of dealloyed gold nanosponge particles. *Nano Lett.* **2016**, *16*, 7248–7253. [CrossRef] [PubMed]
23. Jiang, B.; Li, C.; Tang, J.; Takei, T.; Kim, J.H.; Ide, Y.; Henzie, J.; Tominaka, S.; Yamauchi, Y. Tunable-Sized Polymeric Micelles and Their Assembly for the Preparation of Large Mesoporous Platinum Nanoparticles. *Angew. Chem. Int. Ed.* **2016**, *55*, 10037–10041. [CrossRef]
24. Jiang, B.; Li, C.; Imura, M.; Tang, J.; Yamauchi, Y. Multimetallic mesoporous spheres through surfactant?directed synthesis. *Adv. Sci.* **2015**, *2*, 1500112. [CrossRef]
25. Jiang, B.; Li, C.; Dag, Ö.; Abe, H.; Takei, T.; Imai, T.; Hossain, M.S.A.; Islam, M.T.; Wood, K.; Henzie, J.; Yamauchi, Y. Mesoporous metallic rhodium nanoparticles. *Nat. Commun.* **2017**, *8*, 1–8. [CrossRef]
26. Wang, H.; Jeong, H.Y.; Imura, M.; Wang, L.; Radhakrishnan, L.; Fujita, N.; Castle, T.; Terasaki, O.; Yamauchi, Y. Shape-and size-controlled synthesis in hard templates: sophisticated chemical reduction for mesoporous monocrystalline platinum nanoparticles. *J. Am. Chem. Soc.* **2011**, *133*, 14526–14529. [CrossRef]
27. Kani, K.; Malgras, V.; Jiang, B.; Hossain, M.S.A.; Alshehri, S.M.; Ahamad, T.; Salunkhe, R.R.; Huang, Z.; Yamauchi, Y. Periodically Arranged Arrays of Dendritic Pt Nanospheres Using Cage?Type Mesoporous Silica as a Hard Template. *Chem. Asian J.* **2018**, *13*, 106–110. [CrossRef]
28. Zhang, Q.; Large, N.; Nordlander, P.; Wang, H. Porous Au Nanoparticles with Tunable Plasmon Resonances and Intense Field Enhancements for Single-Particle SERS. *J. Phys. Chem. Lett.* **2014**, *5*, 370–374. [CrossRef]
29. Arnob, M.M.P.; Zhao, F.; Li, J.; Shih, W.C. EBL-based fabrication and different modeling approaches for nanoporous gold nanodisks. *ACS Photonics* **2017**, *4*, 1870–1878. [CrossRef]
30. Kuchmizhak, A.; Gurbatov, S.; Vitrik, O.; Kulchin, Y.; Milichko, V.; Makarov, S.; Kudryashov, S. Ion-beam assisted laser fabrication of sensing plasmonic nanostructures. *Sci. Rep.* **2016**, *6*, 19410. [CrossRef]
31. Babar, S.; Weaver, J. Optical constants of Cu, Ag, and Au revisited. *Appl. Opt.* **2015**, *54*, 477–481. [CrossRef]
32. Meshcheryakov, Y.P.; Bulgakova, N.M. Thermoelastic modeling of microbump and nanojet formation on nanosize gold films under femtosecond laser irradiation. *Appl. Phys. A* **2006**, *82*, 363. [CrossRef]

33. Wang, X.W.; Kuchmizhak, A.A.; Li, X.; Juodkazis, S.; Vitrik, O.B.; Kulchin, Y.N.; Zhakhovsky, V.V.; Danilov, P.A.; Ionin, A.A.; Kudryashov, S.I.; et al. Laser-induced translative hydrodynamic mass snapshots: noninvasive characterization and predictive modeling via mapping at nanoscale. *Phys. Rev. Appl.* **2017**, *8*, 044016. [CrossRef]
34. Liu, J.M. Simple technique for measurements of pulsed Gaussian-beam spot sizes. *Opt. Lett.* **1982**, *7*, 196–198. [CrossRef] [PubMed]
35. Garcia-Lechuga, M.; Gebrayel El Reaidy, G.; Ning, H.; Delaporte, P.; Grojo, D. Assessing the limits of determinism and precision in ultrafast laser ablation. *Appl. Phys. Lett.* **2020**, *117*, 171604. [CrossRef]
36. Fang, Z.; Zhen, Y.R.; Neumann, O.; Polman, A.; García de Abajo, F.J.; Nordlander, P.; Halas, N.J. Evolution of Light-Induced Vapor Generation at a Liquid-Immersed Metallic Nanoparticle. *Nano Lett.* **2013**, *13*, 1736–1742. [CrossRef] [PubMed]
37. Wang, Y.; Zaytsev, M.E.; The, H.L.; Eijkel, J.C.T.; Zandvliet, H.J.W.; Zhang, X.; Lohse, D. Vapor and Gas-Bubble Growth Dynamics around Laser-Irradiated, Water-Immersed Plasmonic Nanoparticles. *ACS Nano* **2017**, *11*, 2045–2051. [CrossRef] [PubMed]
38. Iida, T.; Guthrie, R.I.L. *The Physical Properties of Liquid Metals*; Oxford University Press: New York, NY, USA, 1993.
39. Kneier, F.; Geldhauser, T.; Scheer, E.; Leiderer, P.; Boneberg, J. Nanosecond laser pulse induced vertical movement of thin gold films on silicon determined by a modified Michelson interferometer. *Appl. Phys. A* **2013**, *110*, 321–327. [CrossRef]
40. Kulchin, Y.N.; Vitrik, O.B.; Kuchmizhak, A.A.; Nepomnyashchii, A.V.; Savchuk, A.G.; Ionin, A.A.; Kudryashov, S.I.; Makarov, S.V. Through nanohole formation in thin metallic film by single nanosecond laser pulses using optical dielectric apertureless probe. *Opt. Lett.* **2013**, *38*, 1452–1454. [CrossRef]
41. Šiller, L.; Hunt, M.R.C.; Brown, J.W.; Coquel, J.M.; Rudolf, P. Nitrogen ion irradiation of Au (110): formation of gold nitride. *Surf. Sci.* **2002**, *513*, 78–82. [CrossRef]
42. Caricato, A.P.; Fernandez, M.; Leggieri, G.; Luches, A.; Martino, M.; Romano, F.; Tunno, T.; Valerini, D.; Verdyan, A.; Soifer, Y.M.; et al. Reactive pulsed laser deposition of gold nitride thin films. *Appl. Surf. Sci.* **2007**, *253*, 8037–8040. [CrossRef]
43. Devia, A.; Castillo, H.A.; Benavides, V.J.; Arango, Y.C.; Quintero, J.H. Growth and characterization of AuN films through the pulsed arc technique. *Mater. Charact.* **2008**, *59*, 105–107. [CrossRef]
44. Šiller, L.; Peltekis, N.; Krishnamurthy, S.; Chao, Y.; Bull, S.J.; Hunt, M.R.C. Gold film with gold nitride—A conductor but harder than gold. *Appl. Phys. Lett.* **2005**, *86*, 221912. [CrossRef]
45. Krishnamurthy, S.; Montalti, M.; Wardle, M.G.; Shaw, M.J.; Briddon, P.R.; Svensson, K.; Hunt, M.R.C.; Šiller, L. Nitrogen ion irradiation of Au (110): Photoemission spectroscopy and possible crystal structures of gold nitride. *Phys. Rev. B* **2004**, *70*, 045414. [CrossRef]
46. Pavliuk, G.; Pavlov, D.; Mitsai, E.; Vitrik, O.; Mironenko, A.; Zakharenko, A.; Kulinich, S.A.; Juodkazis, S.; Bratskaya, S.; Zhizhchenko, A.; et al. Ultrasensitive SERS-Based Plasmonic Sensor with Analyte Enrichment System Produced by Direct Laser Writing. *Nanomaterials* **2020**, *10*, 49. [CrossRef]
47. Naghilou, A.; He, M.; Schubert, J.S.; Zhigilei, L.V.; Kautek, W. Femtosecond laser generation of microbumps and nanojets on single and bilayer Cu/Ag thin films. *Phys. Chem. Chem. Phys.* **2019**, *21*, 11846–11860. [CrossRef]
48. Zhang, Y.; Grady, N.K.; Ayala-Orozco, C.; Halas, N.J. Three-dimensional nanostructures as highly efficient generators of second harmonic light. *Nano Lett.* **2011**, *11*, 5519–5523. [CrossRef]
49. Butet, J.; Brevet, P.F.; Martin, O.J.F. Optical second harmonic generation in plasmonic nanostructures: From fundamental principles to advanced applications. *ACS Nano* **2015**, *9*, 10545–10562. [CrossRef]
50. Makarov, S.V.; Zalogina, A.S.; Tajik, M.; Zuev, D.A.; Rybin, M.V.; Kuchmizhak, A.A.; Juodkazis, S.; Kivshar, Y. Light-Induced Tuning and Reconfiguration of Nanophotonic Structures. *Laser Photonics Rev.* **2017**, *11*, 1700108. [CrossRef]
51. Cherepakhin, A.B.; Pavlov, D.V.; Shishkin, I.I.; Voroshilov, P.M.; Juodkazis, S.; Makarov, S.V.; Kuchmizhak, A.A. Laser-printed hollow nanostructures for nonlinear plasmonics. *Appl. Phys. Lett.* **2020**, *117*, 041108. [CrossRef]
52. Baffou, G.; Cichos, F.; Quidant, R. Applications and challenges of thermoplasmonics. *Nat. Mater.* **2020**, *19*, 946–958. [CrossRef]

53. Yi, J.M.; Wang, D.; Schwarz, F.; Zhong, J.; Chimeh, A.; Korte, A.; Zhan, J.; Schaaf, P.; Runge, E.; Lienau, C. Doubly Resonant Plasmonic Hot Spot—Exciton Coupling Enhances Second Harmonic Generation from Au/ZnO Hybrid Porous Nanosponges. *ACS Photonics* **2019**, *6*, 2779–2787. [CrossRef]
54. Larin, A.O.; Nominé, A.; Ageev, E.I.; Ghanbaja, J.; Kolotova, L.N.; Starikov, S.V.; Bruyère, S.; Belmonte, T.; Makarov, S.V.; Zuev, D.A. Plasmonic nanosponges filled with silicon for enhanced white light emission. *Nanoscale* **2020**, *12*, 1013–1021. [CrossRef]
55. Zhong, J.H.; Vogelsang, J.; Yi, J.M.; Wang, D.; Wittenbecher, L.; Mikaelsson, S.; Korte, A.; Chimeh, A.; Arnold, C.L.; Schaaf, P.; et al. Nonlinear plasmon-exciton coupling enhances sum-frequency generation from a hybrid metal/semiconductor nanostructure. *Nat. Commun.* **2020**, *11*, 1–10. [CrossRef]
56. Hashimoto, Y.; Seniutinas, G.; Balčytis, A.; Juodkazis, S.; Nishijima, Y. Au-Ag-Cu nano-alloys: Tailoring of permittivity. *Sci. Rep.* **2016**, *6*, 25010. [CrossRef]
57. Takenaka, M.; Hashimoto, Y.; Iwasa, T.; Taketsugu, T.; Seniutinas, G.; Balčytis, A.; Juodkazis, S.; Nishijima, Y. First Principles Calculations Toward Understanding SERS of 2, 2′-Bipyridyl Adsorbed on Au, Ag, and Au–Ag Nanoalloy. *J. Comput. Chem.* **2019**, *40*, 925–932. [CrossRef]
58. George, E.P.; Raabe, D.; Ritchie, R.O. High-entropy alloys. *Nat. Rev. Mater.* **2019**, *4*, 515–534. [CrossRef]

Publisher's Note: MDPI stays neutral with regard to jurisdictional claims in published maps and institutional affiliations.

© 2020 by the authors. Licensee MDPI, Basel, Switzerland. This article is an open access article distributed under the terms and conditions of the Creative Commons Attribution (CC BY) license (http://creativecommons.org/licenses/by/4.0/).

Article

Synthesis of Oxide Iron Nanoparticles Using Laser Ablation for Possible Hyperthermia Applications

María J. Rivera-Chaverra [1], Elisabeth Restrepo-Parra [1,*], Carlos D. Acosta-Medina [1], Alexandre. Mello [2] and Rogelio. Ospina [3]

1 Laboratorio de Física del Plasma, Department Physics, Universidad Nacional de Colombia, Manizales 170003, Colombia; majriverach@unal.edu.co (M.J.R.-C.); cdacostam@unal.edu.co (C.D.A.-M.)
2 Centro Brasileiro de Pesquisas Físicas, Río de Janeiro 22050-000, Brazil; mello@cbpf.br
3 Universidad Industrial de Santander, Bucaramanga 680001, Colombia; ROSPINAO@uis.edu.co
* Correspondence: erestrepopa@unal.edu.co; Tel.: +57-321-700-4351

Received: 24 July 2020; Accepted: 3 September 2020; Published: 23 October 2020

Abstract: In this work, iron oxide nanoparticles produced using the laser ablation technique were studied in order to determine the characteristics of these nanoparticles as a function of the laser energy for the possible application in magnetic hyperthermia. Nanoparticles were obtained by varying the power of the laser considering values of 90, 173, 279 and 370 mJ. The morphology of these nanoparticles was determined using the dynamic light scattering (DLS) and scattering transmission electron microscopy (STEM) techniques, confirming that the size of the particles was in the order of nanometers. A great influence of the laser power on the particle size was also observed, caused by the competition between the energy and the temperature. The composition was determined by X-ray diffraction and Raman spectroscopy, showing the presence of magnetite, maghemite and hematite. The hyperthermia measurements showed that the temperature rise of the iron oxide nanoparticles was not greatly influenced by the energy change, the heating capacity of magnetic NPs is quantified by the specific absorption rate (SAR), that tends to decrease with increasing energy, which indicates a dependence of these values on the nanoparticles concentration.

Keywords: laser power; XRD; maghemite; hematite

1. Introduction

In recent years, many researchers have studied the synthesis of nanoparticles for several applications. The importance of these nanostructures lies in the characteristics that the nanoparticles possess, different from the characteristics of the bulk materials of the same composition, which is mainly due to the size effects. For instance, magnetic nanomaterials have exceptional properties because their size is comparable to that of magnetic domains [1]. Furthermore, the magnetic and electronic properties are strongly influenced by surface phenomena as the size is reduced [2,3].

For this reason, magnetic nanoparticles exhibit unique physicochemical properties, such as superparamagnetism, high surface/volume ratio, strong magnetic response, and low toxicity [4], depending on their size and shape. This special behavior makes them suitable candidates for a wide variety of applications in areas such as magnetic recording [5], and biomedicine [6,7] among others. Magnetic nanoparticles can be used in different fields of application such as nanotechnology, bioenvironmental, physical medicine, and engineering, among others [2]. More specifically, these types of nanostructures have been used to support diagnosis in magnetic resonance imaging, administration of drugs and their targeted delivery, in addition to environmental remediation, plant growth, catalysis, etc. [8,9].

Iron oxide nanoparticles such as magnetite (Fe_3O_4) or its oxidized form of hematite (α-Fe_2O_3) are the most commonly used nanoparticles for biomedical applications. This is mainly due to the fact that other highly magnetic materials such as cobalt and nickel are susceptible to oxidation and can be toxic, which has made them of little interest in biomedical applications [2].

Magnetite has an inverse structure to the spinel, (Fe^{3+}) (Fe^{2+}, Fe^{3+}) O_4 and is derived from a cubic packing of oxygen with cations at the interstitial tetrahedral and octahedral sites [10], where Fe (II) and Fe (III) are disordered at the octahedral sites, while the tetrahedrons are completely occupied by the Fe (III) cation. From the point of view of magnetic properties, the material is ferrimagnetic up to the Curie temperature (T_C = 858 K). Upon exposure to the ambient atmosphere, the surfaces of Fe_3O_4 crystallites are often covered with Fe_2O_3 layers [11]. Nevertheless, magnetite nanoparticles with diameters less than 30 nm exhibit superparamagnetic behavior, which means that in the absence of an external magnetic field these particles have zero magnetization and less tendency to agglomerate [12]. Additionally, they present a higher performance in terms of chemical stability and biocompatibility compared to metallic nanoparticles [13].

It is also important to consider that the method of nanomaterials synthesis, and in this case, the method used for magnetic nanoparticles production represents one of the most important challenges that will determine the shape, size distribution, particle size, and surface chemistry of the particles and, consequently, the characteristics for their application [14]. Unfortunately, the preparation process can increase the environmental impact and the cost of production [15]. A fundamental drawback when producing nanoparticles, and especially when searching for a specific application, is the agglomeration. This agglomeration is due to the nanoparticles surface is highly reactive which makes them highly unstable. For this reason, it is generally required that they be functionalized by adding a stabilizing agent, being almost always produced as a colloid [16].

Some of the most used techniques to produce these types of nanoparticles are hydrothermal path [17,18] green synthesis [9], co-precipitation [19,20] and laser ablation, amongst others. The laser ablation technique has aroused the interest of the scientific community as a synthesis method. For example, Yang, in his review article [21] mentioned that liquid laser ablation has been shown to be a chemically simple and clean synthesis, without requiring extreme environmental conditions of temperature and pressure. These advantages allow the combination of solid and liquid phases to manufacture composite nanostructures with the desired functions.

However, the good performance of this method strongly depends on the synthesis parameters. In the case of laser ablation, one of the most important parameters is the production energy of the material. This energy can influence the structure, stoichiometry, size, and the concentration, in addition to the properties required for the specific application.

In the literature, there are reports of magnetite nanoparticles produced by laser ablation, varying different parameters for obtaining them, as in the case of Santillán et al. [22] who analyzed the characteristics of the iron oxide nanoparticles immersed in four different solutions, finding several phases of iron oxide and iron carbide. Svetlichnyi et al. [23] and Ismail et al. [24] synthesized iron oxide nanoparticles varying the media. They concluded that several species, such as iron oxides and iron nitrides were obtained and interesting structural characteristics were found, being suitable for application in different fields.

In recent years, attention has been focused in nanomaterial research around thermal therapy, especially in thermo-magnetic therapy known as magnetic hyperthermia. This intracellular treatment, which has as its main attraction the controlled and localized generation of heat in biological targets such as tumors, has found greater efficiency compared to standard treatments.

The first results obtained in the field of magnetic hyperthermia were used by Jordan et al. [25] with nanoparticles of iron oxide contrast tests against a brain tumor. More recently, Yasemian et al. [26] evaluated the effect of the reaction temperature from the magnetic hyperthermia measurements of iron oxide nanoparticles obtained by the co-precipitation method. The size, structure and magnetic properties of the NPs were also studied.

Despite previous reports, not much information was found regarding the effect of energy variation on the production of iron oxide nanoparticles in water.

The objective of this work is to produce nanoparticles of iron oxide by the laser ablation method and later to evaluate the compositional and morphological properties depending on the energy of the laser. These properties play a decisive role in the practical use of nanoparticles in possible biomedical applications. For this reason, a study showing the hyperthermia behavior of nanoparticles was included in order to correlate the dependency between the SAR (specific absorption rate) and the laser energy; this study may guide the authors towards a future application in magnetic hyperthermia

2. Materials and Methods

The iron oxide nanoparticles were synthesized by the laser ablation method, placing a high purity iron target (99.99%) in a beaker with 20 mL of type I water (Milli Q) with water column height of 19.74 mm. The laser beam was focused perpendicular to the surface of the target. A laser system Quantel Q-smart 850 Nd: YAG (Luminbird, Bozeman, MT, USA) was used at a working wavelength of 532 nm (pulse duration of 8ns, repetition rate of 10 Hz). Before synthesis, the iron target was mechanically polished and washed with deionized water and was subsequently washed in an ultrasonic tank for 10 min in isopropyl alcohol, followed by washing with deionized water.

The ablation time for each sample was 10 min, without rotation. All colloids were then transferred to a lidded container. Each sample was produced at different values of laser energy as follows: 370, 279, 173 and 90 mJ.

This method consists of focusing a high-power pulsed laser on a metallic bulk target, submerged in the solvent where the suspension is to be generated. The energy of the laser pulse is absorbed by the target, producing a shock wave that travels in all directions from the point of incidence of the laser, together with a plume of plasma containing the ablated material (top-down process), this shock wave generally propagates at a speed of about 1500 m/s in water [21].

The expansion of the plume in the surrounding liquid produces a decrease in the temperature of the plasma that, together with the generated cavitation bubble, acts as a reactor for the formation of NPs through the condensation of the atoms expelled from the metallic bulk [27,28] (bottom-up process). In this sense, laser ablation turns out to be a hybrid technique between the top-down and bottom-up processes. The NPs generated by this type of technique turned out to be spherical, being able to exhibit a structure without coating (simple) or with coating (core-shell). The irradiation area depends on the laser energy as follows: 0.02217 cm^2 (90 mJ), 0.02835 cm^2 (173 mJ), 0.04599 cm^2 (279 mJ) and 0.0475 cm^2 (and 370 mJ). The ablation threshold of the material can be considered the latent heat of fusion of Iron that is 13.8 kJ/mol.

Composition analysis was determined with an X-ray photoelectron spectroscopy using a SPECS PHOIBOS 100/150 X-ray spectrometer (SPECS, Berlin, Germany), with a hemispherical analyzer and a 1486.6 eV Al-Kα line. XRD patterns were performed with a Rigaku diffractometer (Panalytical, Almelo, The Netherlands), operating with a Cu-Kα radiation source. In addition, this was collected in a step scanning mode, between 2θ = 10 and 80°, with a step of 0.03° and 10 s/step. For STEM measurements, a FEG (Field Emission Gun) Scanning Electron Microscope (Zeiss, Waltham, MA, USA) was used. The images were taken with the following characteristics: high vacuum, 25 kV acceleration voltage, and a STEM I XT detector was used. For the chemical analysis, an EDAX APOLO X detector (STEM I XT, Waltham, MA, USA) with a resolution of 126.1 eV (in. Mn Kα), and an acceleration voltage of 25 kV were used to perform EDS (Energy-Dispersive Spectroscopy) analysis. The results were analyzed using EDX Genesis software (version 3.6, Waltham, MA, USA).

The colloid was deposited in a polystyrene cuvette RefDTS0012 (Malvern Instrument, Worcestershire, United Kingdom). DLS (Dynamic Light Scattering) measurements were performed on a Zetasizer Nano Series equipment brand (Malvern Instrument, Worcestershire, United Kingdom). For the measurement, water with a refractive index 1.33 was used as dispersant and the number of scans of 10–100.

Absorbance spectra were obtained using a UV-Vis UV2600 spectrophotometer (North Lakewood Boulevard, Long Beach, CA, USA) with a spectral range of 200 to 850 nm, equipped with a double beam and an integrating sphere. Spectra were measured in the 200 to 700 nm region.

Finally, a 200 Gauss AC magnetic field was applied to the iron oxide nanoparticles produced at different energies with NanoScale Biomagnetics magnetic hyperthermia equipment (nanoScale Biomagnetics, Zaragoza, Spain).

3. Results and Discussion

In order to determine the average size of the nanoparticles, the DLS technique and images obtained with STEM were used. Table 1 shows the results of these measurements, including the standard deviation of size. The STEM values were obtained by making an average of the sizes measured in the images, for each energy, three images with the same resolution were obtained at different points of the sample, each image with a different number of nanoparticles measured. Figure 1 shows a representative image of each sample. According to these results, the particles are found to have nanometric sizes.

Table 1. Average particle size measured from DLS (Dynamic Light Scattering) and STEM (scanning transmission electron microscopy).

Energy (mJ)	DLS Measurements		STEM Measurements	
	Average Size (nm)	Standard Deviation (nm)	Average Size (nm)	Standard Deviation (nm)
370	25.868	4.189	16.827	6.044
279	65.363	6.680	15.695	4.854
173	42.176	25.585	14.870	8.347
90	25.900	0.761	18.719	12.825

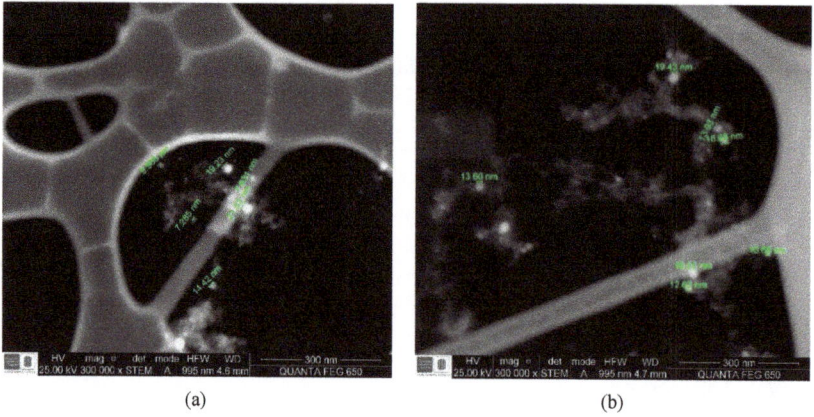

(a)   (b)

Figure 1. *Cont.*

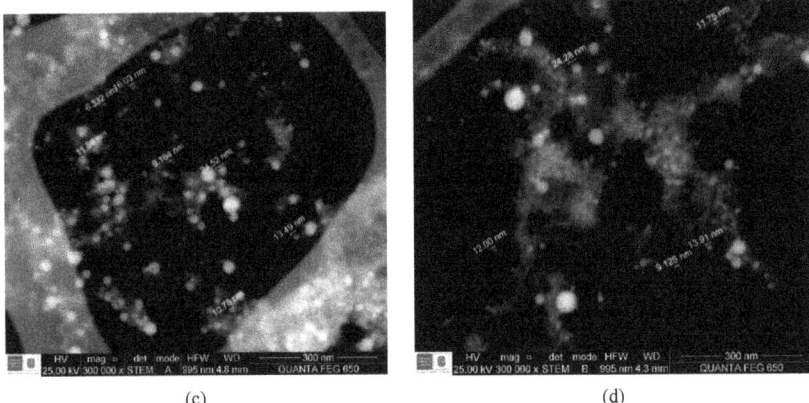

Figure 1. STEM Images for nanoparticles synthesized at (**a**) 370 mJ, (**b**) 279 mJ, (**c**) 173 mJ and (**d**) 90 mJ.

It can be seen for both results that the size of the nanoparticles was not strongly influenced by the energy of the laser, since there was no relationship between the energy and the size of the nanoparticles, however the standard deviation indicates that particles were obtained in a wide size distribution. An explanation for this may be the competition of phenomena that occurs during the ablation process. The energy and temperature produced during the process can generate particles of different sizes, since the energy gives off particles of a certain size, but when temperature is involved, these particles can be divided and have a smaller size, because they have a greater surface area to absorb this thermal energy and use it for dividing [29].

In addition, a high size distribution is due to possible agglomerations of nanoparticles in the water. As their concentration increases, especially in the results obtained by DLS—since the size is measured hydrodynamically—then, it can be concluded that the sizes obtained by this technique are an initial approximation [30]. This can be observed when making a comparison between the values obtained by the two techniques, since they showed very different results, although they indicate that particles in the nanometric range were obtained.

Figure 2 shows the high resolution XPS (X-Ray photoelectron spectrometry) measurements of iron and oxygen for the sample produced at 173 mJ. These spectra are representative of all samples. Figure 2a shows the Fe2p doublet in which two oxidation states are presented, Fe^{2+} and Fe^{3+}, the former coordinated octahedrally and the latter distributed at the octahedral and tetrahedral sites [11]. The spectrum can be successfully adjusted using two main peaks and one satellite peak in the $2p_{3/2}$ region, with a repeated pattern. The lowest binding energy peak at 710.2 eV is attributed to Fe^{2+}, the Fe^{3+} tetrahedral species has a binding energy of 713.3 eV and the satellite peak was identified at 716.0 eV. These values are comparable to others found in the literature [31]. From these results it can be assured that the production of the proposed iron oxide was achieved. Figure 2b shows the high-resolution spectrum O1s, where binding energies, corresponding to the formation of Fe_3O_4 and Fe_2O_3 are identified. However, higher intensity peaks, corresponding to bonds with hydrogen and oxygen are identified. The high amount of these bonds may be due to the nanoparticles being produced in water, which makes them maintain a high humidity. In addition, the carbon, that appears in the spectra, is attributed to the contamination present in the XPS chamber. Carbon is the material which is always present in atmosphere [32].

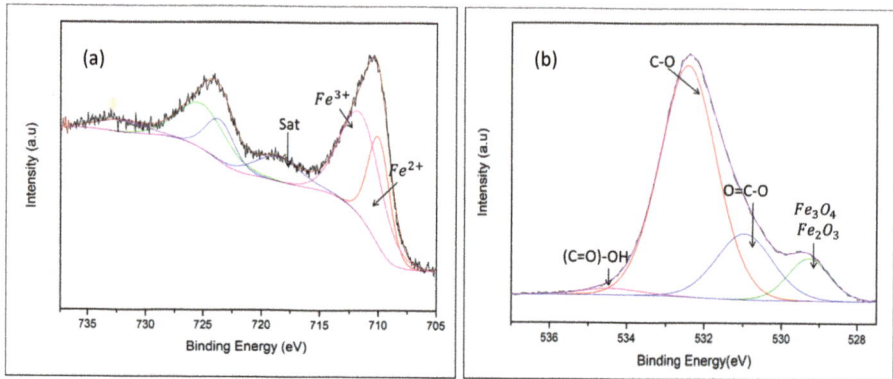

Figure 2. High resolution spectra and their decomposition for nanoparticles produced at 173 mJ. (a) Fe2p and (b) O1s.

In Figure 3, the diffractograms of the samples produced by varying the laser energy are presented. In general, the diffractograms show peaks with wide and low intensities, which is typical of nanometer-sized materials, with low coherence lengths [33]. Furthermore, it is possible to identify peaks in the crystallographic planes (220), (311), (400), (511) and (440) and angles 28.6°, 37.8°, 43.8°, 57.9°, 62.6° respectively, which correspond to Fe_3O_4, and (024), (116) in 54.7° and 46° for Fe_2O_3. The sample produced at 170 mJ has a peak of greater intensity than the others, around 55°, which may be due to the fact that it is the sample with the largest particle size, and therefore, the greatest number of crystallographic planes to produce diffraction; furthermore, the peak positions of Fe_2O_3 and Fe_3O_4 shift to lower 2θ angles; this shifting is due to changes in the surface energy that can compress the nanoparticle and in this way, producing a shifting of the peaks to the right or generating certain possibility that structural efforts be released.

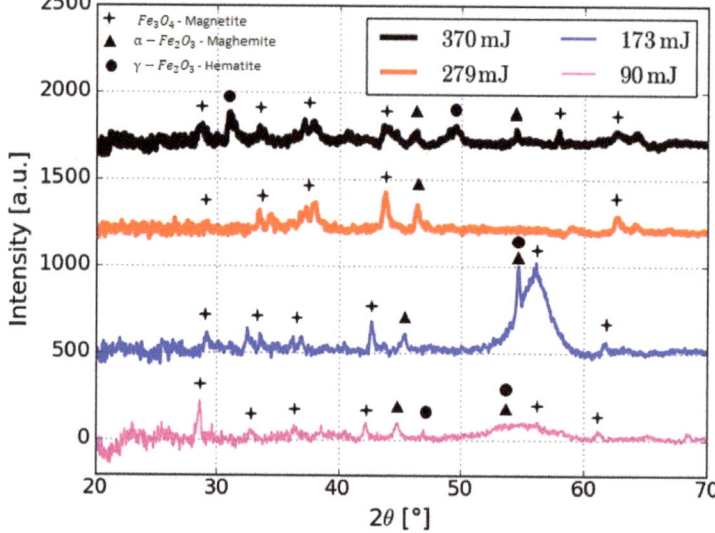

Figure 3. Diffractograms of nanoparticles produced varying the laser energy, where magnetite, maghemite and hematite can be identified.

In order to determine variations in iron and oxygen concentrations in the samples, analyses were performed using EDS. Table 2 shows the atomic percentage values for the different samples. A high concentration of oxygen is observed. This is in accordance with the XPS results in which oxygen not bound to iron is observed, caused by the aqueous medium in which the nanoparticles were produced. No appreciable variation in the concentration of iron and oxygen is observed, indicating that the laser power has no major influence on the stoichiometry of the nanoparticles.

Table 2. Atomic percentage of nanoparticles produced varying the laser energy, calculated from EDS analysis.

Sample	At% (EDS)	
	Fe	O
90 mJ	5.2	94.8
173 mJ	9.3	90.7
279 mJ	10	90
370 mJ	9.9	90.1

Raman spectra of nanoparticles produced at different laser energies were obtained, as shown in Figure 4, where general spectra and the different regions of these spectra are observed. Although Raman spectroscopy is a technique for studying atomic and molecular bonds in the chemistry and physics of condensed matter, care must be taken when applying it to the study of iron oxides. Magnetite bands are not shown as in the diffraction patterns and in the XPS spectra, which may be due to the phase change from magnetite to hematite because excessive exposure of an iron oxide sample to laser radiation has been shown to generate hematite, indicating that some peaks with Fe_2O_3 phase could be attributed to this phenomenon [22]. The reported laser energy threshold values for hematite formation differ widely, depending on experimental conditions such as wavelength, exposure time, and sample surface characteristics. In the spectra presented in Figure 4, the same peaks were identified for all the samples in the fingerprint region and were related in Table 3. In this table it can be seen that the intensities do not have appreciable variations depending on the energy of the laser.

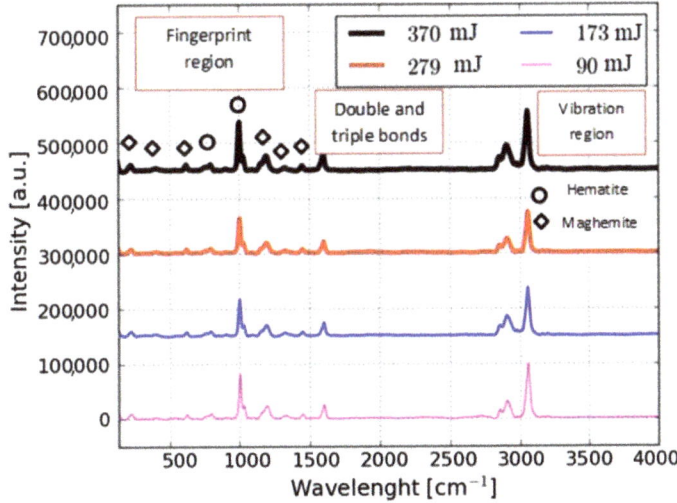

Figure 4. Raman spectra of nanoparticles of iron oxide synthesized by varying the energy.

Table 3. Identification of peaks obtained from Raman analysis.

Peak	Assignation	Area under the Curve (cm^{-1})			
		90 mJ	173 mJ	279 mJ	370 mJ
224.83	Hematite	1.6	2.47	1.72	1.85
410.90	Hematite	0.47	3.87	0.67	1.04
626.73	Hematite	1.13	1.85	1.05	1.19
797.91	Maghemite	2.27	4.01	2.49	2.33
1004.82	Maghemite	12.99	14.19	12.53	11.56
1190.88	Hematite	8.17	9.18	7.87	8.16
1317.40	Hematite	1.71	3.14	1.32	1.12
1451.38	Hematite	1.29	1.84	1.46	1.17

In Figure 5, the absorbance spectra of the colloids obtained for samples produced at different energies can be seen. In these spectra, a decrease in the slope between 200 and 400 nm is identified. This means a decrease in absorbance as the pulse energy used to manufacture it decreases which can be directly related to the decrease in nanoparticle concentrations [22]. This fact is qualitatively supported by the decrease in the color of each sample observed when producing them.

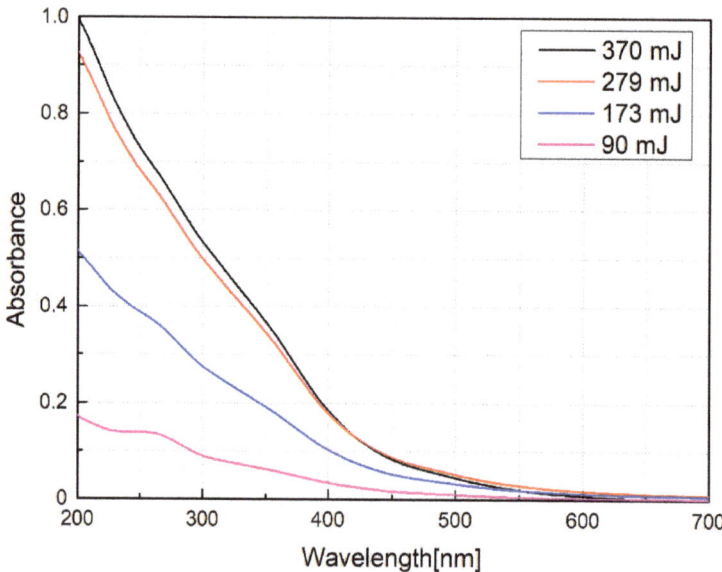

Figure 5. UV-Vis absorption spectra for nanoparticles produced varying the laser energy.

Iron oxide nanoparticles are commonly used as heating mediators for the treatment of magnetic hyperthermia cancer due to their ability to release heat when an AC magnetic field of sufficient intensity and frequency is applied.

Figure 6 shows the temperature change versus time curves in the samples of the iron oxide nanoparticles for different laser energies. Here, the temperature is shown to rise for all samples. Although the 90 mJ sample exhibits an increase of 1 °C, more than the other samples, it is not a significant variation, which could be due to the temperature increase of iron oxide NPs and was not greatly influenced by the energy change [34]. Muller et al. [35] reported that a wide size distribution can negatively influence the magnetic properties due to the statistical orientation of the particles. This is consistent with the DLS and STEM results where a wide distribution can be observed. For having a better understanding of the influence of the laser energy on the hyperthermia behavior of iron oxide nanoparticles, the specific absorption rate (SAR) must be calculated.

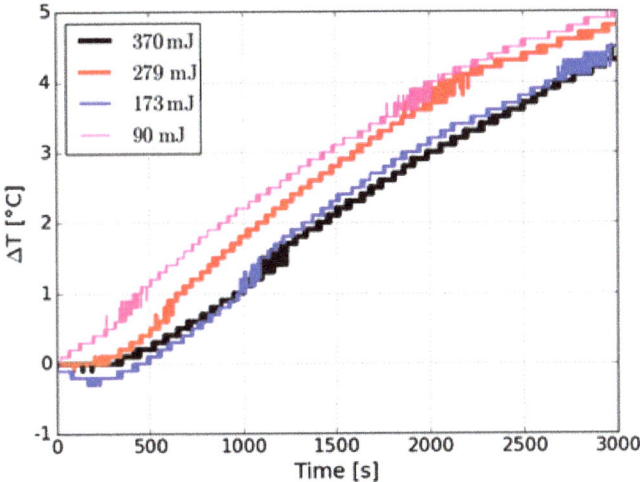

Figure 6. Magnetic hyperthermia measurements for nanoparticles produced varying the laser energy.

Magnetic energy dissipation in a ferrofluid sample is measured in terms of SAR. These values are obtained from two graphical linear fit methods. For the first method, the initial linear slope of the ΔT-time curves was obtained, which was substituted in the second method, values a and b were obtained with the Box Lucas Fitting method of the ΔT-time curves [36]. These were substituted in Equation (2).

$$SAR_{linear} = \frac{M_s}{M_n} C \frac{\Delta T}{\Delta t} \quad (1)$$

$$SAR_{B.L} = \frac{M_s}{M_n} C\,(a.b) \quad (2)$$

where M_s is the mass of suspension including distilled water and NPs, M_n is the mass of NPs, and C is the specific heat capacity of distilled water.

The obtained SAR values of samples with linear fitting and Box Lucas Fitting are shown in Figure 7. SAR values decreased with increased energy, which could indicate a dependence of these values on concentration, and then, on the laser energy. As the laser energy is increased, the nanoparticles concentration in the colloid also increased, since more quantity of material is extracted so more nanoparticles are obtained for converting the magnetic field into heat.

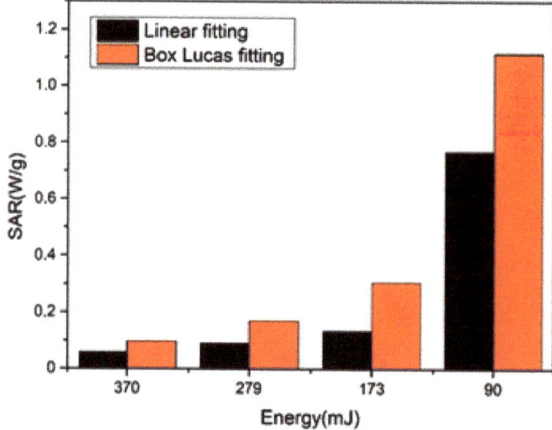

Figure 7. SAR (specific absorption rate) of the iron oxide nanoparticles as a function of laser energy, using linear fitting and Box Lucas fitting.

The difference between the methods is due to linear fitting being just an approximation of the Box Lucas Fitting, that, according to the literature, is the most accurate method [37].

4. Conclusions

Nanoparticles of iron oxide were obtained using the laser ablation method varying the power of the laser. The morphology was determined by DLS and STEM techniques. From this analysis, it was observed that there exists a strong competition between the energy of the laser and the temperature, not only on the particle size, but also on the standard deviation, since, as the energy increased, more of the material was ablated, which increased the concentration and the particle size, while the temperature produced divisions of the large sized particles. On the other hand, the composition was determined using XRD which showed wide peaks with small intensities, indicating nanosized domains. The sample exhibiting the greater particle sizes showed a peak with greater intensity, possibly because there is a greater quantity of crystallographic planes. The Fe3p and O1s peaks were identified in all samples confirming the iron oxide formation. Raman spectroscopy allows the identification of peaks belonging to hematite and maghemite. The temperature rise of iron oxide NPs was not greatly influenced by the energy change in magnetic hyperthermia measurements. Results show that, for hyperthermia applications, low laser energy is better because the SAR exhibited the greater value

Author Contributions: Conceptualization, M.J.R.-C. and R.O.; methodology, M.J.R.-C. and R.O.; validation, E.R.-P., C.D.A.-M. and A.M.; formal analysis, M.J.R.-C. and E.R.-P.; investigation, M.J.R.-C., R.O., C.D.A.-M. and A.M.; resources, R.O. and C.D.A.-M.; data curation, M.J.R.-C.; writing—original draft preparation, M.J.R.-C. and E.R.-P.; writing—review and editing, E.R.-P., C.D.A.-M. and A.M.; visualization, R.O.; supervision, E.R.-P.; project administration, C.D.A.-M.; funding acquisition, C.D.A.-M. All authors have read and agreed to the published version of the manuscript.

Funding: Financied by the Colombia Científica call, contract No FP44842-213-2018.

Acknowledgments: This paper is a result o the work developed under the Project PROGRAMA DE INVESTIGACIÓN RECONSTRUCCIÓN DEL TEJIDO SOCIAL EN ZONAS DE POS-CONFLICTO EN COLOMBIA Code SIGP: 57579 with the reserach "COMPETENCIAS EMPRESARIALES Y DE INNOVACIÓN PARA EL DESARROLLO ECONÓMICO Y LA INCLUSIÓN PRODUCTIVA DE LAS REGIONES AFECTADAS POR EL CONFLICTO COLOMBIANO" Código SIGP: 58907. Financed by the Colombia Científica call, contract No FP44842-213-2018.

Conflicts of Interest: The authors declare no conflict of interest.

References

1. Frey, N.A.; Peng, S.; Cheng, K.; Sun, S. Magnetic nanoparticles: Synthesis, functionalization, and applications in bioimaging and magnetic energy storage. *Chem. Soc. Rev.* **2009**, *38*, 2532–2542. [CrossRef] [PubMed]
2. Schrand, A.M.; Rahman, M.F.; Hussain, S.M.; Schlager, J.J.; Smith, D.A.; Ali, S.F. Metal-based nanoparticles and their toxicity assessment. *Wiley Interdiscip. Rev. Nanomed. Nanobiotechnol.* **2010**, *2*, 544–568. [CrossRef] [PubMed]
3. Huang, S.-H.; Juang, R.-S. Biochemical and biomedical applications of multifunctional magnetic nanoparticles: A review. *J. Nanoparticle Res.* **2011**, *13*, 4411–4430. [CrossRef]
4. Lassoued, A.; Lassoued, M.S.; Dkhil, B.; Ammar, S.; Gadri, A. Synthesis, photoluminescence and Magnetic properties of iron oxide (α-Fe_2O_3) nanoparticles through precipitation or hydrothermal methods. *Phys. E Low-Dimensional Syst. Nanostruct.* **2018**, *101*, 212–219. [CrossRef]
5. Benz, M. Superparamagnetism: Theory and Applications. *Superparamagn. Theory Appl.* **2012**, *22*, 1–27.
6. Akbarzadeh, A.; Samiei, M.; Davaran, S. Magnetic nanoparticles: Preparation, physical properties, and applications in biomedicine. *Nanoscale Res. Lett.* **2012**, *7*, 144. [CrossRef]
7. Simonsen, G.; Strand, M.; Øye, G. Potential applications of magnetic nanoparticles within separation in the petroleum industry. *J. Pet. Sci. Eng.* **2018**, *165*, 488–495. [CrossRef]
8. Chomoucka, J.; Drbohlavova, J.; Huska, D.; Adam, V.; Kizek, R.; Hubalek, J. Magnetic nanoparticles and targeted drug delivering. *Pharmacol. Res.* **2010**, *62*, 144–149. [CrossRef]
9. Karade, V.C.; Dongale, T.D.; Sahoo, S.C.; Kollu, P.; Chougale, A.; Patil, P.S.; Patil, P. Effect of reaction time on structural and magnetic properties of green-synthesized magnetic nanoparticles. *J. Phys. Chem. Solids* **2018**, *120*, 161–166. [CrossRef]
10. Levy, D.; Giustetto, R.; Hoser, A. Structure of magnetite (Fe_3O_4) above the Curie temperature: A cation ordering study. *Phys. Chem. Miner.* **2011**, *39*, 169–176. [CrossRef]
11. Wilson, D.; Langell, M. XPS analysis of oleylamine/oleic acid capped Fe_3O_4 nanoparticles as a function of temperature. *Appl. Surf. Sci.* **2014**, *303*, 6–13. [CrossRef]
12. Barcena, C.; Sra, A.K.; Gao, J. Applications of Magnetic Nanoparticles in Biomedicine. *Nanoscale Magn. Mater. Appl.* **2009**, *167*, 591–626. [CrossRef]
13. Gregorio-Jáuregui, K. "Item 1025/494 | Repositorio CIQA", Ciqa.repositorioinstitucional.mx. 2020. Available online: http://ciqa.repositorioinstitucional.mx/jspui/handle/1025/494 (accessed on 12 September 2020).
14. Lassoued, A.; Lassoued, M.S.; Dkhil, B.; Ammar, S.; Gadri, A. Synthesis, structural, morphological, optical and magnetic characterization of iron oxide (α-Fe_2O_3) nanoparticles by precipitation method: Effect of varying the nature of precursor. *Phys. E Low-Dimensional Syst. Nanostruct.* **2018**, *97*, 328–334. [CrossRef]
15. Marathe, K.; Doshi, P. Magnetic nanoparticles: Preparation, physical properties, and applications in biomedicine. *IEEE Int. Conf. Intell. Robot. Syst.* **2015**, *2015*, 2550–2555.
16. Bantz, C.; Koshkina, O.; Lang, T.; Galla, H.-J.; Kirkpatrick, C.J.; Stauber, R.H.; Maskos, M. The surface properties of nanoparticles determine the agglomeration state and the size of the particles under physiological conditions. *Beilstein J. Nanotechnol.* **2014**, *5*, 1774–1786. [CrossRef]
17. Wan, J.; Yao, Y.; Tang, G. Controlled-synthesis, characterization, and magnetic properties of Fe_3O_4 nanostructures. *Appl. Phys. A Mater. Sci. Process.* **2007**, *89*, 529–532. [CrossRef]
18. Zolghadr, S.; Kimiagar, S.; Davarpanah, A.M. Magnetic property of α-Fe_2O_3-GO nanocomposite. *IEEE Trans. Magn.* **2017**, *53*, 1–6. [CrossRef]
19. Nagarjuna, R.; Challagulla, S.; Ganesan, R.; Roy, S. High rates of Cr (VI) photoreduction with magnetically recoverable nano-Fe_3O_4@Fe_2O_3/Al_2O_3 catalyst under visible light. *Chem. Eng. J.* **2017**, *308*, 59–66. [CrossRef]
20. Cirtoaje, C.; Petrescu, E.; Stan, C.; Creanga, D. Ferromagnetic nanoparticles suspensions in twisted nematic. *Phys. E Low-Dimensional Syst. Nanostruct.* **2016**, *79*, 38–43. [CrossRef]
21. Yang, G.W. Laser Ablation in Liquids: Applications in the Synthesis of Nanocrystals. *Prog. Mater. Sci.* **2007**, *38*, 648–698. [CrossRef]
22. Santillán, J.M.J.; Arboleda, D.M.; Coral, D.F.; Van Raap, M.F.; Muraca, D.; Schinca, D.C.; Scaffardi, L.B. Optical and Magnetic Properties of Fe Nanoparticles Fabricated by Femtosecond Laser Ablation in Organic and Inorganic Solvents. *ChemPhysChem* **2017**, *18*, 1192–1209. [CrossRef] [PubMed]

23. Svetlichnyi, V.A.; Shabalina, A.V.; Lapin, I.N.; Goncharova, D.A.; Kharlamova, T.S.; Stadnichenko, A.I. Comparative study of magnetite nanoparticles obtained by pulsed laser ablation in water and air. *Appl. Surf. Sci.* **2019**, *468*, 402–410. [CrossRef]
24. Ismail, R.A.; Sulaiman, G.M.; Abdulrahman, S.A.; Marzoog, T.R. Antibacterial activity of magnetic iron oxide nanoparticles synthesized by laser ablation in liquid. *Mater. Sci. Eng. C* **2015**, *53*, 286–297. [CrossRef] [PubMed]
25. Jordan, A.; Maier-Hauff, K. Magnetic Nanoparticles for Intracranial Thermotherapy. *J. Nanosci. Nanotechnol.* **2007**, *7*, 4604–4606. [CrossRef] [PubMed]
26. Yasemian, A.R.; Almasi-Kashi, M.; Ramazani, A. Surfactant-free synthesis and magnetic hyperthermia investigation of iron oxide (Fe_3O_4) nanoparticles at different reaction temperatures. *Mater. Chem. Phys.* **2019**, *230*, 9–16. [CrossRef]
27. Zhang, D.; Gökce, B.; Barcikowski, S. Laser Synthesis and Processing of Colloids: Fundamentals and Applications. *Chem. Rev.* **2017**, *117*, 3990–4103. [CrossRef]
28. Plata, L. Caracterización de suspensiones coloidales de nanopartículas metálicas sintetizadas por ablación láser de pulsos ultracortos. Ph.D. Thesis, Facultad de Ciencias Exactas, La Plata, Argentina, 18 February 2018.
29. Riabinina, D.; Chaker, M.; Margot, J. Dependence of gold nanoparticle production on pulse duration by laser ablation in liquid media. *Nanotechnol.* **2012**, *23*, 135603. [CrossRef]
30. Amendola, V.; Meneghetti, M. What controls the composition and the structure of nanomaterials generated by laser ablation in liquid solution? *Phys. Chem. Chem. Phys.* **2013**, *15*, 3027–3046. [CrossRef]
31. Poulin, S.; França, R.; Moreau-Bélanger, L.; Sacher, E. Confirmation of X-ray Photoelectron Spectroscopy Peak Attributions of Nanoparticulate Iron Oxides, Using Symmetric Peak Component Line Shapes. *J. Phys. Chem. C* **2010**, *114*, 10711–10718. [CrossRef]
32. Sharma, H. Re: What are the Major Sources of Carbon Peaks in XPS Spectra of Non-Carbonous Samples? 2016. Available online: https://www.researchgate.net/post/What_are_the_major_sources_of_carbon_peaks_in_XPS_spectra_of_non-carbonous_samples/58555ceb615e276a3f466071/citation/download (accessed on 19 August 2020).
33. García-Benjume, M.L.; Espitia-Cabrera, M.I.; Contreras-García, M.E. Hierarchical macro-mesoporous structures in the system TiO_2-Al_2O_3, obtained by hydrothermal synthesis using Tween-20® as a directing agent. *Mater. Charact.* **2009**, *60*, 1482–1488. [CrossRef]
34. Hedayatnasab, Z.; Abnisa, F.; Daud, W.M.A.W. Review on magnetic nanoparticles for magnetic nanofluid hyperthermia application. *Mater. Des.* **2017**, *123*, 174–196. [CrossRef]
35. Dutz, S.; Hergt, R. Magnetic nanoparticle heating and heat transfer on a microscale: Basic principles, realities and physical limitations of hyperthermia for tumour therapy. *Int. J. Hyperth.* **2013**, *29*, 790–800. [CrossRef] [PubMed]
36. Ebrahimisadr, S.; Aslibeiki, B.; Asadi, R. Magnetic hyperthermia properties of iron oxide nanoparticles: The effect of concentration. *Phys. C Supercond.* **2018**, *549*, 119–121. [CrossRef]
37. Lanier, O.; Korotych, O.I.; Monsalve, A.G.; Wable, D.; Savliwala, S.; Grooms, N.W.F.; Nacea, C.; Tuitt, O.R.; Dobson, J. Evaluation of magnetic nanoparticles for magnetic fluid hyperthermia. *Int. J. Hyperth.* **2019**, *36*, 686–700. [CrossRef] [PubMed]

© 2020 by the authors. Licensee MDPI, Basel, Switzerland. This article is an open access article distributed under the terms and conditions of the Creative Commons Attribution (CC BY) license (http://creativecommons.org/licenses/by/4.0/).

Article

Photocatalytic Properties of Graphene/Gold and Graphene Oxide/Gold Nanocomposites Synthesized by Pulsed Laser Induced Photolysis

Li-Hsiou Chen [1], Huan-Ting Shen [1], Wen-Hsin Chang [2], Ibrahim Khalil [3,4], Su-Yu Liao [5], Wageeh A. Yehye [4], Shih-Chuan Liu [6], Chih-Chien Chu [2,7,*] and Vincent K. S. Hsiao [3,*]

[1] Department of Pulmonary Medicine, Taichung Tzu Chi Hospital, Buddhist Tzu Chi Medical Foundation, Taichung 427, Taiwan; chenindy@tzuchi.com.tw (L.-H.C.); ryenhat@tzuchi.com.tw (H.-T.S.)
[2] Department of Medical Applied Chemistry, Chung Shan Medical University, Taichung 40201, Taiwan; dadada5202008@gmail.com
[3] Department of Applied Materials and Optoelectronic Engineering, National Chi Nan University, Nantou 54561, Taiwan; ikhalilcu@gmail.com
[4] Nanotechnology & Catalysis Research Centre (NANOCAT), Institute for Advanced Studies, University of Malaya, Kuala Lumpur 50603, Malaysia; wdadboub@um.edu.my
[5] Department of Electrical Engineering, National Chi Nan University, Nantou 54561, Taiwan; suyu@ncnu.edu.tw
[6] Department of Health Diet and Industry Management, Chung Shan Medical University, Taichung 40201, Taiwan; liou@csmu.edu.tw
[7] Department of Medical Education, Chung Shan Medical University Hospital, Taichung 40201, Taiwan
* Correspondence: jrchu@csmu.edu.tw (C.-C.C.); kshsiao@ncnu.edu.tw (V.K.S.H.)

Received: 7 September 2020; Accepted: 3 October 2020; Published: 7 October 2020

Abstract: Graphene (Gr)/gold (Au) and graphene-oxide (GO)/Au nanocomposites (NCPs) were synthesized by performing pulsed-laser-induced photolysis (PLIP) on hydrogen peroxide and chloroauric acid ($HAuCl_4$) that coexisted with Gr or GO in an aqueous solution. A 3-month-long aqueous solution stability was observed in the NCPs synthesized without using surfactants and additional processing. The synthesized NCPs were characterized using absorption spectroscopy, transmission electron microscopy, Raman spectroscopy, energy dispersive spectroscopy, and X-ray diffraction to prove the existence of hybrid Gr/Au or GO/Au NCPs. The synthesized NCPs were further evaluated using the photocatalytic reaction of methylene blue (MB), a synthetic dye, under UV radiation, visible light (central wavelength of 470 nm), and full spectrum of solar light. Both Gr/Au and GO/Au NCPs exhibited photocatalytic degradation of MB under solar light illumination with removal efficiencies of 92.1% and 94.5%, respectively.

Keywords: graphene; graphene oxide; gold; nanocomposite; photolysis; photocatalysis

---

1. Introduction

With the rapid progress in industrialization, the wastewaters containing synthetic dyes, such as methylene blue (MB), are not easily degradable and potentially toxic. Hence, these dyes cause adverse effects to the aquatic organisms and humans, even when present in minute concentrations [1,2]. Hence, an effective treatment that enables the removal or degradation of synthetic dyes present in industrial effluents before discharging the effluents into the environment should be developed urgently for minimizing environmental pollution. To date, various techniques, such as adsorption, coagulation, aerobic oxidation, membrane filtration, chemical precipitation, ozonization, and photocatalysis have been employed for the removal of dyes from industrial wastewater [3–7]. Of these techniques, photocatalysis has emerged as a promising green technique and been widely used for the degradation of dyes because of its easy operation process, low cost, and energy conservation ability. Moreover, under the irradiation of light, particularly solar light, the photocatalytic treatment yields CO_2, H_2O, and other nontoxic compounds. A complete degradation of the dyes occurs at a large extent under ambient conditions of temperature and pressure [8–10]. However, the photocatalyst itself remains unchanged during the treatment process. Moreover, the process does not require any consumable chemicals, and landfill is not required for sludge disposal because no residue of the original material remains [9,11].

Of the different types of heterogeneous photocatalysts, TiO_2 has been the most studied [12,13]. Other semiconductor materials, such as ZnO, CuO, CeO_2, SnO_2, CdS, and ZnS, have also been used for the photocatalytic degradation and removal of organic dyes [14,15]. Semiconductor materials exhibit an excellent photoinduced redox reaction because of their specific electronic structure (filled valence and empty conduction bands). Hence, semiconductor materials are materials of choice for the photocatalytic removal of organic and synthetic dyes [16]. However, suitable matrix materials, such as zeolite, silica, activated carbon, carbon nanotube, graphene (Gr), and graphene oxide (GO), have been reported to enhance the photocatalytic degradation efficiency [13]. Gr has a two-dimensional planar structure comprising single-layer sp^2-bonded carbon atoms arranged in a honeycomb lattice structure, zero band-gap semiconductor properties with a large surface area, high charge carrier mobility, high adsorption capacity, and excellent electron transfer rate [3,17–19].

Gr and GO have extraordinary properties and have been widely used in various applications since their discovery in 2004. Moreover, Gr-based nanocomposites (NCPs), such as Gr-metal and Gr-metal-oxide NCPs, have been tested for exhibiting more advantageous properties than those of the individual material alone to explore a higher number of applications and to enhance the expected outcome or efficiency. Gr/gold (Au) and GO/Au NCPs are the most commonly used hybrid NCPs among the different hybrid NCPs because of the remarkable features of Au nanoparticles (NPs), such as higher chemical stability, catalytic activity, easy surface functionalization, and biocompatibility [20]. Gr reveals the improved activities as photocatalysis to remove the dyes. For example, Yang et al. synthesized a porous TiO_2/Gr composite that presented a higher degradation rate (6.5 times) of MB in commercial P25 [2]. Similarly, the TiO_2/reduced-GO (rGO) NCPs were prepared using different amounts of Gr (1–20%) through two different synthesis routes. When 10% Gr was added in TiO_2/rGO NCPs during the fabrication process, the highest photocatalytic activity was observed toward MB. This activity attained a value of 93% for the NCPs prepared using the sol-gel method that is followed by the hydrothermal treatment and a value of 82% for the NCPs prepared using the hydrothermal route only [13]. Some other Gr-based binary NCPs, such as the GO/TiO_2 hydrogel [12], TiO_2-doped calcined mussel shell [11], Gr/SnO_2 [15], Gr/ZnO [21], ZnO-decorated GO [22], Gr/CeO_2 [14], and Bi_2MoO_6/rGO aerogel NCPs [9] were evaluated for the photocatalytic degradation of MB under visible light illumination. Moreover, ternary composites, such as rGO/Fe_3O_4/TiO_2 [3], Fe_2O_3/Gr/CuO [5], GO/mesoporous TiO_2/Au [10], and MoO_3/Fe_2O_3/rGO NCPs [1] were investigated. However, few discussions on photocatalytic properties of Gr/Au or GO/Au NCPs were proposed [23].

Thus far, many fabrication methods, such as some green synthesis approaches, have been reported for the synthesis of Gr/Au or GO/Au NCPs [24,25]. The most common approach is the chemical reduction method that use different chemical reductants [26,27]. A major disadvantage of using a chemical reducing agent is the presence of this reducing agent in the final composite even after washing multiple times, thus considerably limiting their application. By contrast, Gr/metal or GO/metal NCP synthesis performed using photochemical or photolysis reaction has many advantages. For instance, avoiding the use of toxic chemical reducing agents, which are generated from reducing or capping agents, could provide intended applications, could prevent negative influences, and can present better control of Au over Gr or GO sheets without requiring a high temperature [28]. Pulsed laser-induced photolysis (PLIP) has been used for fabricating pure Au NPs [29], Au NP micelles [30,31], and the Au/Ag NP alloy [32]. Instead of using an Au target with pulsed laser ablation to fabricate Au NP, NCPs or the Au NP hybrid NCPs [33–36], chloroauric acid ($HAuCl_4$) can be used as a precursor and photodecomposing agent to obtain Au^{3+}. This ion is later reduced by the photon, and Au NPs are generated under the high-energy pulsed laser [29]. Because of the unique properties of ultra-short pulse duration and ultra-high peak power intensity, the pulsed-laser-induced synthesis technique, such as pulsed laser ablation and PLIP, is considered a clean and prompt technique to reduce metal ions into NPs without using any other chemical reagent. However, the long-term stability of synthesized Au NPs hinders their practical application. In this study, we reported the clean synthesis of Gr/Au and GO/Au NCPs by using the photolysis technique induced using a nanosecond pulsed laser operated at a 532 nm wavelength, as shown in Figure 1. A one-pot synthesis was conducted in which all precursors, $HAuCl_4$, hydrogen peroxide (H_2O_2) and Gr or GO, were dissolved or dispersed in an aqueous solution. Both H_2O_2 and $HAuCl_4$ undergo the photolysis process and generate Au^{3+} and HO–O$^\bullet$. The photolysis-induced Au^{3+} was further reduced by HO–O$^\bullet$, which is an effective one-electron reducing agent [37]. Gr/Au and GO/Au NCPs with long-term stability in aqueous solution could be synthesized by adding Gr or GO into the precursor solution. Different H_2O_2 and $HAuCl_4$ amounts were used to optimize the stability and photocatalytic properties of Gr/Au and GO/Au NCPs under different light exposure conditions, dark, UV, visible, and solar light.

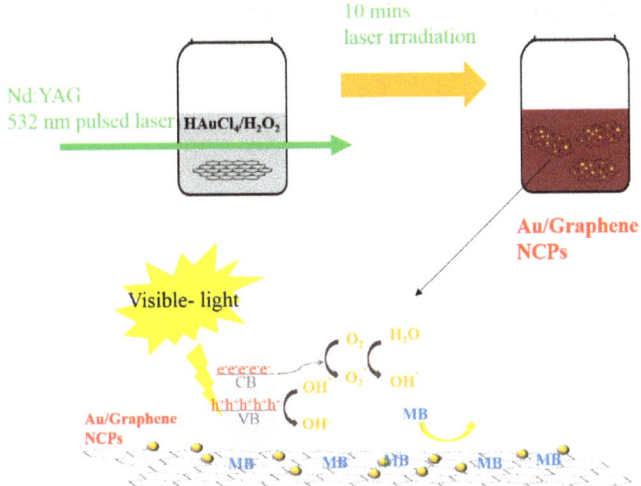

Figure 1. Schematic of the pulsed-laser-induced photolysis (PLIP) method to fabricate Au/Graphene nanocomposites (NCPs) used in visible photocatalysis.

2. Materials and Methods

2.1. Materials

Gr and GO powders were purchased from Ritedia Corporation (Hsinchu, Taiwan) and Tokyo Chemical Industry (Tokyo, Japan), respectively. MB ($C_{16}H_{13}N_3SCl$) was obtained from Katayama Chemical Company (Osaka, Japan). Tetrachloroauric (III) acid trihydrate ($HAuCl_4 \cdot 3H_2O$) and H_2O_2 (35 wt.% solution in water) were obtained from Acros Organics (Geel, Belgium). Milli-Q water (18.2 MΩ cm) was used as the aqueous solution throughout the study and was prepared in house.

2.2. Characterization

X-ray diffraction (XRD) patterns of all the synthesized samples were obtained in the 2θ range of 0–90° on a high-resolution X-ray diffractometer (Bruker AXS Gmbh, Karlsruhe, Germany) The elemental analysis of the samples was performed through X-ray energy dispersive spectrometry (EDS) (Bruker AXS Gmbh, Karlsruhe, Germany). The morphology, particle size, and distribution of the Au

NPs on Gr or GO sheets were investigated through transmission electron microscopy (TEM) on a JEM-2100 transmission electron microscope (JEOL, Tokyo, Japan). UV/Vis spectrophotometry was performed using a UV/Vis spectrophotometer (GENESYS 10S; Thermo Scientific, Waltham, MH, USA) to justify GO/Au NP and Gr/Au NP NCP syntheses and evaluate degradation percentage of all samples under different conditions. Raman spectra were obtained using a microRaman system (LabRAM HR800; HORIBA Jobin Yvon, Northampton, UK) with a helium–neon laser as the excitation source operating at a 633-nm wavelength and a 40× objective lens. The other instruments are as follows: delta ultrasonic cleaner (DC200H; Delta, Taipei, Taiwan), bench-top centrifuge (Velocity 14; Dynamica, Hong Kong), 2996 photodiode-array detector (PDA; Waters, MA, USA), solar simulator (Prosper OptoElectronics, Hualien, Taiwan), and universal centrifuges (Hermle LaborTechnik, Wehingen, Germany).

2.3. Gr/Au and GO/Au NCP Syntheses

Gr/Au and GO/Au NCPs were synthesized using pulsed-laser-induced photolysis. For synthesis, different concentrations and amounts of $HAuCl_4 \cdot 3H_2O$ were added to a 7 mL transparent glass bottle. Then, a different amount of H_2O_2 at fixed concentration of 10 mM and fixed 1 mL of Gr or GO suspension (0.1 mg/mL) was added to the bottle. The details of each sample fabricated using different experimental conditions are shown in Table S1. The aqueous suspension was then irradiated for 10 min with a pulsed Q-switch Nd:YAG laser (LS-2137U; LOTIS TII, Minsk, Belarus) with a wavelength of 532 nm, pulse duration of 6–7 ns, pulse repetition rate of 10 Hz, and fluence of approximately 37 mJ/cm^2. The laser beam was delivered in the middle of the precursor solution to ensure a homogenous light exposure to the sample. The precursor solution changed from light grey to reddish purple after 10 min. The pulsed-laser-treated solutions containing NCPs which have long-time stability were used for photocatalytic measurement without additional treatment.

2.4. Photocatalytic Activity Test

The photocatalytic activity of the synthesized binary NCPs was evaluated based on the degradation of MB in a homemade dark room setup, under UV light emitting diode (LED) (365 nm wavelength, 100 mW/cm^2), under laser diode (LD) (470 nm wavelength, 500 mW/cm^2), and under solar simulator (Xenon lamp, 50 W/cm^2). For the evaluation of photocatalytic degradation, 50 mL of 10 mg/L MB solution was transferred into a sample bottle. Then, 10 mg of the respective synthesized sample composite were added into the bottle, and the sample composite was uniformly dispersed by magnetic stirring at 550 rpm for 5–10 min. Each sample suspension was then tested for photocatalytic activity by placing in a dark room and under UV LED, visible LD, and solar light sources. The absorption spectra were recorded at different time intervals. The photocatalytic degradation rate and removal efficiency of MB were calculated by measuring the absorbance maximum value of the treated solutions at 664 nm by using the following equation: $(A_0 - A_t)/A_0 \times 100\%$, where A_0 and A_t are the initial and final absorbance values of MB solution at the specified time, respectively.

3. Results and Discussion

3.1. Laser-Induced Photolysis and Formation of Au NPs

Gr/Au and GO/Au NCPs were synthesized by performing Nd:YAG laser-induced photolysis of $HAuCl_4$ in the presence of aqueous H_2O_2 as a reducing agent and with the addition of Gr or GO. In the reaction mechanism, on laser excitation, $HAuCl_4$ dissociates into $AuCl_3^-$ and H_2O_2 undergoes the photolytic process and generates an effective one-electron reducing agent HO–O$^\bullet$ [37]. $AuCl_3^-$ is further reduced to Au$^\circ$ in the presence of metal ions through the following three steps. Finally, Au atoms aggregated to form Au NPs supported on the Gr or GO nanosheets. Hence, the reaction protocol is a fast and clean approach to generate NCPs by using H_2O_2 as a reducing agent in an aqueous solution.

Step 1: Au³⁺ Formation

$$HAuCl_4 \rightarrow H^+ + AuCl_4^- \tag{1}$$

$$AuCl_4^- \xrightarrow{h\nu} Au^{3+}Cl_3^- + Cl^\bullet \tag{2}$$

Step 2: H₂O₂ Photolysis

$$H_2O_2 \xrightarrow{h\nu} 2HO^\bullet \tag{3}$$

$$HO^\bullet + H_2O_2 \rightarrow HO-O^\bullet + H_2O \tag{4}$$

Step 3: Au NP Generation

$$HO-O^\bullet + Au^{3+} \rightarrow Au^{2+} + O_2 + H^+ \tag{5}$$

$$HO-O^\bullet + Au^{2+} \rightarrow Au^+ + O_2 + H^+ \tag{6}$$

$$HO-O^\bullet + Au^+ \rightarrow Au^0 + O_2 + H^+ \tag{7}$$

$$nAu^0 \rightarrow AuNPs \tag{8}$$

3.2. Stability of Synthesized NCP

UV/Vis spectroscopy was conducted to justify the formation of Au NPs either in the NP or NCP form. Upon laser excitation, the HAuCl₄ solution co-existing with H₂O₂ was converted to Au NPs. This conversion was indicated by the color change to a reddish-purple color and the shifting of the UV/Vis absorption peak from 290 nm (HAuCl₄ solution) to 526 nm (Figure S1). Here we used different combination of the concentration and the amount of HAuCl₄ and H₂O₂ to fabricate Gr/Au and GO/Au NCPs. Some experimental conditions failed to fabricate NCPs. Table S1 shows the experimental results indicating higher concentration of HAuCl₄ is not suitable to fabricate NCPs, for example, the sample number AuG6, AuG7, AuGO6 and AuGO7 of 4 mM concentration. The interesting result from sample number AuG3 and AuGO3 shows the total volume of precursors has to be controlled in a small volume (<3.5 mL) to achieve the success fabrication of NCPs. This result may be attributed to the small exposure area from the pulsed laser.

Similarly, the Gr/Au and GO/Au NCPs were characterized using the UV/Vis absorption spectra to identify the existence of Au NPs with the characteristic absorption peak at 526 and 535 nm, respectively (Figure 2a,b). All samples show red absorption peaks, and the GO/Au NCPs (AuGO1 and AuGO4) show extra absorption at around 700 nm wavelength. The absorbance from the sample solutions was measured continuously to determine the stability of the synthesized Au NPs, Gr/Au NCPs, and GO/Au NCPs, as shown in Figures S2 and S3, separately. The Gr/Au and GO/Au NCPs reveal high stability after 48 h compared with the unstable Au NPs in aqueous solution. Gr/Au NCPs (AuG1) reveal optimal stability with an absorbance value of 2.88 at a wavelength of 537 nm after 48 h, as shown in Figure 2c. This value decreased to 2.46 after 30 days with 2-nm blue shifting (Figure S2). The synthesized Gr/Au NCPs maintain high stability even after 30 days (AuG1 and AuG5); thus, Gr/Au NCPs are considered a better candidate for photocatalysis because of the high stability for a long duration. GO/Au NCPs reveal high stability in 1 week (Figure S3) by observing the characteristic absorbance located in the 537 nm, the same resonance peak as Gr/Au NCP, with only decreasing the absorbance intensity. However, the peak absorbance observed from GO/Au NCPs became even, as shown in Figure 2d, thus indicating a decrease in the amount of Au remaining inside the GO nanosheets.

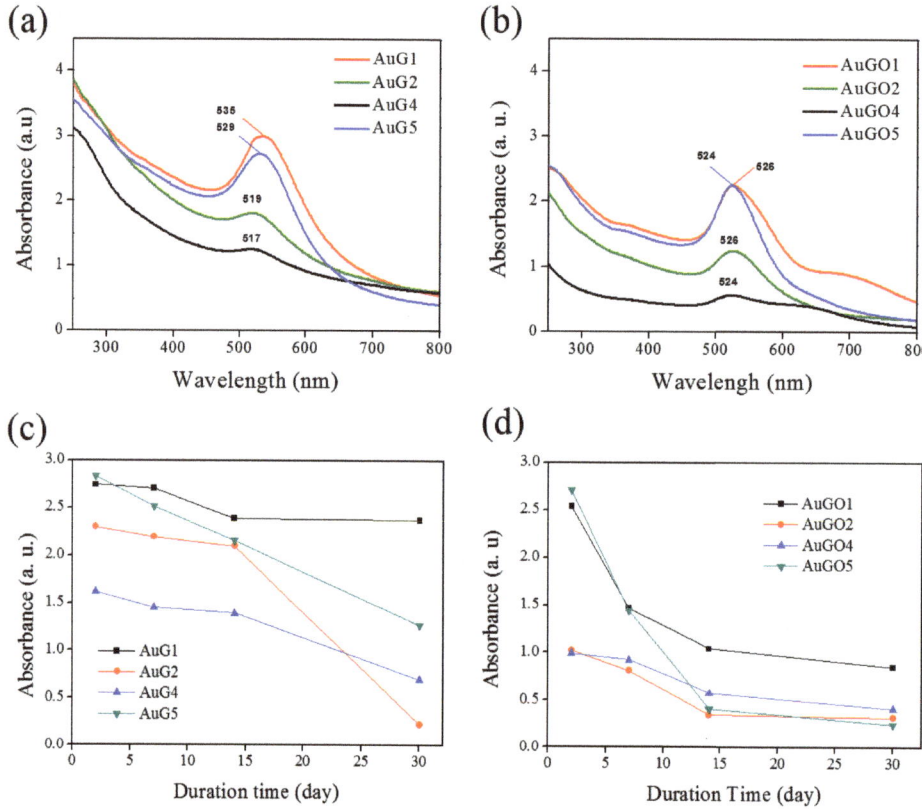

Figure 2. Characteristic absorption spectra of (**a**) Gr/Au NCPs and (**b**) GO/Au NCPs. Stability evaluation of (**c**) Gr/Au NCPs, and (**d**) GO/Au NCPs by recording the peak absorption of samples at different time intervals at room temperature.

3.3. Characterization of Synthesized NCPs

The XRD spectra of the synthesized Gr/Au NCPs that is presented in Figure 3a revealed four major diffraction peaks at 2θ of 38.22°, 44.44°, 64.64°, and 77.66° that correspond to (111), (200), (220), and (311) crystal planes of Au NPs [38], respectively. Moreover, Gr/Au NCPs exhibited another diffraction peak at the 2θ of 24° that corresponds to the (002) plane of Gr nanosheets, thus indicating the presence of Gr in the synthesized NCPs [5]. Similarly, XRD spectra of GO/Au NCPs, as shown in Figure 3b, presented the same diffraction peaks with minor red-shifting to 2θ of 39.1°, 45.3°, 65.52°, and 78.5°. GO/Au NCPs exhibited a broadened diffraction peak at a 2θ of 24° that corresponded to the (002) plane assigned to the Gr material, thus indicating that the starting material GO was reduced to Gr by using a Nd:YAG-induced pulsed laser irradiation at a wavelength of 532 nm [39]. Moreover, the intense peak at 38.22° indicates the preferential growth of Au NPs in the (111) plane. The observed XRD spectra of the synthesized NCPs reveal the successful fabrication of Au NPs over the Gr and GO nanosheets.

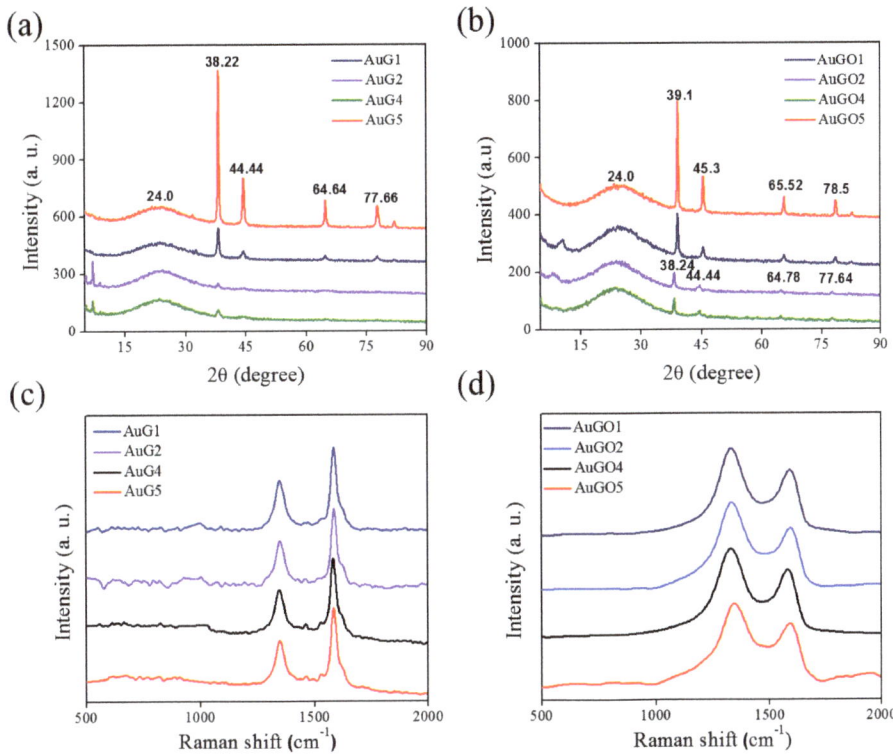

Figure 3. XRD spectra of (**a**) Gr/Au and (**b**) GO/Au NCPs, and Raman spectra of (**c**) Gr/Au and (**d**) GO/Au NCPs.

The Raman spectra of the Gr/Au NCPs were characterized using a sharp D band at approximately 1350 cm^{-1} and G band at approximately 1585 cm^{-1} (Figure 3c). The D band is attributable to the defect induced features resulting from vibrations of carbon atoms with dangling bonds and corresponds to the breathing mode of k-point phonons of the A_{1g} symmetry. However, the G band refers to the in-plane phonon vibration of sp2 carbon atoms with the E_{2g} symmetry. By contrast, GO/Au NCPs represent the typical Raman spectra with a strong and broad D band [39] at approximately 1330 cm^{-1} and G band at approximately 1590 cm^{-1} (Figure 3d). The observation of Raman spectra also proves the existence of the precursors (graphene, graphene oxide, and chloroauric acid) originally added into the aqueous solution.

The size, morphology, distribution, and elemental compositions of the Gr/Au and GO/Au NCPs were evaluated through TEM and EDS, as shown in Figures S4 and S5. Those findings clearly demonstrate that the NCPs are a combination of Au NPs and Gr or GO materials. Figure 4 shows the typical TEM images and corresponding size distribution of NCPs observed from AuG1 sample. Elemental analysis reveals that the samples comprise gold, carbon, and oxygen. We also evaluate the Au loading by theoretically calculating the amount of added HAuCl$_4$ and experimentally observed results from EDS, as shown in Table S2. The evaluations are matched in the fabricated samples using Gr as Au loading materials; however, the use of GO shows interesting results that both the amounts of HAuCl$_4$ and H$_2$O$_2$ have to be the same to get the maximum loading of Au NPs. All TEM images reveal that the Au NPs were distributed uniformly. Moreover, in few cases under agglomerated conditions, the Au NPs were distributed on smooth, almost transparent single- or few-layered wrinkled Gr or GO nanosheets. Hence, TEM images provide a direct evidence of the decoration of Au NPs on planar

Gr or GO sheets, which provide high active surface areas for the intended photocatalytic degradation of the dye.

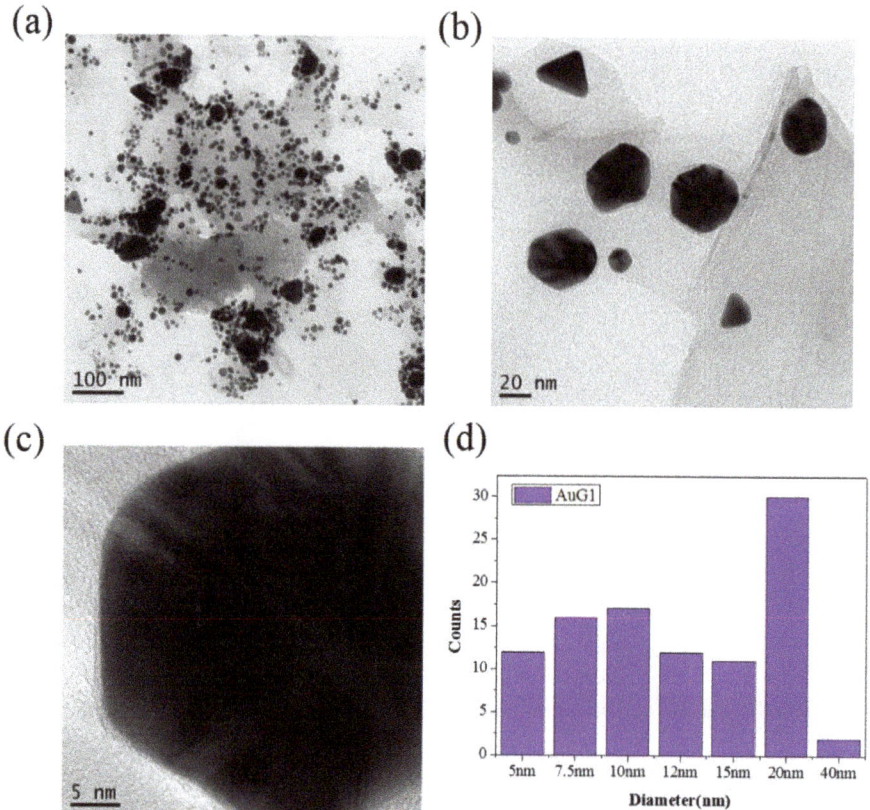

Figure 4. TEM image of different magnification (**a**–**c**) and (**d**) corresponding size distribution of Gr/Au NCPs (Sample AuG1).

3.4. Photocatalytic Activity of Synthesized Materials

The photocatalytic performance of the synthesized and selected Gr/Au and GO/Au NCPs were evaluated through the degradation ability of MB in the aqueous solution under four different conditions—in a darkroom, under UV radiation (wavelength, 365 nm), visible light (wavelength, 470 nm), and simulated solar light (wavelength, 350–915 nm). Aqueous MB solution (10 mg/L) was used as the control. The MB degradation activity of the samples treated in the dark and under UV and visible light was evaluated at an interval of 8 h for a total of 64 h. The samples irradiated with solar light were checked at an interval of 2 h for a total 10 h period. Both composites revealed slight degradation of the dye in the dark-room condition even after storage for a long period of 64 h. The degradation percentages for Gr/Au and GO/Au NCPs were 6.55% and 7.98%, respectively (Figures S6a and S7a, respectively). Hence, the dark-room condition confirmed that the absorption of dye from Gr or GO has no obvious catalytic role in dyes because the degradation ratio is significantly small. The absorption spectra of MB in a dark room, as shown in Figure 5 (black dot), at a gradual time interval did not reveal any change, thus indicating that MB cannot easily degrade by itself. However, the samples treated under UV radiation revealed a gradual increase in the degradation activity with increase in the treatment time and attained a value of 28.18% by using Gr/Au NCPs (Figure S6b) and 21.63% by using GO/Au NCPs (Figure S7b) after 64 h. By contrast, the degradation rate under visible illumination was really convincing and much greater than that under UV radiation. The NCPs under visible light (470 nm) illumination reveal a gradual increase in the degradation activity with increase in the exposure period. The lowest absorption peak was observed at 664 nm after a 64-h treatment period. Hence, the highest degradation percentage was 90.51% for Gr/Au NCPs (Figure S6c) and 86.09% by the GO/Au NCPs (Figure S7c).

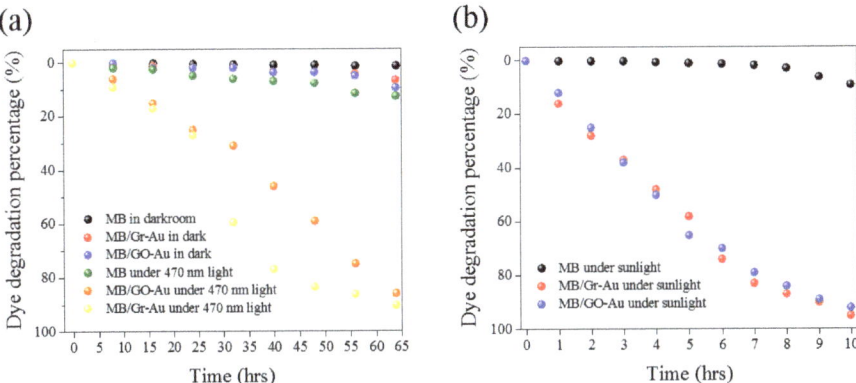

Figure 5. Photocatalytic activity of Gr/Au and GO/Au NCPs in MB solution as a function of the exposure time under (**a**) visible and (**b**) solar light.

The most potential photocatalytic degradation mechanism was achieved by treating the MB solution with NCPs under solar light. The degradation efficiency of the samples was checked at a 1 h interval when placed under a 100 W solar simulator. Both NCPs revealed strong MB degradation activities under solar light with gradual dye removal, thus presenting a declining trend of absorbance maxima at 664 nm with the progress of the exposure period. At the end of the 10 h exposure period, 94.5% of the dye was removed by the activity of Gr/Au NCPs (Figure S6d). However, the MB removal percentage was 92.1% when GO/Au NCPs were used (Figure S7d). With increase in the exposure period, the UV spectra also slightly shifted to the left from 664 nm. Throughout the study, the photocatalytic degradation of the MB, Gr/Au NP showed better degradation ability than GO/Au NP (Figure 5). Comparing the previous studies [23,24] of using similar NCPs to measure their photocatalytic ability, our NCPs required more time to achieve 100% degradation; however, other studies have no discussion regarding the long-term stability of NCPs. Therefore, our synthesized NCPs still show potential use in a hash environment where long-term stability is necessary to achieve photocatalytic ability.

Au NPs were formed and anchored immediately onto the Gr/GO nanosheets by using pulsed-laser-induced photolysis. Hence, the stable, binary NCPs with uniform dispersion of Au NPs were suspended inside the sample solution and actively participated in the MB degradation activity. A Schottky barrier was formed at the interface of Gr and Au particles because of the higher work function of Au than that of Gr. Thus, the degradation rate considerably increased under visible light irradiation [40–43]. The electrons injected on the Gr surface subsequently moved to Au NPs, which spatially separated MB^+ and electrons. Thus, the recombination process was delayed. The decrease in electron accumulation on the Gr surface evidently enhanced the continuous electron transfer from $MB^{\bullet}$ to Gr. Moreover, surface adsorbed O_2 could easily trap the electrons from Au particles to form various reactive oxidative species, such as $HO^{\bullet}$, because Au is an efficient electron donor, thus greatly improving MB degradation. Because of differences between the work functions of Gr (−4.42 eV) and Au NPs (−4.75 eV), direct electron transfer from $MB^{\bullet}$ to the semiconducting Gr or GO enables MB photodegradation [23]. Hence, the possible degradation mechanism of MB by the synthesized NCPs under solar radiation involves the excitation of the dye MB•, followed by electron transfer from MB• to Gr or GO. The electrons subsequently move to Au NPs and are trapped using O_2 to produce HO•. The MB•+ finally degrades by itself and/or is degraded by HO•. Thus, Gr/Au or GO/Au NCPs synthesized through a simple, straightforward PLIP method without requiring a potential reducing agent can be promising agents in the photocatalysis field.

4. Conclusions

Gr/Au and GO/Au NCP hybrid photocatalysts were successfully synthesized through a simple, fast, and clean technique by using PLIP. H_2O_2, a volatile reductant, was used in this technique during pulsed laser irradiation in the presence of $HAuCl_4$ that coexisted with aqueous Gr or GO suspension to generate NCPs within a few minutes with H_2O and O_2 generation as the final byproducts. Both Gr/Au and GO/Au NCPs were characterized by the uniform dispersion of Au NPs over GO and Gr sheets. Hence, a planar and large surface area is provided for the photocatalysis of MB. The photocatalytic performances of the synthesized NCPs were evaluated under UV, visible, and solar light. Both Gr/Au and GO/Au NCPs exhibited optimal photocatalytic performance under solar light and exhibited degradation efficiencies of 94.5% and 92.1% after a 10-h exposure period. Therefore, the current study outcome suggests that the NCPs obtained using the green synthesis approach are excellent and stable photocatalysts for the photocatalytic degradation of MB, a common water polluting agent. Hence, these NCPs have a large potential in the practical treatment of dye-contaminated wastewater through an ecofriendly fabrication process.

Supplementary Materials: The following are available online at http://www.mdpi.com/2079-4991/10/10/1985/s1, Figure S1: Characteristic absorption spectra of $HAuCl_4$ when used as an aqueous solution and the corresponding Au NPs, Gr/Au NCPs, and GO/Au NCPs. Figure S2: Characteristic absorption spectra of the corresponding Gr/Au NCPs by observing the absorption spectra of samples at different time intervals at room temperature. The inset shows the photo of the solution after 30 day storage at dark room. Figure S3: Characteristic absorption spectra of the corresponding GO/Au NCPs by observing the absorption spectra of samples at different time intervals at room temperature. The inset shows the photo of the solution after 30 day storage at dark room. Figure S4: Elementary analysis of Gr/Au and GO/Au NCPs. Figure S5: TEM image and corresponding size distribution of Gr/Au and GO/Au NCPs. Figure S6: Change in absorbance of MB when photocatalytic degradation is performed (a) in a darkroom and under (b) UV, (c) visible, and (d) solar light in the presence of Gr/Au NCPs. Figure S7: Change in the absorbance of MB when photocatalytic degradation is performed (a) in a darkroom and under (b) UV, (c) visible, and (d) solar lights in the presence of GO/Au NCPs. Table S1: Experimental conditions of nanocomposites fabricated by PLIP method. Table S2: Theoretical and experimental evaluation of Au loading.

Author Contributions: Conceptualization, L.-H.C., C.-C.C. and V.K.S.H.; methodology, H.-T.S., W.-H.C., S.-Y.L. and S.-C.L.; formal analysis, L.-H.C. and W.-H.C.; writing—original draft preparation, L.-H.C., H.-T.S. and I.K.; writing—review and editing, C.-C.C. and V.K.S.H.; supervision, C.-C.C., W.A.Y. and V.K.S.H; All authors have read and agreed to the published version of the manuscript.

Funding: This research was funded by Ministry of Science and Technology (MOST) of Taiwan under the project number MOST107-2113-M-040-003 and MOST107-2221-E-260-016-MY3.

Conflicts of Interest: The authors declare no conflict of interest.

References

1. Anjaneyulu, R.B.; Mohan, B.S.; Naidu, G.P.; Muralikrishna, R. Visible light enhanced photocatalytic degradation of methylene blue by ternary nanocomposite, MoO_3/Fe_2O_3/rGO. *J. Asian Ceram. Soc.* **2018**, *6*, 183–195. [CrossRef]
2. Yang, Y.; Xu, L.; Wang, H.; Wang, W.; Zhang, L. TiO_2/graphene porous composite and its photocatalytic degradation of methylene blue. *Mater. Des.* **2016**, *108*, 632–639. [CrossRef]
3. Benjwal, P.; Kumar, M.; Chamoli, P.; Kar, K.K. Enhanced photocatalytic degradation of methylene blue and adsorption of arsenic (iii) by reduced graphene oxide (rGO)–metal oxide (TiO_2/Fe_3O_4) based nanocomposites. *RSC Adv.* **2015**, *5*, 73249–73260. [CrossRef]
4. Roushani, M.; Mavaei, M.; Daneshfar, A.; Rajabi, H.R. Application of graphene quantum dots as green homogenous nanophotocatalyst in the visible-light-driven photolytic process. *J. Mater. Sci. Mater. Electron.* **2017**, *28*, 5135–5143. [CrossRef]

5. Nuengmatcha, P.; Porrawatkul, P.; Chanthai, S.; Sricharoen, P.; Limchoowong, N. Enhanced photocatalytic degradation of methylene blue using Fe_2O_3/graphene/CuO nanocomposites under visible light. *J. Environ. Chem. Eng.* **2019**, *7*, 103438. [CrossRef]
6. LimaBeluci, N.; Mateus, G.A.P.; Miyashiro, C.S.; Homem, N.C.; Gomes, R.G.; Fagundes-Klen, M.R.; Bergamasco, R.; Vieira, A.M.S. Hybrid treatment of coagulation/flocculation process followed by ultrafiltration in TiO_2-modified membranes to improve the removal of reactive black 5 dye. *Sci. Total Environ.* **2019**, *664*, 222–229.
7. Fayoud, N.; Tahiri, S.; Alami Younssi, S.; Albizane, A.; Gallart- Mateu, D.; Cervera, M.L.; de la Guardia, M. Kinetic, isotherm and thermodynamic studies of the adsorption of methylene blue dye onto agro-based cellulosic materials. *Desalin. Water Treat.* **2016**, *57*, 16611–16625. [CrossRef]
8. Nguyen, C.H.; Fu, C.-C.; Juang, R.-S. Degradation of methylene blue and methyl orange by palladium-doped TiO2 photocatalysis for water reuse: Efficiency and degradation pathways. *J. Clean. Prod.* **2018**, *202*, 413–427. [CrossRef]
9. Liu, X.; Wang, J.; Dong, Y.; Li, H.; Xia, Y.; Wang, H. One-step synthesis of Bi_2MoO_6/reduced graphene oxide aerogel composite with enhanced adsorption and photocatalytic degradation performance for methylene blue. *Mater. Sci. Semicon. Proc.* **2018**, *88*, 214–223. [CrossRef]
10. Yang, Y.; Ma, Z.; Xu, L.; Wang, H.; Fu, N. Preparation of reduced graphene oxide/meso-TiO_2/AuNPs ternary composites and their visible-light-induced photocatalytic degradation n of methylene blue. *Appl. Surf. Sci.* **2016**, *369*, 576–583. [CrossRef]
11. Echabbi, F.; Hamlich, M.; Harkati, S.; Jouali, A.; Tahiri, S.; Lazar, S.; Lakhmiri, R.; Safi, M. Photocatalytic degradation of methylene blue by the use of titanium-doped Calcined Mussel Shells CMS/TiO_2. *J. Environ. Chem. Eng.* **2019**, *7*, 103293. [CrossRef]
12. Fu, Z.; Zhang, S.; Fu, Z. Preparation of Multicycle GO/TiO2 Composite Photocatalyst and Study on Degradation of Methylene Blue Synthetic Wastewater. *Appl. Sci.* **2019**, *9*, 3282. [CrossRef]
13. Atout, H.; Álvarez, M.G.; Chebli, D.; Bouguettoucha, A.; Tichit, D.; Llorca, J.; Medina, F. Enhanced photocatalytic degradation of methylene blue: Preparation of TiO_2/reduced graphene oxide nanocomposites by direct sol-gel and hydrothermal methods. *Mater. Sci. Bull.* **2019**, *95*, 578–587. [CrossRef]
14. Kim, Y.; Cho, B.; Ko, W. Photocatalytic degradation of methylene blue by graphene impregnated with CeO_2 nanoparticles under ultrasonic irradiation. *Asian J. Chem.* **2013**, *25*, 8178–8180. [CrossRef]
15. Seema, H.; Kemp, K.C.; Chandra, V.; Kim, K.S. Graphene–SnO_2 composites for highly efficient photocatalytic degradation of methylene blue under sunlight. *Nanotechnology* **2012**, *23*, 355705. [CrossRef]
16. Khoa, N.T.; Kim, S.W.; Yoo, D.; Cho, S.; Kim, E.J.; Hahn, S.H. Fabrication of Au/graphene-wrapped ZnO-nanoparticle-assembled hollow spheres with effective photoinduced charge transfer for photocatalysis. *ACS Appl. Mater. Interfaces* **2015**, *7*, 3524–3531. [CrossRef]
17. Khalil, I.; Rahmati, S.; Julkapli, N.M.; Yehye, W.A. Graphene metal nanocomposites—Recent progress in electrochemical biosensing applications. *J. Indust. Eng. Chem.* **2018**, *59*, 425–439. [CrossRef]
18. Allen, M.J.; Tung, V.C.; Kaner, R.B. Honeycomb carbon: A review of graphene. *Chem. Rev.* **2010**, *110*, 132–145. [CrossRef]
19. Geim, A.K. Graphene: Status and prospects. *Science* **2009**, *324*, 1530–1534. [CrossRef]
20. Khalil, I.; Julkapli, N.M.; Yehye, W.A.; Basirun, W.J.; Bhargava, S.K. Graphene–gold nanoparticles hybrid—synthesis, functionalization, and application in a electrochemical and surface-enhanced raman scattering biosensor. *Materials* **2016**, *9*, 406. [CrossRef]
21. Fan, H.; Zhao, X.; Yang, J.; Shan, X.; Yang, L.; Zhang, Y.; Li, X.; Gao, M. ZnO–graphene composite for photocatalytic degradation of methylene blue dye. *Catal. Comm.* **2012**, *29*, 29–34. [CrossRef]
22. Atchudan, R.; Immanuel Edison, T.N.J.; Perumal, S.; Karthikeyan, D.; Lee, Y.R. Facile synthesis of zinc oxide nanoparticles decorated graphene oxide composite via simple solvothermal route and their photocatalytic activity on methylene blue degradation. *J. Photochem. Photobiol. B Biol.* **2016**, *162*, 500–510. [CrossRef] [PubMed]
23. Xiong, Z.; Zhang, L.L.; Ma, J.; Zhao, X.S. Photocatalytic degradation of dyes over graphene–gold nanocomposites under visible light irradiation. *Chem. Commun.* **2010**, *46*, 6099–6101. [CrossRef] [PubMed]
24. Šimšíková, M.; Bartoš, M.; Keša, P.; Šikola, T. Green approach for preparation of reduced graphene oxide decorated with gold nanoparticles and its optical and catalytic properties. *Mater. Chem. Phys.* **2016**, *177*, 339–345. [CrossRef]

25. Hurtado, R.B.; Cortez-Valadez, M.; Aragon-Guajardo, J.R.; Cruz-Rivera, J.J.; Martínez-Suárez, F.; Flores-Acosta, M. One-step synthesis of reduced graphene oxide/gold nanoparticles under ambient conditions. *Arabian J. Chem.* **2020**, *13*, 1633–1640. [CrossRef]
26. Goncalves, G.; Marques, P.A.A.P.; Granadeiro, C.M.; Nogueira, H.I.S.; Singh, M.K.; Grácio, J. Surface modification of graphene nanosheets with gold nanoparticles: The role of oxygen moieties at graphene surface on gold nucleation and growth. *Chem. Mater.* **2009**, *21*, 4796–4802. [CrossRef]
27. Chuang, M.-K.; Lin, S.-W.; Chen, F.-C.; Chu, C.-W.; Hsu, C.-S. Gold nanoparticle-decorated graphene oxides for plasmonic-enhanced polymer photovoltaic devices. *Nanoscale* **2014**, *6*, 1573–1579. [CrossRef]
28. Moussa, S.; Atkinson, G.; El-Shall, M.S.; Shehata, A.; AbouZeid, K.M.; Mohamed, M.B. Laser assisted photocatalytic reduction of metal ions by graphene oxide. *J. Mater. Chem.* **2011**, *21*, 9608–9619. [CrossRef]
29. Zhao, C.; Qu, S.; Qiu, J.; Zhu, C. Photoinduced formation of colloidal Au by a near-infrared femtosecond laser. *J. Mater. Res.* **2003**, *18*, 1710–1714. [CrossRef]
30. Kurihara, K.; Kizling, J.; Stenius, P.; Fendler, J.H. Laser and pulse radiolytically induced colloidal gold formation in water and in water-in-oil microemulsions. *J. Am. Chem. Soc.* **1983**, *105*, 2574–2579. [CrossRef]
31. Bronstein, L.; Chernyshov, D.; Valetsky, P.; Tkachenko, N.; Lemmetyinen, H.; Hartmann, J.; Förster, S. Laser photolysis formation of gold colloids in block copolymer micelles. *Langmuir* **1999**, *15*, 83–91. [CrossRef]
32. Kuladeep, R.; Jyothi, L.; Shadak Alee, K.; Deepak, K.L.N.; Narayana Rao, D. Laser-assisted synthesis of Au-Ag alloy nanoparticles with tunable surface plasmon resonance frequency. *Opt. Mater. Express* **2012**, *2*, 161–172. [CrossRef]
33. Shirk, M.D.; Molian, P.A. A review of ultrashort pulsed laser ablation of materials. *J. Laser Appl.* **1998**, *10*, 18–28. [CrossRef]
34. Zeng, H.; Du, X.-W.; Singh, S.C.; Kulinich, S.A.; Yang, S.; He, J.; Cai, W. Nanomaterials via laser ablation/irradiation in liquid: A review. *Adv. Funct. Mater.* **2012**, *22*, 1333–1353. [CrossRef]
35. Kubiliūtė, R.; Maximova, K.A.; Lajevardipour, A.; Yong, J.; Hartley, J.S.; Mohsin, A.S.M.; Blandin, P.; Chon, J.; Sentis, M.; Stoddart, P.R.; et al. Ultra-pure, water-dispersed Au nanoparticles produced by femtosecond laser ablation and fragmentation. *Int. J. Nanomed.* **2013**, *8*, 2601–2611.
36. Yu, Y.; Yan, L.; Si, J.; Xu, Y.; Hou, X. Femtosecond laser assisted synthesis of gold nanorod and graphene hybrids and its photothermal property in the near-infrared region. *J. Phys. Chem. Solids* **2019**, *132*, 116–120. [CrossRef]
37. McGilvray, K.L.; Granger, J.; Correia, M.; Banks, J.T.; Scaiano, J.C. Opportunistic use of tetrachloroaurate photolysis in the generation of reductive species for the production of gold nanostructures. *Phys. Chem. Chem. Phys.* **2011**, *13*, 11914–11918. [CrossRef] [PubMed]
38. Khalil, I.; Chou, C.-M.; Tsai, K.-L.; Hsu, S.; Yehye, W.A.; Hsiao, V.K.S. Gold Nanofilm-Coated Porous Silicon as Surface-Enhanced Raman Scattering Substrate. *Appl. Sci.* **2019**, *9*, 4806. [CrossRef]
39. Ghasemi, F.; Razi, S.; Madanipour, K. Single-step laser-assisted graphene oxide reduction and nonlinear optical properties exploration via CW laser excitation. *J. Electron. Mater.* **2018**, *47*, 2871–2879. [CrossRef]
40. Li, X.; Zhu, J.; Wei, B. Hybrid nanostructures of metal/two-dimensional nanomaterials for plasmon-enhanced applications. *Chem. Soc. Rev.* **2016**, *45*, 3145–3187. [CrossRef]
41. Ren, R.; Li, S.; Li, J.; Ma, J.; Liu, H.; Ma, J. Enhanced catalytic activity of Au nanoparticles self-assembled on thiophenol functionalized graphene. *Catal. Sci. Technol.* **2015**, *5*, 2149–2156. [CrossRef]
42. Biroju, R.K.; Choudhury, B.; Giri, P.K. Plasmon-enhanced strong visible light photocatalysis by defect engineered CVD graphene and graphene oxide physically functionalized with Au nanoparticles. *Catal. Sci. Technol.* **2016**, *6*, 7101–7112. [CrossRef]
43. Movahed, S.K.; Fakharian, M.; Dabiri, M.; Bazgir, A. Gold nanoparticle decorated reduced graphene oxide sheets with high catalytic activity for ullmann homocoupling. *RSC Adv.* **2014**, *4*, 5243–5247. [CrossRef]

© 2020 by the authors. Licensee MDPI, Basel, Switzerland. This article is an open access article distributed under the terms and conditions of the Creative Commons Attribution (CC BY) license (http://creativecommons.org/licenses/by/4.0/).

Article

ZnO Nanoparticle/Graphene Hybrid Photodetectors via Laser Fragmentation in Liquid

Kristin Charipar *, Heungsoo Kim, Alberto Piqué and Nicholas Charipar

U.S. Naval Research Laboratory, 4555 Overlook Ave., SW, Washington, DC 20375, USA;
heungsoo.kim@nrl.navy.mil (H.K.); alberto.pique@nrl.navy.mil (A.P.); nicholas.charipar@nrl.navy.mil (N.C.)
* Correspondence: kristin.charipar@nrl.navy.mil

Received: 28 July 2020; Accepted: 16 August 2020; Published: 21 August 2020

Abstract: By combining the enhanced photosensitive properties of zinc oxide nanoparticles and the excellent transport characteristics of graphene, UV-sensitive, solar-blind hybrid optoelectronic devices have been demonstrated. These hybrid devices offer high responsivity and gain, making them well suited for photodetector applications. Here, we report a hybrid ZnO nanoparticle/graphene phototransistor that exhibits a responsivity up to 4×10^4 AW^{-1} and gain of up to 1.3×10^5 with high UV wavelength selectivity. ZnO nanoparticles were synthesized by pulsed laser fragmentation in liquid to attain a simple, efficient, ligand-free method for nanoparticle fabrication. By combining simple fabrication processes with a promising device architecture, highly sensitive ZnO nanoparticle/graphene UV photodetectors were successfully demonstrated.

Keywords: graphene; laser fragmentation; laser processing; nanoparticles; ultraviolet photodetection; zinc oxide

1. Introduction

Optoelectronic devices utilizing graphene have been studied extensively over the past decade, paving the way for the fabrication of thin, lightweight, highly efficient devices. The advantages of graphene in sensor applications are numerous, including high mobility (>10^4 cm^2 V^{1}s^{-1}) [1], optical transparency (~2.3% for monolayer) [2,3], excellent mechanical and chemical stability, and an inherently ultrathin, flexible form factor [3,4]. While a reported absorption of ~2.3% is quite large for monolayer materials [2], it is insufficient for high quantum efficiency optoelectronic devices. Additionally, because the ultrafast exciton lifetime of graphene leads to fast carrier recombination times [5,6], photocurrent development is hindered making graphene alone not ideal for photoconductor applications [6,7]. Nonetheless, by combining photosensitive nanostructures, such as metal oxide semiconductors, with graphene as a transport layer, many enhanced effects are observed [8]. These photodetectors can be tailored to operate in specific spectral ranges depending on the bandgap of the material (e.g., ZnO for ultraviolet detection, PbS for near-infrared detection [7], and Ti$_2$O$_3$ for mid-infrared detection [9]).

Because of its wide band gap (~3.3 eV) and high exciton binding energy (~60 meV), zinc oxide (ZnO) is a promising candidate for ultraviolet (UV)-sensitive hybrid photodetectors [8]. In addition, ZnO is radiation-resistant and non-toxic making it an attractive material for wearable sensor technologies. While the inherent mobility of bulk crystalline ZnO is not high (~200 cm^2 V^{-1}s^{-1} at room temperature) [10], its combination with graphene offers an efficient charge transport pathway due to the high mobility of graphene resulting in significant photoconductive gain. Moreover, ZnO has enhanced wavelength selectivity in the UV range, while graphene provides broadband optical transparency, thus enabling solar-blind photodetectors. As a result, photodetectors that combine ZnO and graphene have been studied recently by many research groups. A wide assortment of ZnO structures has been studied for UV photodetection, including nanoparticles [11–15], nanowires [16–19], and thin films [20–23]. Because of the unique properties afforded by nanoscale structures, an optimal size is achieved when the ZnO nanostructures approach the Debye length, which is on the order of ~18 nm [24]. At this size scale, the surface depletion effect is maximized, shortening the carrier transit time, leading to photoconductive gain. In addition, by using nanoparticles instead of bulk ZnO thin films, the high surface-to-volume ratio provides a high density of hole trap states for charge transfer into the underlying graphene layer [14].

Many different techniques have been utilized to fabricate ZnO nanostructures for hybrid graphene photodetector applications, such as hydrolysis methods for nanoparticle fabrication [15] and hydrothermal [18] and chemical vapor deposition [19] methods for nanowire synthesis. To simplify the nanostructure fabrication process, we demonstrate the use of pulsed laser fragmentation in liquid (PLFL) as an alternative for nanoparticle generation. This technique relies on ultrafast laser pulses to generate nanoparticles in solution via various physicochemical processes. It offers the advantages of simple experimental set-up, control over size distribution and particle morphology, and the potential to maintain the stoichiometry of the original particle [25]. PLFL of Ag nanoclusters was first demonstrated over two decades ago by Kamat et al. [26], followed by many other research efforts focused primarily on noble metal nanoparticles [27,28]. In recent years, the use of PLFL has extended well beyond Au and Ag to other metals [29], alloys [30], and semiconductors [31], including indium tin oxide [32] and ZnO [33,34].

Here, we have demonstrated PLFL for the synthesis of ZnO nanoparticles with a bimodal size distribution (~18 nm and 46 nm). These nanoparticles were integrated into graphene-based hybrid phototransistors, which were then characterized to determine the optical and electrical performance, including wavelength selectivity and responsivity.

2. Materials and Methods

Phototransistor devices were fabricated using standard wet transfer [35–38] microfabrication processing techniques. Highly-doped (0.001 $\Omega\cdot$cm–0.005 $\Omega\cdot$cm) Si wafers with a 285 nm thermal oxide layer were laser-diced to 2 cm × 2 cm. The SiO$_2$ on the backside of the Si wafer was laser-micromachined to expose the highly conductive Si for device back-gating. Graphene on Cu foils (Graphene Supermarket, Ronkonkoma, NY, USA) were spin-coated with poly(methyl methacrylate), or PMMA (Kayaku Advanced Materials, Westborough, MA, USA, 495 PMMA A2) resulting in a ~600 nm thick layer. The foils were then baked at 100 °C in air for 2 min on a hot plate. The graphene on the backside of the Cu foil was etched by floating the foil on a 10% HNO$_3$ solution for 3 min followed by rinsing with deionized water. The Cu was removed by etching in a ferric chloride solution for 2 h. The remaining PMMA/graphene film was then floated on a dilute 2% HCl solution to remove any particulates introduced during the Cu etching process. The film was rinsed in deionized water before wet transfer. The PMMA/graphene film was then transferred onto the SiO$_2$/Si substrate and allowed to air dry. A small droplet of PMMA was drop casted onto the surface of the PMMA/graphene to encourage flattening of the film, followed by air drying. The PMMA was then removed with acetone, followed by rinsing in isopropanol and then water. After processing the graphene, source and drain electrodes (5 nm Ti/150 nm Au) were deposited via electron beam evaporation using a shadow mask that was laser-micromachined from a thin (75 μm) polyetherimide sheet. After electrode deposition, isolation lines were laser-micromachined around each device on the chip, yielding active device areas of 2 mm × 1 mm.

ZnO nanoparticles were produced using pulsed laser fragmentation in liquid (PLFL) [25]. ZnO powders were used as received (Millipore Sigma, St. Louis, MO, USA 140 nm avg. diameter) and dispersed in deionized water at 0.1 wt%. PLFL was performed using a pulsed femtosecond laser system (Light Conversion Ltd, Vilnius, Lithuania, Pharos Yb:KGW laser, λ = 1030 nm, 10 kHz, pulse duration ~200 fs). The ZnO particle solution was laser-treated for 1 h at a laser pulse energy of 17 μJ. The laser was focused with a 10 cm focal length lens into a quartz cuvette containing the

ZnO/water solution. Because the laser spot was focused into a cuvette containing the ZnO solution, it was difficult to determine an exact fluence as the laser light was absorbed and scattered by the ZnO particles as the beam converged into focus. After PLFL, the water was exchanged for ethanol via centrifugation and decanting. The final solution was sonicated to re-disperse the nanoparticles and break apart any agglomerates. The final ZnO/graphene devices were fabricated by drop-casting the ZnO nanoparticle ethanol solution onto the active graphene area of the previously fabricated phototransistors. The ethanol was allowed to evaporate in air resulting in a film of ZnO nanoparticles across the entire device. ZnO nanoparticles fabricated by PLFL were characterized via scanning electron microscopy (JEOL USA Inc., Peabody, MA, USA, JSM7001F), particle analysis, and photoluminescence measurements (343 nm excitation source) to determine final particle size, distribution, and quality, respectively. Additionally, the optical absorption spectra of ZnO nanoparticle solutions were collected using a UV/Vis spectrophotometer (JASCO Inc., Easton, MD, USA, V670). A schematic of the final phototransistor device with ZnO nanoparticles dispersed on the surface is shown in Figure 1.

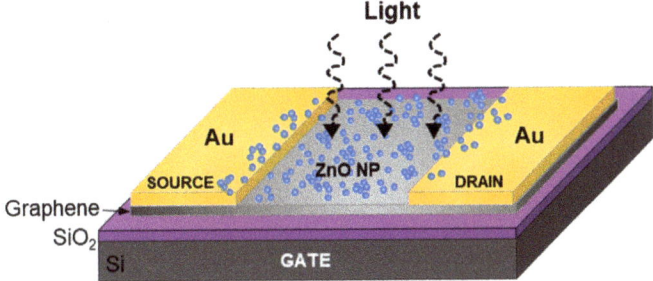

Figure 1. Schematic of the ZnO nanoparticle/graphene phototransistor architecture (not to scale).

The optical properties of the fabricated graphene transistors before nanoparticle deposition were characterized via Raman spectroscopy (WITec Instruments Corp., Knoxville, TN, USA, alpha300 RAS) which revealed the quality of the graphene layer. Optoelectronic characterization was performed using UV illumination that was fiber coupled from a monochromator into a 10× objective, mounted on a probe station. The light intensity was adjusted by a computer-controlled attenuator, maintaining a uniform spot size of ~2 mm for all experiments. Electrical characterization, including the drain and gate sweeps as well as temporal measurements, was conducted using a semiconductor characterization system (Keithley Instruments, Solon, OH, USA, 4200SCS).

3. Results and Discussion

3.1. Pulsed Laser Fragmentation in Liquid (PLFL)

Because of the unique properties afforded by nanoscale materials, different methods have been developed for simple and efficient fabrication, including wet chemical synthesis [12], sol gel [39], thermal vaporization [40], and pyrolytic reactions [41]. Chemical synthesis methods are often time-consuming, complex multi-step processes, involving a variety of potentially hazardous materials and solvents. Additionally, these chemical synthesis methods often require the use of ligands either during or after fabrication of nanoparticles [42], which can affect nanoparticle packing and electrical transport, ultimately impacting device performance. Alternatively, a ligand-free synthesis technique that has been widely studied is pulsed laser ablation in liquid (PLAL), which relies on laser–matter interactions for the generation of nanostructures typically from bulk materials [43]. There has been much research conducted on the generation of ZnO nanoparticles via PLAL; however, typical experiments involve the use of either a solid Zn or a ZnO target submerged in a liquid medium [43–47]. In this work, we begin with a ZnO particle powder dispersed in water and use PLFL to create smaller, more uniform nanoparticles. While PLAL is performed using a solid target material, PLFL relies on micro- or nano-sized particles suspended in liquid, which is shown schematically in Figure 2. Similar to PLAL, the resulting size and shape of the particles produced by PLFL can be controlled via pulse energy, pulse duration, and the initial material properties of the target material.

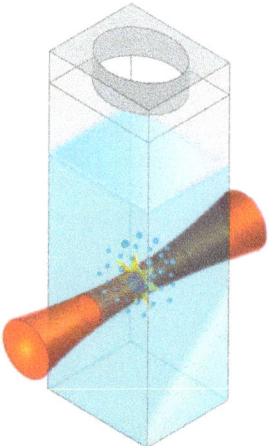

Figure 2. Schematic of the pulsed laser fragmentation in liquid (PLFL) process, where an aqueous solution of ZnO particles is irradiated with a femtosecond IR laser to synthesize smaller ZnO nanoparticles (not to scale).

The mechanisms responsible for nanoparticle formation via PLFL and the effect that initial size, concentration and material have on the resulting particle size have been studied extensively [25–34]. However, the exact mechanisms responsible for nanoparticle formation via PLFL are not entirely understood. Nonetheless, two mechanisms are often used to explain the formation of smaller particles, including photothermal evaporation and Coulombic explosion [25]. During photothermal evaporation, the laser energy is absorbed by the particle, causing surface evaporation when the boiling point of the material is exceeded [28,31]. When the vaporized species cool, they condense into smaller particles. During Coulombic explosion, electrons are ejected from the original particle, generating ionized nanoparticles. These particles then undergo additional fragmentation because of electrical charge repulsion [48]. These two mechanisms can occur independently or can compete depending on material properties and operating conditions (laser pulse duration and laser fluence, to name a few). Additionally, PLFL can often be accompanied with some degree of simultaneous laser melting. During the PLFL process, laser attenuation in the liquid can create a fluence gradient, where a portion of the liquid experiences a low fluence regime that results in laser melting of the particles. Thus, the resulting nanoparticle size is often a complex balance between the laser fragmentation process which reduces particle size with a laser melting process which can cause the produced nanoparticles to coalesce and grow [25].

There are several advantages to this technique, including simple experimental set-up and the ability to maintain complex stoichiometries with narrow particle size distributions [49]. The mechanism of particle formation allows for nanoparticle surfaces that are ligand-free [25]. While traditional solution-based nanoparticle synthesis methods often involve the use of ligands either during or after synthesis [42,50], PLFL offers a ligand-free fabrication method [25]. Solution-based chemistry techniques can leave insulating surface chemistries on the nanoparticle that can be time consuming to remove before device integration. This is important because ligands can interfere with optoelectronic device performance and efficiency by inhibiting charge transport and preventing the close-packing of particles [51,52]. Additionally, PLFL offers a method to produce bimodal size distributions, further enhancing particle packing which can improve performance [53].

Both the original ZnO powder particles before PLFL and the resulting smaller nanoparticles after PLFL can be seen in Figure 3a,b, respectively. The largest particles observed after laser processing were ~65 nm and represent a very small fraction (<1%) of the overall particle count. During laser processing, the completion of the fragmentation process was determined via optical scattering. The laser-processed nanoparticles resulted in a bimodal size distribution (Figure 3c), where a portion of the particles was ~18 nm and another portion was ~46 nm, while the original ZnO particles showed a uniform distribution centered around ~140 nm. This bimodal distribution of nanoparticles as a result of PLFL [33] is of interest because the smaller particles are close to the Debye length for ZnO, which is on the order of ~18 nm. At this length scale, the depletion layer on the ZnO nanoparticle surface is enhanced, which minimizes the photodetector response time while maximizing responsivity. The bimodal distribution ($r_{small}/r_{large} \approx 0.42$) is potentially advantageous for the final device design because a more efficient packing factor becomes possible compared to the unimodal particle distribution. For bimodal spheres, ideal packing can be efficiently achieved up to a particle size ratio of ~0.41 [53], where the smaller particle simply fits into the interstitial spaces between the larger particles. Additionally, it has been shown that UV absorption in ZnO nanoparticles is dependent on nanoparticle diameter, where absorption increases as the particle diameter increases, peaking at 40 nm and then decreasing as the particle size increases beyond 40 nm [54].

Figure 3. SEM micrographs of ZnO nanoparticles (NPs) (**a**) before and (**b**) after pulsed laser fragmentation in liquid (PLFL) processing; (**c**) ZnO nanoparticle size distribution both before and after PLFL.

To give insight into the quality and defect content of the generated nanoparticles, photoluminescence measurements (λ = 343 nm) were conducted. The photoluminescence of ZnO nanoparticles has been studied extensively and typically reveals two distinct emission bands, one in the UV and one in the visible spectrum. The emission peak observed in the UV region is a result of near-band-edge emission which is mediated through exciton–exciton interactions. As the particle size decreases, the fluorescence is blue-shifted due to an increase in transition energy [55,56]. The second photoluminescence peak for ZnO is commonly observed as a green emission and is most likely due to deep level emission in the band gap through electron-hole recombination. This green emission peak is often broad and weak compared to the UV peak, with emissions lines reported from 510–583 nm [57]; however, other visible emission has been observed, including blue, yellow, violet, and red. The cause of these different visible emission peaks is still controversial, but is attributed to intrinsic defects such as Zn interstitials, oxygen vacancies, and the formation of free carriers [56,58,59]. It has been shown that green emission can be suppressed by coating the ZnO nanoparticle surface with surfactants, suggesting that surface defects are responsible [57,60]. Specifically, the mechanism responsible for green emission is often partially attributed to single ionized oxygen vacancies [56,60].

The as-received ZnO powders, which can be seen in Figure 4a, show a strong, narrow UV emission band at 375 nm and no discernible green emission, indicating high quality, low surface defect particles. A second UV emission peak is then observed in the photoluminescence spectra of ZnO nanoparticles generated via PLFL, which can also be seen in Figure 4a. The size of these particles is unlikely to directly affect the UV peak emission position, as quantum confinement effects are not observed at these scales because the Bohr radius of ZnO is significantly smaller at ~2.34 nm [57]. This second UV peak, observed at ~388 nm, can be attributed to either band-edge exciton emission or energy transitions involving Zn interstitials [57]. These UV emission peaks are both strong and narrow, but a broad, weaker peak seen at 571 nm indicates defect states on the ZnO nanoparticle surfaces. The emission characteristics of ZnO typically exhibit stronger UV peaks with structures of larger size with better crystalline quality, while smaller, more defective surface states show higher visible emission. While the exact origin of the green defect emission in ZnO remains contentious and poorly understood, there are several processing parameters that can be adjusted to control this defect emission, including solvent choice [61,62].

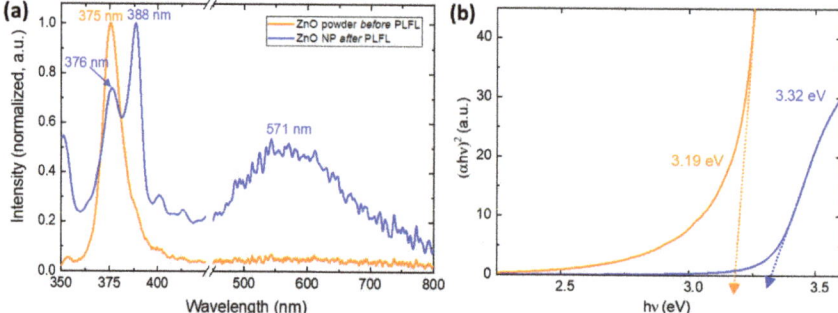

Figure 4. (a) Photoluminescence spectra and (b) Tauc plots for the ZnO nanoparticles both before (orange lines) and after (blue lines) PLFL processing.

In addition to the insight into the quality and defect density induced in the nanoparticles during laser processing, the optical transmission spectra of the ZnO nanoparticle solutions were collected to help understand the effect of laser-processing on the particles optical bandgap, which is shown in Figure 4b before and after PLFL.

3.2. ZnO Nanoparticle/Graphene Phototransistors

The mechanism of photoconduction in ZnO nanoparticle/graphene phototransistors is schematically illustrated in Figure 5. In the absence of UV light, oxygen molecules adsorb onto the ZnO nanoparticle surface and capture free electrons which form oxygen ions, creating a low conductivity depletion zone on the surface of the ZnO nanoparticles. When illuminated with UV light with energy higher than the bandgap of ZnO (~3.3 eV), electron-hole pairs are generated with the holes crossing the depletion layer and traveling to the surface of the ZnO nanoparticle. These holes recombine with negatively charged oxygen ions which results in the desorption of neutral oxygen molecules [15]. The remaining unpaired electrons in the conduction band of the ZnO nanoparticles transfer to the graphene layer, where they move to the drain electrode as a result of an applied source–drain voltage potential, resulting in a change in channel resistance. It is known that the size of the nanoparticle affects the performance of the phototransistor, where ZnO nanoparticles, close to or smaller than the Debye length (~18 nm), allow for a high density of trapped hole states on the surface, providing substantial photoconductive gain. Thus, by combining the advantageous surface depletion zone achieved with ZnO nanoparticles and a high mobility of graphene layers, enhanced responsivity and gain can be achieved in a hybrid photodetector.

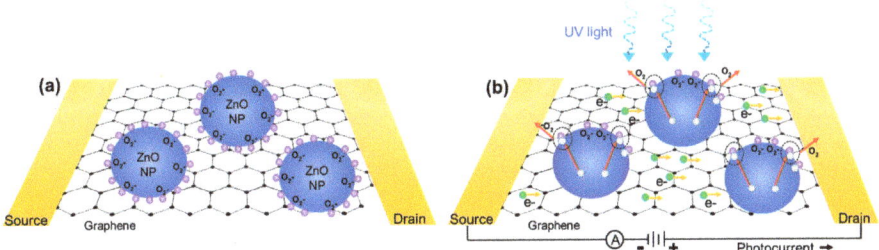

Figure 5. Schematic of photoconduction mechanism for ZnO nanoparticle/graphene photodetectors (a) without and (b) with UV illumination (not to scale).

To understand the photoresponse of the devices, electrical transport properties were measured with UV illumination (λ = 365 nm), from 34 µW/cm² up to 1.4 mW/cm². Drain current as a function of drain voltage can be seen in Figure 6a, where different illumination conditions are plotted. The I–V characterization shows a bipolar behavior as a function of drain voltage. By applying a gate voltage to the transistor, an electric field is produced which can enhance the device response. Additionally, drain current as a function of gate voltage can be seen in Figure 6b.

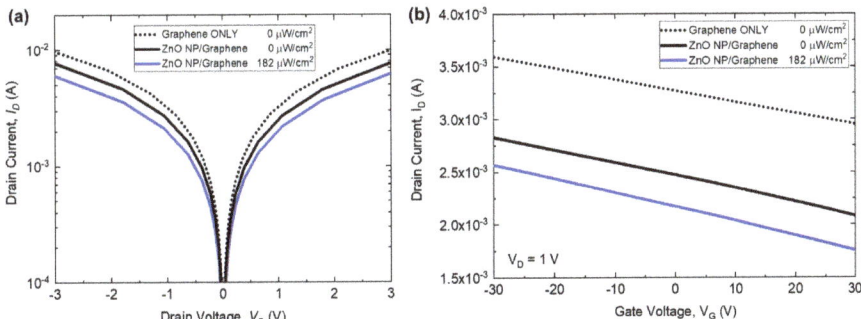

Figure 6. Electrical transport properties of both the ZnO/graphene phototransistors and the graphene-only phototransistors. Photocurrent as a function of (a) drain voltage and (b) the applied gate voltage at a drain voltage, V_D of 1V.

The responsivity and gain as a function of incident UV power were measured to determine the performance of the phototransistors. The responsivity of the devices is the ratio of the measured photocurrent to the UV illumination, where:

$$R = \frac{I_{ph} - I_{dark}}{P} = \frac{\Delta I}{P} \tag{1}$$

Here, I_{ph} and I_{dark} are the induced photocurrent under UV illumination and the dark current, respectively. The responsivity at a drain current of V_D = 5 V with no gate voltage applied can be seen in Figure 7a. As the incident power approaches zero, the responsivity can be extrapolated and rises to a maximum of ~4 × 10⁴ AW⁻¹, where experimentally at P = 34 µW/cm², the responsivity is measured to be 2 × 10³ AW⁻¹. The photoconductive gain G can be described as the ratio between the number of electrons collected per unit time and the number of absorbed photons per unit time [12,19]:

$$G = \frac{\Delta I}{qF} = \frac{\Delta I}{P} \cdot \frac{hc}{e\lambda} = R \cdot \frac{hc}{e\lambda} \tag{2}$$

where h is Planck's constant, c is the speed of light, e is electron charge, and λ is incident wavelength. By fitting the available data, the maximum gain achieved as P approaches zero is $G_{max} \approx 1.3 \times 10^5$ at $V_D = 5$ V with no applied gate voltage. The gain of the ZnO nanoparticle/graphene phototransistor is a function of the applied optical intensity. The number of electron-hole pairs generated is directly related to the optical intensity applied, where with a higher intensity, more hole trap states are filled at the surface eventually reaching a surface saturation state. Once this occurs, the electron-hole pairs generated do not aid in charge transfer into the graphene layer, thus limiting the efficiency of the hybrid device [15].

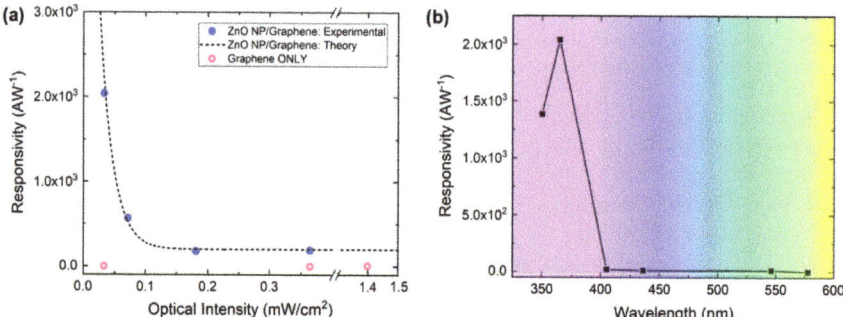

Figure 7. (a) Responsivity as a function of optical intensity for both ZnO nanoparticle/graphene devices (solid blue circles) and graphene only devices (open pink circles). Dashed black line represents a simulated fit. (b) Responsivity as a function of incident wavelength for ZnO nanoparticle/graphene devices.

In addition to responsivity and gain, the wavelength selectivity of the photodetector is critical because it determines the spectral range. The spectral dependence of responsivity for ZnO nanoparticle/graphene devices can be seen in Figure 7b, where the photoresponse of the phototransistors was measured at various wavelengths, chosen to correspond with the experimental UV source (Hg lamp) emission lines. Since the bandgap of the ZnO nanoparticles was measured to be ~3.32 eV, large photocurrent generation at a wavelength of 365 nm was expected, whereas visible light does not provide enough energy to cause electron excitation to the conduction band [18]. Above ~400 nm, there is no discernable photoresponse in the visible range, making these detectors solar-blind.

In order to understand the time-varying behavior of the phototransistors, the photocurrent response was measured as a function of time as the devices were exposed to UV illumination and then subsequently turned off. These temporal measurements were conducted at a drain voltage of $V_D = 1$ V, with no applied gate voltage ($V_G = 0$ V). The graphene-only phototransistors show no photoresponse when illuminated at 365 µW/cm² over a time scale of several hundred seconds, which can be seen in Figure 8a. Other research efforts have demonstrated graphene photoresponse, but the time scale for these changes typically occurs on the order of tens of minutes, so any photoresponse observed in the ZnO nanoparticle/graphene phototransistors can be attributed to photon absorption by the ZnO nanoparticles [14]. The temporal response of the ZnO nanoparticle/graphene detectors at both 182 and 365 µW/cm² can also be seen in Figure 8a, where the change in drain current increases from 500 µA to greater than 1.3 mA. The photoresponse behavior at 182 µW/cm² can be seen in Figure 8b where the UV illumination was turned on and the device was allowed to equilibrate, followed by a recovery time after the illumination was turned off (at ~175 s). A sharp increase in photocurrent was observed once the UV light is turned on, followed by recovery on the order of tens of seconds.

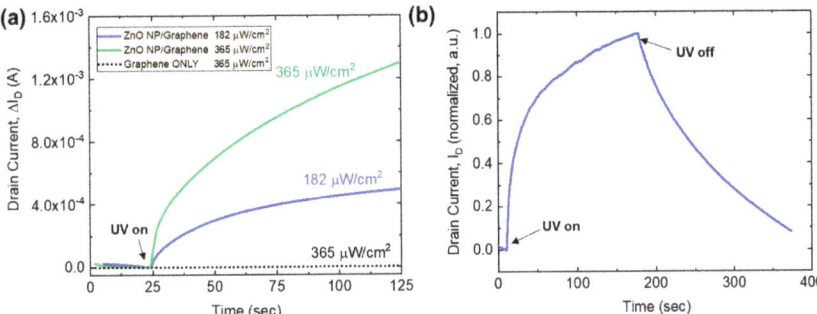

Figure 8. Temporal photoresponse of ZnO nanoparticle/graphene devices: (**a**) ZnO nanoparticle/graphene device compared with the graphene only device at 182 µW/cm^2 and 365 µW/cm^2; (**b**) optical intensity of 182 µW/cm^2 showing the rise and fall as intensity is turned both on and off, respectively.

Both the responsivity and the response times of these ZnO nanoparticle/graphene photodetectors can be optimized by adjusting several experimental parameters. It is well known that the nanoparticle layer is strongly dependent on particle packing in its ability to absorb light and efficiently transfer electrons into the underlying graphene layer [63]. Thus, future studies should focus on controlling the nanoparticle packing factor in an effort to increase the photoresponse and reduce switching speeds. Additionally, the transistor design could be optimized, including making the active area narrower (< mm) in size, which would reduce the possibility for defects introduced during the fabrication process. Because the PLFL process can be controlled via operating parameters, the synthesis of nanoparticles could be optimized to yield even smaller particles.

4. Conclusions

Hybrid ZnO nanoparticle/graphene phototransistors were demonstrated, exhibiting a responsivity of up to 4×10^4 AW^{-1} with a maximum gain of 1.3×10^5 and superior spectral selectivity below 400 nm, making them ideal solar-blind UV photodetectors. We have demonstrated the use of pulsed laser fragmentation in liquid (PLFL) as a simple, ligand-free alternative to traditional nanoparticle synthesis techniques for the fabrication of ZnO nanoparticles.

Author Contributions: Conceptualization, K.C., A.P., and N.C.; investigation, K.C., H.K., and N.C.; writing—original draft preparation, K.C.; writing—review and editing, K.C., H.K., A.P., and N.C. All authors have read and agreed to the published version of the manuscript.

Funding: This work was funded by the Office of Naval Research (ONR) through the Naval Research Laboratory Basic Research Program.

Conflicts of Interest: The authors declare no conflict of interest.

References

1. Novoselov, K.S.; Geim, A.K.; Morozov, S.V.; Jiang, D.; Zhang, Y.; Dubonos, S.V.; Grigorieva, I.V.; Firsov, A.A. Electric field effect in atomically thin carbon films. *Science* **2004**, *306*, 666–669. [CrossRef]
2. Nair, R.R.; Blake, P.; Grigorenko, A.N.; Novoselov, K.S.; Booth, T.J.; Stauber, T.; Peres, N.M.R.; Geim, A.K. Fine Structure Constant Defines Visual Transparency of Graphene. *Science* **2008**, *320*, 1308. [CrossRef] [PubMed]
3. Bonaccorso, F.; Sun, Z.; Hasan, T.; Ferrari, A.C. Graphene photonics and optoelectronics. *Nat. Phys.* **2010**, *4*, 611–622. [CrossRef]
4. Geim, A.K.; Novoselov, K.S. The rise of graphene. *Nat. Mater.* **2007**, *6*, 183–191. [CrossRef] [PubMed]
5. Dawlaty, J.M.; Shivaraman, S.; Chandrashekhar, M.; Rana, F.; Spencer, M.G. Measurement of ultrafast carrier dynamics in epitaxial graphene. *Appl. Phys. Lett.* **2008**, *92*, 042116. [CrossRef]
6. Liu, C.-H.; Chang, Y.-C.; Norris, T.B.; Zhong, Z. Graphene photodetectors with ultra-broadband and high responsivity at room temperature. *Nat. Nanotechnol.* **2014**, *9*, 273–278. [CrossRef]

7. Konstantatos, G.; Badioli, M.; Gaudreau, L.; Osmond, K.; Bernechea, M.; de Arquer, F.P.G.; Gatti, F.; Koppens, F.H.L. Hybrid graphene-quantum dot phototransistors with ultrahigh gain. *Nat. Nanotechnol.* **2012**, *7*, 363–368. [CrossRef]
8. Liang, F.-X.; Gao, Y.; Xie, C.; Tong, X.-W.; Li, Z.-J.; Luo, L.-B. Recent advances in the fabrication of graphene-ZnO heterojunctions for optoelectronic device applications. *J. Mater. Chem. C* **2018**, *6*, 3815–3833. [CrossRef]
9. Yu, X.; Li, Y.; Hu, X.; Zhang, D.; Tao, Y.; Liu, Z.; He, Y.; Haque, A.; Liu, Z.; Wu, T.; et al. Narrow bandgap oxide nanoparticles coupled with graphene for high performance mid-infrared photodetection. *Nat. Commun.* **2018**, *9*, 4299. [CrossRef]
10. Look, D.C.; Reynolds, D.C.; Sizelove, J.R.; Jones, R.L.; Litton, C.W.; Cantwell, G.; Harsh, W.C. Electrical properties of bulk ZnO. *Solid State Commun.* **1998**, *105*, 399–401. [CrossRef]
11. Jun, J.H.; Seong, H.; Cho, K.; Moon, B.-M.; Kim, S. Ultraviolet photodetectors based on ZnO nanoparticles. *Ceram. Int.* **2009**, *35*, 2797–2801. [CrossRef]
12. Gong, M.; Liu, Q.; Cook, B.; Kattel, B.; Wang, T.; Chan, W.-L.; Ewing, D.; Casper, M.; Stramel, A.; Wu, J.Z. All-Printable ZnO Quantum Dots/Graphene van der Waals Heterostructures for Ultrasensitive Detection of Ultraviolet Light. *ACS Nano* **2017**, *11*, 4114–4123. [CrossRef] [PubMed]
13. Liu, A.Q.; Gong, M.; Cook, B.; Ewing, D.; Casper, M.; Stramel, A.; Wu, J. Transfer-free and printable graphene/ZnO-nanoparticle nanohybrid photodetectors with high performance. *J. Mater. Chem. C* **2017**, *5*, 6427–6432. [CrossRef]
14. Guo, W.; Xu, S.; Wu, Z.; Wang, N.; Loy, M.M.T.; Du, S. Oxygen-Assisted Charge Transfer Between ZnO Quantum Dots and Graphene. *Small* **2013**, *9*, 3031–3036. [CrossRef] [PubMed]
15. Shao, D.; Gao, J.; Chow, P.; Sun, H.; Xin, G.; Sharma, P.; Lian, J.; Koratkar, N.A.; Sawyer, S. Organic-Inorganic Heterointerfaces for Ultrasensitive Detection of Ultraviolet Light. *Nano Lett.* **2015**, *15*, 3787–3792. [CrossRef]
16. Xu, Q.; Cheng, Q.; Zhong, J.; Cai, W.; Zhang, Z.; Wu, Z.; Zhang, F. A metal-semiconductor-metal detector based on ZnO nanowires grown on a graphene layer. *Nanotechnology* **2014**, *25*, 055501. [CrossRef]
17. Dang, V.Q.; Trung, T.Q.; Duy, L.T.; Kim, B.-Y.; Siddiqui, S.; Lee, W.; Lee, N.-E. High-performance flexible ultraviolet (UV) phototransistor using hybrid channel of vertical ZnO nanorods and graphene. *ACS Appl. Mater. Interface* **2015**, *7*, 11032–11040. [CrossRef]
18. Dang, V.Q.; Trung, T.Q.; Kim, D.-I.; Duy, L.T.; Hwang, B.-U.; Lee, D.-W.; Kim, B.-Y.; Toan, L.D.; Lee, N.-E. Ultrahigh responsivity in graphene-ZnO nanorod hybrid UV photodetector. *Small* **2015**, *11*, 3054–3065. [CrossRef]
19. Soci, C.; Zhang, A.; Xiang, B.; Dayeh, S.A.; Aplin, D.P.R.; Park, J.; Bao, X.Y.; Lo, Y.H.; Wang, D. ZnO Nanowire UV Photodetectors with High Internal Gain. *Nano Lett.* **2007**, *7*, 1003–1009. [CrossRef]
20. Lee, Y.; Kim, D.Y.; Lee, S. Low-power graphene/ZnO Schottky UV photodiodes with enhanced lateral schottky barrier homogeneity. *Nanomaterials* **2019**, *9*, 799. [CrossRef]
21. Lee, S.; Lee, Y.; Kim, D.Y.; Song, E.B.; Kim, S.M. Back-gate tuning of Schottky barrier height in graphene/zinc-oxide photodiodes. *Appl. Phys. Lett.* **2013**, *102*, 242114. [CrossRef]
22. Lee, H.; An, N.; Jeong, S.; Kang, S.; Kwon, S.; Lee, J.; Lee, Y.; Kim, D.Y.; Lee, S. Strong dependence of photocurrent on illumination-light colors for ZnO/graphene Schottky diode. *Curr. Appl. Phys.* **2017**, *17*, 552–556. [CrossRef]
23. Cheng, C.-C.; Zhan, J.-Y.; Liao, Y.-M.; Lin, T.-Y.; Hsieh, Y.-P.; Chen, Y.-F. Self-powered and broadband photodetectors based on graphene/ZnO/silicon triple junctions. *Appl. Phys. Lett.* **2016**, *109*, 053501. [CrossRef]
24. Hongsith, N.; Wongrat, E.; Kerdcharoen, T.; Choopun, S. Sensor response formula for sensor based on ZnO nanostructures. *Sens. Actuators B Chem.* **2010**, *144*, 67–72. [CrossRef]
25. Zhang, D.; Gökcev, B.; Barcikowski, S. Laser Synthesis and Processing of Colloids: Fundamentals and Applications. *Chem. Rev.* **2017**, *117*, 3990–4103. [CrossRef]
26. Kamat, P.V.; Flumiani, M.; Hartland, G.V. Picosecond Dynamics of Silver Nanoclusters. Photoejection of Electrons and Fragmentation. *J. Phys. Chem. B* **1998**, *102*, 3123–3128. [CrossRef]
27. Werner, D.; Furube, A.; Okamoto, T.; Hashimoto, S. Femtosecond Laser-Induced Size Reduction of Aqueous Gold Nanoparticles: In Situ and Pump-Probe Spectroscopy Investigations Revealing Coulomb Explosion. *J. Phys. Chem. C* **2011**, *115*, 8503–8512. [CrossRef]
28. Takami, A.; Kurita, J.; Koda, S. Laser-Induced Size Reduction of Noble Metal Particles. *J. Phys. Chem. B* **1999**, *103*, 1226–1232. [CrossRef]
29. Kuzmin, P.G.; Shafeev, G.A.; Serkov, A.A.; Kirichenko, N.A.; Shcherbina, M.E. Laser-assisted fragmentation of Al particles suspended in liquid. *Appl. Surf. Sci.* **2014**, *294*, 15–19. [CrossRef]

30. Chubilleau, C.; Lenoir, B.; Migot, S.; Dauscher, A. Laser fragmentation in liquid medium: A new way for the synthesis of PbTe nanoparticles. *J. Colloid Interface Sci.* **2011**, *357*, 13–17. [CrossRef]
31. Singh, S.C.; Mishra, S.K.; Srivastava, R.K.; Gopal, R. Optical Properties of Selenium Quantum Dots Produced with Laser Irradation of Water Suspended Se Nanoparticles. *J. Phys. Chem. C* **2010**, *114*, 17374–17384. [CrossRef]
32. Usui, H.; Sasaki, T.; Koshizaki, N. Optical Transmittance of Indium Tin Oxide Nanoparticles Prepared by Laser-Induced Fragmentation in Water. *J. Phys. Chem. B* **2006**, *110*, 12890–12895. [CrossRef] [PubMed]
33. Lau, M.; Barcikowski, S. Quantification of mass-specific laser energy input converted into particle properties during picosecond pulsed laser fragmentation of zinc oxide and boron carbide in liquids. *Appl. Surf. Sci.* **2015**, *348*, 22–29. [CrossRef]
34. Zeng, H.; Yang, S.; Cai, W. Reshaping Formation and Luminescence Evolution of ZnO Quantum Dots by Laser-Induced Fragmentation in Liquid. *J. Phys. Chem. C* **2011**, *115*, 5038–5043. [CrossRef]
35. Suk, J.W.; Kitt, A.; Magnuson, C.W.; Hao, Y.; Ahmed, S.; An, J.; Swan, A.K.; Goldberg, B.B.; Ruoff, R.S. Transfer of CVD-grown monolayer graphene onto arbitrary substrates. *ACS Nano* **2011**, *5*, 6916–6924. [CrossRef]
36. Liang, X.; Sperling, B.A.; Calizo, I.; Cheng, G.; Hacker, C.A.; Zhang, Q.; Obeng, Y.; Yan, K.; Peng, H.; Li, Q.; et al. Toward clean and crackless transfer of graphene. *ACS Nano* **2011**, *5*, 9144–9153. [CrossRef]
37. Li, X.; Zhu, Y.; Cai, W.; Borysiak, M.; Han, B.; Chen, D.; Piner, R.D.; Colombo, L.; Ruoff, R.S. Transfer of large-area graphene films for high-performance transparent conductive electrodes. *Nano Lett.* **2009**, *9*, 4359–4363. [CrossRef]
38. Li, X.; Cai, W.; An, J.; Kim, S.; Nah, J.; Yang, D.; Piner, R.; Velamakanni, A.; Jung, I.; Tutuc, E.; et al. Large-area synthesis of high-quality and uniform graphene films on copper films. *Science* **2009**, *324*, 1312–1314. [CrossRef]
39. Spanhel, L.; Anderson, M.A. Semiconductor cluster in the sol-gel process: Quantized aggregation, gelation, and crystal growth in concentrated ZnO colloids. *J. Am. Chem. Soc.* **1991**, *113*, 2826–2833. [CrossRef]
40. Wang, Z.L. Zinc oxide nanostructures: Growth, properties and applications. *J. Phys. Condens. Matter* **2004**, *16*, R829–R858. [CrossRef]
41. Wang, Z.; Zhang, H.; Zhang, L.; Yuan, J.; Yan, S.; Wang, C. Low-temperature synthesis of ZnO nanoparticles by solid-state pyrolytic reaction. *Nanotechnology* **2003**, *14*, 11–15. [CrossRef]
42. Heuer-Jungemann, A.; Feliu, N.; Bakaimi, I.; Hamaly, M.; Alkilany, A.; Chakraborty, I.; Masood, A.; Casula, M.F.; Kostopoulou, A.; Oh, E.; et al. The Role of Ligands in the Chemical Synthesis and Applications of Inorganic Nanoparticles. *Chem. Rev.* **2019**, *119*, 4819–4880. [CrossRef]
43. Yan, Z.; Chrisey, D.B. Pulsed laser ablation in liquid for micro-/nanostructure generation. *J. Photochem. Photobiol. C Photochem. Rev.* **2012**, *13*, 204–223. [CrossRef]
44. Tan, D.; Zhou, S.; Qiu, J.; Khusro, N. Preparation of functional nanomaterials with femtosecond laser ablation in solution. *J. Photochem. Photobiol. C Photochem. Rev.* **2013**, *17*, 50–68. [CrossRef]
45. Singh, S.C.; Gopal, R. Synthesis of colloidal zinc oxide nanoparticles by pulsed laser ablation in aqueous media. *Phys. E Low Dimens. Syst. Nanostruct.* **2008**, *40*, 724–730. [CrossRef]
46. Al-Nassar, S.I.; Hussein, F.I.; Mahmoud, A.K. The effect of laser pulse energy on ZnO nanoparticles formation by liquid phase pulse laser ablation. *J. Mater. Res. Technol.* **2019**, *8*, 4026–4031. [CrossRef]
47. Huang, H.; Lai, J.; Lu, J.; Li, Z. Pulsed laser ablation of bulk target and particle products in liquid for nanomaterial fabrication. *AIP Adv.* **2019**, *9*, 015307. [CrossRef]
48. Naher, U.; Bjornholm, S.; Frauendorf, S.; Garcias, F.; Guet, C. Fission of Metal Clusters. *Phys. Rep.* **1997**, *285*, 245–320. [CrossRef]
49. Siebeneicher, S.; Waag, F.; Castillo, M.E.; Shvartsman, V.V.; Lupascu, D.C.; Gökce, B. Laser Fragmentation Synthesis of Colloidal Bismuth Ferrite Particles. *Nanomaterials* **2020**, *10*, 359. [CrossRef]
50. Ling, D.; Hackett, M.J.; Hyeon, T. Surface ligands in synthesis, modification, assembly and biomedical applications of nanoparticles. *Nanotoday* **2014**, *9*, 457–477. [CrossRef]
51. Richter, T.V.; Stelzl, F.; Schulz-Gericke, J.; Kerscher, B.; Würfel, U.; Niggermann, M.; Ludwigs, S. Room temperature vacuum-induced ligand removal and patterning of ZnO nanoparticles: From semiconducting films towards printed electronics. *J. Mater. Chem.* **2010**, *20*, 874–879. [CrossRef]
52. Luo, J.; Dai, X.; Bai, S.; Jin, Y.; Ye, Z.; Guo, X. Ligand Exchange of Colloidal ZnO Nanocrystals from the High Temperatuer and Nonaqueous Approach. *Nano Micro Lett.* **2013**, *5*, 274–280. [CrossRef]
53. O'Toole, P.I.; Hudson, T.S. New High-Density Packings of Similarly Sized Binary Spheres. *J. Phys. Chem. C* **2011**, *115*, 19037–19040. [CrossRef]

54. Goh, E.G.; Xu, X.; McCormick, P.G. Effect of particle size on the UV absorbance of zinc oxide nanoparticles. *Scr. Mater.* **2014**, *78–79*, 49–52. [CrossRef]
55. Thareja, R.K.; Shukla, S. Synthesis and characterization of zinc oxide nanoparticles by laser ablation of zinc in liquid. *Appl. Surf. Sci.* **2007**, *253*, 8889–8895. [CrossRef]
56. Zhang, L.; Yin, L.; Wang, C.; Iun, N.; Qi, Y.; Xiang, D. Origin of visible photoluminescence of ZnO quantum dots: Defect-dependent and size-dependent. *J. Phys. Chem. C* **2010**, *114*, 9651–9658. [CrossRef]
57. Djurisic, A.B.; Leung, Y.H. Optical Properties of ZnO nanostructures. *Small* **2006**, *2*, 944–961. [CrossRef]
58. Ischenko, V.; Polarz, S.; Grote, D.; Stavarache, V.; Fink, K.; Driess, M. Zinc Oxide Nanoparticles with Defects. *Adv. Funct. Mater.* **2005**, *15*, 1945–1954. [CrossRef]
59. Zhao, J.-H.; Lu, C.-J.; Lv, Z.-H. Photoluminescence of ZnO nanoparticles and nanorods. *Optik* **2016**, *127*, 1421–1423. [CrossRef]
60. Djurisic, A.B.; Choy, W.C.H.; Roy, V.A.L.; Leung, Y.H.; Kwong, C.Y.; Cheah, K.W.; Rao, T.K.G.; Chan, W.K.; Lui, H.F.; Surya, C. Photoluminescence and Electron Paramagnetic Resonance of ZnO Tetrapod Structures. *Adv. Funct. Mater.* **2004**, *14*, 856–864. [CrossRef]
61. Raoufi, D. Synthesis and photoluminescence characterization of ZnO nanoparticles. *J. Lumin.* **2013**, *134*, 213–219. [CrossRef]
62. Xu, C.H.; Si, J.; Xu, Y.; Hou, X. Femtosecond Laser-Assisted Synthesis of ZnO Nanoparticles in Solvent with Visible Emission for Temperature Sensing. *Nano Brief Rep. Rev.* **2019**, *14*, 1950054. [CrossRef]
63. Liu, Q.; Gong, M.; Cook, B.; Ewing, D.; Casper, M.; Stramel, A.; Wu, J. Fused nanojunctions of electron-depleted ZnO Nanoparticles for extraordinary performance in ultraviolet detection. *Adv. Mater. Interface* **2017**, *4*, 1601064. [CrossRef]

© 2020 by the authors. Licensee MDPI, Basel, Switzerland. This article is an open access article distributed under the terms and conditions of the Creative Commons Attribution (CC BY) license (http://creativecommons.org/licenses/by/4.0/).

Communication

Hybrid Orthorhombic Carbon Flakes Intercalated with Bimetallic Au-Ag Nanoclusters: Influence of Synthesis Parameters on Optical Properties

Muhammad Abdullah Butt [1,2,3], Daria Mamonova [4], Yuri Petrov [5], Alexandra Proklova [4], Ilya Kritchenkov [4], Alina Manshina [4,*], Peter Banzer [1,2,3,*] and Gerd Leuchs [1,2,3]

1. Emeritus Group Leuchs, Max Planck Institute for the Science of Light, 91058 Erlangen, Germany; muhammad-abdullah.butt@mpl.mpg.de (M.A.B.); gerd.leuchs@mpl.mpg.de (G.L.)
2. Institute of Optics, Information and Photonics, University Erlangen-Nuremberg, 91058 Erlangen, Germany
3. School of Advanced Optical Technologies, University Erlangen-Nuremberg, 91052 Erlangen, Germany
4. Institute of Chemistry, St. Petersburg State University, 198504 St. Petersburg, Russia; magwi@mail.ru (D.M.); proklova_97@mail.ru (A.P.); i.s.kritchenkov@spbu.ru (I.K.)
5. Faculty of physics, St. Petersburg State University, 198504 St. Petersburg, Russia; y.petrov@spbu.ru
* Correspondence: a.manshina@spbu.ru (A.M.); peter.banzer@mpl.mpg.de (P.B.)

Received: 18 June 2020; Accepted: 11 July 2020; Published: 15 July 2020

Abstract: Until recently, planar carbonaceous structures such as graphene did not show any birefringence under normal incidence. In contrast, a recently reported novel orthorhombic carbonaceous structure with metal nanoparticle inclusions does show intrinsic birefringence, outperforming other natural orthorhombic crystalline materials. These flake-like structures self-assemble during a laser-induced growth process. In this article, we explore the potential of this novel material and the design freedom during production. We study in particular the dependence of the optical and geometrical properties of these hybrid carbon-metal flakes on the fabrication parameters. The influence of the laser irradiation time, concentration of the supramolecular complex in the solution, and an external electric field applied during the growth process are investigated. In all cases, the self-assembled metamaterial exhibits a strong linear birefringence in the visible spectral range, while the wavelength-dependent attenuation was found to hinge on the concentration of the supramolecular complex in the solution. By varying the fabrication parameters one can steer the shape and size of the flakes. This study provides a route towards fabrication of novel hybrid carbon-metal flakes with tailored optical and geometrical properties.

Keywords: laser-induced deposition; hybrid carbon-metal flake; orthorhombic carbon; metallic nanoparticles; polarization analysis

---

1. Introduction

In recent years, metamaterials and metasurfaces have paved the pathway for technological developments [1–3] from guiding and shaping light [4,5], enhanced optical effects [6,7], all the way to the design and fabrication of flat optics [8,9], countless applications have been suggested and realized. Corresponding metasurfaces are usually fabricated using electron-beam lithography [9,10], ion-beam milling, or chemical vapor deposition [11,12]. The key features of metamaterials and metasurfaces are ruled by the design of their building blocks. The most important components of metamaterials are metal nanostructures that are arranged in a regular or irregular fashion defining the optical response. In most cases reported to date, these nanostructures are either fabricated on dielectric-metal interfaces or embedded homogeneously into a dielectric environment. However, considering a crystalline and/or anisotropic embedding medium would be of interest as well, because it would add another parameter to the system enabling fine control of the optical properties. A special class of metamaterials is based on carbon nanostructures intercalated with metal clusters. They are considered as promising materials for optics, nanophotonics, electrocatalysis, and sensing technologies [13,14]. Most carbon-metal hybrid metamaterials are based on amorphous carbon [15,16]. Fullerenes (molecular form of carbon) and carbon nanotubes can also be decorated with metal NPs, rendering possible interesting applications in catalysis and sensing [17,18]. It should be noted that metamaterials based on regular 2D carbon structures (e.g., graphene), with plasmonic nanoparticles (NPs) on the surface, have led to the observation of interesting modulation effects of the plasmon response in the hybrid system caused by the influence of the host carbon matrix [19,20]. Therefore, the combination of metal NPs with crystalline carbon in a metamaterial is an exciting field. Graphene and thermally expanded graphite are the available options for this purpose. However, the first allows only surface decoration with NPs. Thermally expanded graphite in contrast can be intercalated with metal NPs, but it is of little use for optical studies due to the irregular morphology and structure [21]. The fabrication of novel and versatile crystalline carbon-metal metamaterials is thus of great interest for the field of optics and photonics. The progress in this direction is restrained by the difficulties in the synthesis of regular structures based on crystalline carbon and plasmonic NPs suitable for optical measurements.

A promising method for the fabrication of metal NPs embedded in a carbon matrix has been reported recently [22,23]. It is based on the decomposition of a dissolved supramolecular complex $[Au_{13}Ag_{12}(C_2Ph)_{20}(PPh_2(C_6H_4)_3PPh_2)_3][PF_6]_5$ (herein called SMC), as a result of laser irradiation of a substrate/solution interface. This leads to a controllable formation of a crystalline structure with Au-Ag nanoclusters embedded in the carbon matrix, on a substrate surface caused by the self-organization of the SMC constituents. Further details on SMC preparation and chemicals used can be found as the Supplementary Materials. The laser-induced deposition of the hybrid carbon-metal structures is a straight-forward and easy-to-implement procedure that does not require intense laser irradiation or special equipment. This process results in the formation of NPs [23–25] or other structures such as flakes and flowers [26] of controllable composition, depending on the chosen SMC and the solvent. An interesting type of structures resulting from this laser-induced process are carbon flakes, which are the first reported orthorhombic form of carbon with embedded Au-Ag nanoclusters. Detailed analysis with a transmission electron microscope (TEM) revealed that the average radius of these metallic nanoclusters is $d = 2R = 2.5 \pm 0.9$ nm, with a center-to-center distance of 7.3 ± 1.5 nm [27]. Due to their inherent combination of bimetal Au-Ag nanoclusters with a crystalline carbon material of orthorhombic nature, these flakes possess intriguing optical properties [27], and may find applications in nano-optics and spectroscopy [7,22].

2. Experimental

2.1. Carbon Flakes Fabrication

In this article, we report on how various fabrication parameters affect the properties of the resulting hybrid carbon-metal flakes. This includes the laser irradiation time, the SMC concentration in the solution and the application of an external electric field during the growth process. Figure 1a shows the synthesis system that consists of a cuvette with the SMC solution mentioned above, prepared following a standard procedure [28], a substrate (industrial ITO-coated glass TIX 005 series from TECHINSTRO with a thickness of 1.1 mm, in contact with the solution) and a CW He-Cd Laser with $\lambda = 325$ nm and $I = 0.1$ W/cm^2 (Plasma, JSC Research Institute of Gas Discharge Devices, Tsiolkovskz St., Rzayan, Russian Federation). The laser beam is illuminated on the substrate-solution interface from the side of the substrate.

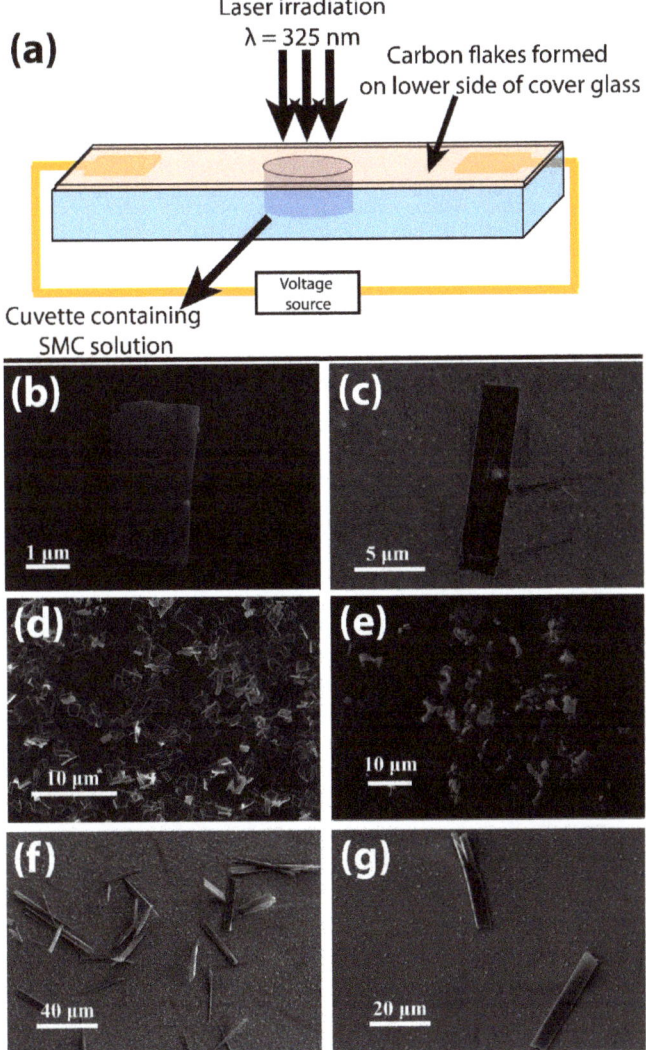

Figure 1. Fabrication of carbon flakes in dependence on various fabrication parameters. The scanning electron micrographs are from samples mentioned in Table 1. (**a**) The sample present in the cuvette with a certain concentration of supramolecular complex (SMC) in the solution is irradiated with laser for a certain amount of time. There is also the possibility to apply an electric field across the substrate to affect the formation of flakes. Flakes are formed by varying the laser irradiation time: (**b**) Sample C-2 with a laser irradiation time of 40 min and (**c**) sample C-4 with a laser irradiation time of 80 min. Varying SMC concentration: (**d**) Sample C-5 with an SMC concentration of 6 g/L and (**e**) sample C2 with an SMC concentration of 2 g/L. Images in (**f**) and (**g**) show examples of flakes from sample C-3 resulting from a fabrication with an external electric field switched on and causing lateral elongation of flakes.

The hybrid nanostructures are formed in the laser-affected zone on the substrate surface. The laser-induced deposition can be also realized with a DC electric field (7 V) applied via two metal electrodes (2 cm apart) deposited onto the ITO-coated glass (sample C3). The construction of the cuvette prevents the metal electrodes from being in contact with the SMC solution.

Table 1. Samples with different fabrication parameters.

Sample	Laser Irradiation Time (min)	SMC Concentration (g/L)	Electric Field (V/m)	Resulting Thickness of Carbon Flakes (nm)
C1 [27]	15	4 *	Off	150–500
C2	40	2	Off	250–750
C3	40	2	On	250–750
C4	80	2	Off	1200–1700
C5	40	6	Off	150–750

* In the earlier experiments, additional phenylacetylene was present in the SMC solution. We did not yet study the effect of phenylacetylene systematically, but that preliminary results indicate that phenylacetylene does not change the composition of the generated flakes, but only changes the number of flakes formed.

We experimentally investigate the effects of the aforementioned control parameters on the optical properties of the resulting carbon flakes. The investigated samples (containing multiple flakes each) with different fabrication parameters are listed in Table 1. Each of the parameters affects the reaction rate that all together lead to the variation of the flakes morphology and dimensions.

2.2. Methods

For the experimental study of the aforementioned carbon flakes, we use a custom method based on the microscopic Mueller matrix technique [26,27]. It allows performing a spatially resolved far-field polarization analysis of micron-sized specimens. The method combines the benefits of k-space microscopy [29,30] with liquid crystal variable retarders to avoid any moving optical elements in the analysis part of the setup [31].

For the optical investigation, we utilize a home-built setup [27] to raster-scan the carbon flakes with lateral dimensions between several micrometers to tens of micrometers, as shown in Figure 2. The incoming light beam of tunable wavelength and polarization is focused onto the sample using a microscope objective (Leica HCX PL FL 100×/0.9, Leica Microsystems GmbH, Wetzlar, Germany) of numerical aperture (NA) of 0.9 (effective NA = 0.5).

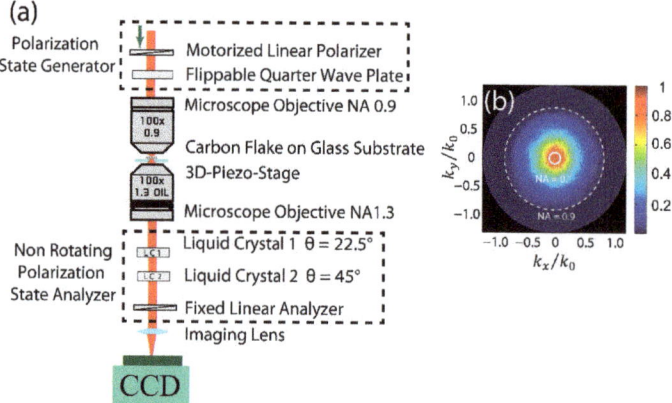

Figure 2. Experimental setup for the Mueller matrix measurement. (a) The polarization of the incident laser beam can be controlled (H, V, +45, −45, right circular polarization (RCP), and left circular polarization (LCP) with a motorized polarizer and a quarter-wave plate. The incident beam is focused onto the sample (flakes on the glass substrate), which is placed on a piezo stage to raster-scan the flakes with respect to the fixed input beam. The transmitted light is collected with another objective and polarization analysis is performed using liquid crystals. The back focal plane of the lower objective is imaged onto the charge-coupled device (CCD) camera and recorded for further analysis. (b) During the analysis of the Fourier space image, only k-vectors corresponding to normal incidence (numerical aperture (NA) region 0.1) are used for extracting the Mueller matrix.

The small focal spot size (on the order of the wavelength) ensures that no edge effects contribute to the results. The light is collected using an oil immersion objective of NA 1.3 (Leica HCX PL FL 100×/1.3 Oil, Leica Microsystems GmbH, Wetzlar, Germany). The collected light passes through a set of two liquid crystal cells and a fixed polarizer to perform a complete Stokes analysis fully characterizing the polarization state of light [32]. The Stokes vector S can be calculated by projecting the incident light onto six polarization states (linear horizontal, vertical, diagonal, anti-diagonal as well as right- and left-handed circular).

$$S = [s_0\ s_1\ s_2\ s_3]^T \quad (1)$$

where $s_0 = I_h + I_v$, $s_1 = I_h - I_v$, $s_2 = I_{p45} - I_{m45}$, $s_3 = I_{RCP} - I_{LCP}$.

For polarimetry we use liquid crystal variable retarders (LCVR) from Thorlabs GmbH, 85232 Bergkirchen, Germany [33]. The retardance of these LCVRs can be controlled via a voltage applied across the cells. This property makes them suitable for the flexible polarization analysis [27,31]. In our technique, the six different polarization states are generated by aligning the optic axis of the liquid crystals at 22.5 and 45°, and by using a combination of different retardances (applied voltages). This way, all six polarization states can be projected onto a linear fixed polarizer. An additional lens is used to image the back focal plane of the lower objective onto a CCD camera with 8-bit dynamic range (DMK 31BU03, The Imaging Source Europe GmbH, Bremen, Germany). This allows us to access momentum space (k-space; see Figure 2b). Incoming and outgoing Stokes vectors are linked via a 4 × 4 matrix, the Mueller matrix. By performing a pseudo-inverse analysis of input and output Stokes vectors, we can extract the Mueller matrix in the following way [34].

$$S_{out} = MS_{in} \quad (2)$$

$$M = S_{out} \times S_{in}^+ = S_{out} \times \left(S_{in}^\dagger S_{in}\right)^{-1} S_{in}^\dagger \quad (3)$$

where S_{in}^+ denotes the pseudo-inverse of S_{in}, which can be calculated using $S_{in}^\dagger$, the conjugate transpose of S_{in}. The Mueller matrix contains information about the optical properties of the carbon flake, such as attenuation, diattenuation, and birefringence, which can be further extracted by using decomposition techniques [34–36].

3. Results and Discussion

We perform a computational analysis of the experimentally recorded data to extract the Mueller matrix of each flake investigated. Here, we analyze two optical properties of carbon flakes. Firstly, the attenuation of the flakes is studied. To compare the results for flakes of different thickness, we normalize the attenuation accordingly (attenuation per 100 nm thickness of the flake). Secondly, we study the birefringence Δn, which is related to the retardance δ at a particular wavelength λ and for a certain thickness d of the carbon flake in the following way:

$$\Delta n = \frac{\delta \lambda}{2\pi d} \quad (4)$$

The thicknesses of the studied flakes were determined by atomic force microscopy (AFM), while the lateral dimensions were analyzed in a scanning electron microscope (SEM). A spatial average across data in the central region of the flake is used to analyze the average value of the aforementioned optical properties. For the three fabrication parameters discussed in this report, we retrieved the following results.

SEM images of the flakes deposited at different experimental parameters are presented in Figure 1b–g (for sample C2, C3, C4, and C5). For all cases, minor fluctuations in the flake's dimensions are observed. However, we can conclude that the dimensions of the flakes are clearly correlated with the deposition conditions. For instance, if a DC electric field is applied across the cuvette during the fabrication process (see Figure 1(f,g-C3)), the lateral dimensions of the flake structures are modified strongly in comparison to the case with the field switched off (see Figure 1(b,e-C2)). An elongation along one axis of the flakes is observed. The typical sizes of such flakes reach 40–60 µm in length and 5–10 µm in width (see Figure 1(f,g-C3)). These results suggest that the lateral dimensions of flakes can be fine-tuned by an externally applied voltage. To single out this effect only the electric field was switched on and off for samples C3 and C2, respectively, while all other parameters were kept unchanged. However, the optical properties, i.e., attenuation (normalized to the thickness) and birefringence are not significantly modified by this fabrication condition (see Figure 3).

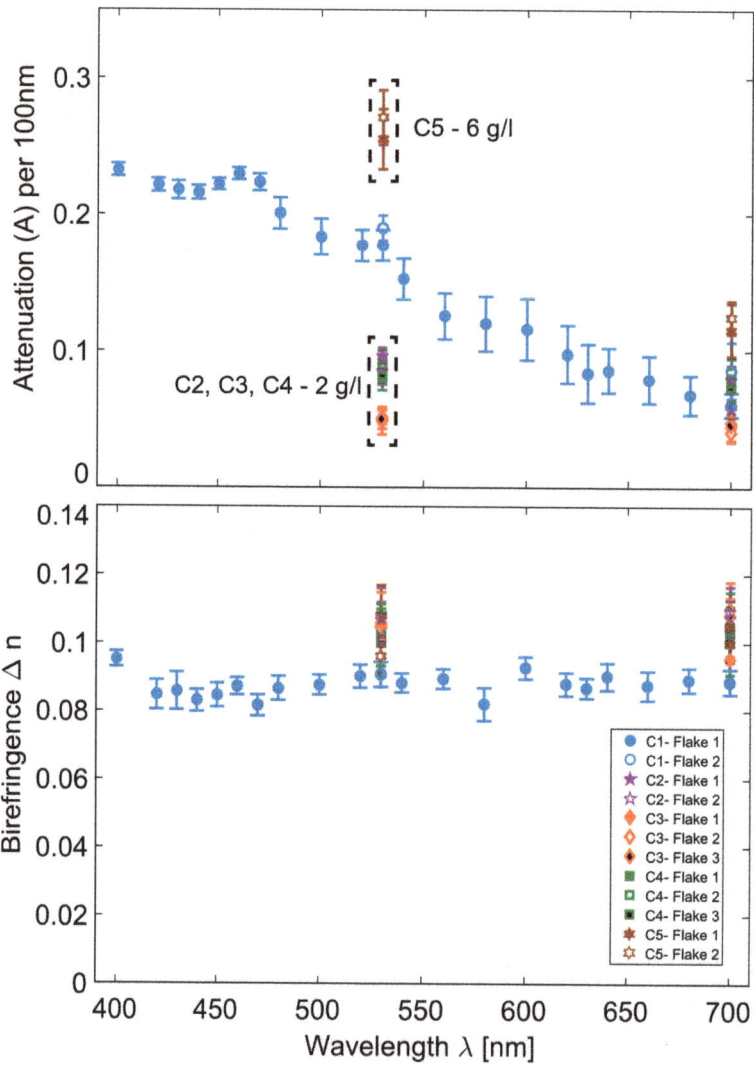

Figure 3. Experimental results for various flakes on different samples. The graphs contain data from our previous work [27] to serve as a reference (C1-Flake 1 and C1-Flake 2). The birefringence for flakes with different fabrications remain constant around $\Delta n = 0.1$. The attenuation is affected by the concentration of the SMC used during the fabrication process.

Furthermore, the laser irradiation time was found to also directly influence the overall growth of the flakes. For samples C2 and C4, only the irradiation time was chosen to be different, keeping all other parameters the same. AFM and SEM scans of the carbon flakes for samples reveal a direct relation between the irradiation time and the thickness of the resulting flakes, with a longer irradiation resulting in thicker flakes. The results of the optical measurements presented in Figure 3 show unambiguously that the attenuation (normalized to the thickness) and birefringence are not significantly changed for different laser irradiation times.

By changing the third parameter studied here, the SMC concentration in the solution, we see clearly in Figure 3 that the attenuation of the individual flakes can be modified. We analyze sample C2 and C5, for which other fabrication parameters were kept constant and only the SMC concentration was changed. We observe a clear change in the attenuation for different SMC concentrations. To serve as a reference, we also plot the attenuation and birefringence (C1-Flake 1 and C1-Flake2) from [27] in Figure 3.

It is worth noting that the linear birefringence remains unchanged for all different SMC densities tested here. In addition, the SMC concentration was also found to affect the number of flakes formed on the substrate (not shown here). The structural parameters of the carbon flakes are found to not be effected by the SMC concentration (see Figure 1d-C5,e-C2).

In the theoretical modeling discussed in detail in our earlier study [27], we found that the attenuation depends sensitively on the embedded bimetallic nanoclusters whereas the linear birefringence is a direct consequence of the orthorhombic carbon lattice. Here, we confirm this result by analyzing the results for flakes fabricated with different parameters, which do not affect the crystalline phase of carbon and, hence, the birefringence does not change significantly. Therefore, we speculate that a lower density per unit volume of bimetallic nanoclusters embedded in the carbon flakes is the cause of the substantially lower attenuation observed for a lower SMC concentration. Some other parameters such as the choice of a substrate, electro-optical effects within the flakes, etc., not discussed here in more detail, could also affect the attenuation of flakes. We summarize in Table 2, all observed dependences of the tested fabrication parameters on the geometrical and optical properties of carbon flakes.

Table 2. Variable fabrication parameters and their effect on carbon flakes.

Varied Fabrication Parameter	Effect on Optical Properties of Carbon Flakes	Effect on Geometrical Properties of Carbon Flakes
Electric field on or off	Optical properties remain unchanged	Lateral elongation of flakes (with field switched on)
Laser irradiation time	Optical properties remain unchanged	Increase in thickness with increasing irradiation time
SMC concentration	Increase in attenuation with the increasing SMC concentration	No observed effect on structural parameters

4. Conclusions

In conclusion, we studied the effect of different concentrations of the SMC in the solution, laser irradiation times, as well as the application of an external DC electric field during the fabrication process. The attenuation of the studied carbon flakes was found to be mainly affected by the SMC concentration. Applying an electric field results in lateral elongation of carbon flakes. The laser irradiation time directly influences the thickness of the resulting carbon flakes. In all cases, the birefringence was found to be wavelength independent and nearly constant around $\Delta n \approx 0.1$. Our study shows that the geometrical parameters of orthorhombic metal-carbon hybrid flakes can be tailored during the fabrication process by controlling various fabrication parameters, such as irradiation time and the application of an external field along the substrate. Furthermore, optical parameters such as the attenuation can be modified as well, while not changing the extraordinarily high linear birefringence of the flakes, which results from the orthorhombic carbon matrix. Hence, the material platform is promising to provide unique hybrid flakes, which can be engineered towards certain applications in opto-electronics and nano-optics.

Supplementary Materials: Details regarding SMC solution preparation and its composition are available online at http://www.mdpi.com/2079-4991/10/7/1376/s1.

Author Contributions: G.L., P.B., and A.M. conceived the general research idea. M.A.B. and P.B. designed and implemented the experimental method. D.M., A.P., I.K. and Y.P. fabricated and characterized the carbon flakes. M.A.B. and D.M. performed the optical experiments. M.A.B. analyzed the data. M.A.B. and P.B. interpreted the data. P.B. and G.L. supervised the project. All authors contributed to writing the manuscript and reviewed it. All authors have read and agreed to the published version of the manuscript.

Funding: This research received no external funding and APC was funded by emeritus group Leuchs at Max Planck institute for the science of Light, Germany.

Acknowledgments: This work was supported by the German Research Foundation (Deutsche Forschungsgemeinschaft, DFG) by funding the Erlangen Graduate School in Advanced Optical Technologies (SAOT) within the German Excellence Initiative. The authors are also grateful to "Interdisciplinary Resource Centre for Nanotechnology" of Saint-Petersburg State University Research Park and the program "Dmitriy Mendeleev" of DAAD (The German Academic Exchange Service).

Conflicts of Interest: The authors declare no conflict of interest.

References

1. Butler, S.Z.; Hollen, S.M.; Cao, L.; Cui, Y.; Gupta, J.A.; Gutiérrez, H.R.; Heinz, T.F.; Hong, S.S.; Huang, J.; Ismach, A.F.; et al. Progress, challenges, and opportunities in two-dimensional materials beyond graphene. *ACS Nano* **2013**, *7*, 2898–2926. [CrossRef] [PubMed]
2. Baron, A.; Aradian, A.; Ponsinet, V.; Barois, P. Self-assembled optical metamaterials. *Opt. Laser Technol.* **2016**, *82*, 94–100. [CrossRef]
3. Neshev, D.; Aharonovich, I. Optical metasurfaces: New generation building blocks for multi-functional optics. *Light Sci. Appl.* **2018**, *7*, 1–5. [CrossRef]
4. Wang, Z.; Li, T.; Soman, A.; Mao, D.; Kananen, T.; Gu, T. On-chip wavefront shaping with dielectric metasurface. *Nat. Commun.* **2019**, *10*, 1–7. [CrossRef] [PubMed]
5. Papaioannou, M.; Plum, E.; Valente, J.; Rogers, W.T.; Zheludev, N.I. Two-dimensional control of light with light on metasurfaces. *Light Sci. Appl.* **2016**, *5*, e16070. [CrossRef]
6. de Leon, I.; Horton, M.J.; Schulz, S.A.; Upham, J.; Banzer, P.; Boyd, R.W. Strong, spectrally-tunable chirality in diffractive metasurfaces. *Sci. Rep.* **2015**, *5*, 13034. [CrossRef]
7. Bashouti, M.Y.; Povolotckaia, A.V.; Povolotskiy, A.V.; Tunik, S.P.; Christ, S.H.; Leuchs, G.; Manshina, A.A. Spatially-controlled laser-induced decoration of 2D and 3D substrates with plasmonic nanoparticles. *RSC Adv.* **2016**, *6*, 75681–75685. [CrossRef]
8. Yu, N.; Capasso, F. Flat optics with designer metasurfaces. *Nat. Mater.* **2014**, *13*, 139–150. [CrossRef]
9. Capasso, F. The future and promise of flat optics: A personal perspective. *Nanophotonics* **2018**, *7*, 953–957. [CrossRef]
10. Su, V.C.; Chu, C.H.; Sun, G.; Tsai, D.P. Advances in optical metasurfaces: Fabrication and applications. *Opt. Express* **2018**, *26*, 13148–13182. [CrossRef]
11. Wang, J.; Du, J. Plasmonic and dielectric metasurfaces: Design, fabrication and applications. *Appl. Sci.* **2016**, *6*, 239. [CrossRef]
12. Park, C.-S.; Shrestha, V.R.; Yue, W.; Gao, S.; Lee, S.S.; Kim, E.S.; Choi, D.Y. Structural color filters enabled by a dielectric metasurface incorporating hydrogenated amorphous silicon nanodisks. *Sci. Rep.* **2017**, *7*, 1–9. [CrossRef]
13. Hong, W.; Bai, H.; Xu, Y.; Yao, Z.; Gu, Z.; Shi, G.J. Preparation of gold nanoparticle/graphene composites with controlled weight contents and their application in biosensors. *Phys. Chem. C* **2010**, *114*, 1822–1826. [CrossRef]
14. Shen, J.; Shi, M.; Li, N.; Yan, B.; Ma, H.; Hu, Y.; Ye, M. Facile synthesis and application of Ag-chemically converted graphene nanocomposite. *Nano Res.* **2010**, *3*, 339–349. [CrossRef]
15. Uslu, B.; Ozkan, S.A. Electroanalytical application of carbon based electrodes to the pharmaceuticals. *Anal. Lett.* **2007**, *40*, 817–853. [CrossRef]

16. Cao, X.; Tan, C.; Sindoro, M.; Zhang, H. Hybrid micro-/nano-structures derived from metal–organic frameworks: Preparation and applications in energy storage and conversion. *Chem. Soc. Rev.* **2017**, *46*, 2660–2677. [CrossRef]
17. Khalil, I.; Julkapli, N.; Yehye, W.; Basirun, W.; Bhargava, S. Graphene–gold nanoparticles hybrid—synthesis, functionalization, and application in a electrochemical and surface-enhanced Raman scattering biosensor. *Materials* **2016**, *9*, 406. [CrossRef]
18. Juang, Z.Y.; Tseng, C.C.; Shi, Y.; Hsieh, W.P.; Ryuzaki, S.; Saito, N.; Hsiung, C.E.; Chang, W.H.; Hernandez, Y.; Han, Y.; et al. Graphene-Au nanoparticle based vertical heterostructures: A novel route towards high-ZT Thermoelectric devices. *Nano Energy* **2017**, *38*, 385–391. [CrossRef]
19. Hajati, Y.; Zanbouri, Z.; Sabaeian, M. Low-loss and high-performance mid-infrared plasmon-phonon in graphene-hexagonal boron nitride waveguide. *J. Opt. Soc. Am. B* **2018**, *35*, 446–453. [CrossRef]
20. Zhang, Z.Y.; Li, D.M.; Zhang, H.; Wang, W.; Zhu, Y.H.; Zhang, S.; Zhang, X.P.; Yi, J.M. Coexistence of two graphene-induced modulation effects on surface plasmons in hybrid graphene plasmonic nanostructures. *Opt. Express* **2019**, *27*, 13503–13515. [CrossRef]
21. Bala, K.; Suriyaprakash, J.; Singh, P.; Chauhan, K.; Villa, A.; Gupta, N. Copper and cobalt nanoparticles embedded in naturally derived graphite electrodes for the sensing of the neurotransmitter epinephrine. *New J. Chem.* **2018**, *42*, 6604–6608. [CrossRef]
22. Povolotckaia, A.; Pankin, D.; Petrov, Y.; Vasileva, A.; Kolesnikov, I.; Sarau, G.; Christiansen, S.; Leuchs, G.; Manshina, A. Plasmonic carbon nanohybrids from laser-induced deposition: Controlled synthesis and SERS properties. *J. Mater. Sci.* **2019**, *54*, 8177–8186. [CrossRef]
23. Manshina, A.A.; Grachova, E.V.; Povolotskiy, A.V.; Povolotckaia, A.V.; Petrov, Y.V.; Koshevoy, I.O.; Makarova, A.A.; Vyalikh, D.V.; Tunik, S.P. Laser-induced transformation of supramolecular complexes: Approach to controlled formation of hybrid multi-yolk-shell Au-Ag@a-C:H nanostructures. *Sci Rep.* **2015**, *5*, 12027. [CrossRef]
24. Manshina, A.; Povolotskaya, A.V.; Petrov, Y.V.; Willinger, E.; Willinger, M.-G.; Banzer, P.; Leuchs, G. Novel 2D carbon allotrope intercalated with Au-Ag nanoclusters: From laser design to functionality. *OSA Adv. Photonics* **2017**. [CrossRef]
25. Bashouti, M.; Manshina, A.; Povolotckaia, A.; Povolotskiy, A.; Kireev, A.; Petrov, Y.; Mačković, M.; Spiecker, E.; Koshevoy, I.; Tunik, S.; et al. Direct laser writing of μ-chips based on hybrid C–Au–Ag nanoparticles for express analysis of hazardous and biological substances. *Lab Chip* **2015**, *15*, 1742–1747. [CrossRef] [PubMed]
26. Butt, A.; Neugebauer, M.; Lesina, A.C.; Ramunno, L.; Berini, P.; Vaccari, A.; Bauer, T.; Manshina, A.A.; Banzer, P.; Leuchs, G. Investigating the optical properties of a novel 3D self-assembled metamaterial made of carbon intercalated with bimetal nanoparticles. *OSA Adv. Photonics* **2018**. [CrossRef]
27. Butt, M.A.; Lesina, A.C.; Neugebauer, M.; Bauer, T.; Ramunno, L.; Vaccari, A.; Berini, P.; Petrov, Y.; Danilov, D.; Manshina, A.; et al. Investigating the optical properties of a laser induced 3D self-assembled carbon–metal hybrid structure. *Small* **2019**, *15*, 1900512. [CrossRef]
28. Koshevoy, I.O.; Karttunen, A.J.; Tunik, S.P.; Haukka, M.; Selivanov, S.I.; Melnikov, A.S.; Serdobintsev, P.Y.; Pakkanen, T.A. Synthesis, characterization, photophysical, and theoretical studies of supramolecular gold(I)–silver(I) alkynyl-phosphine complexes. *Organometallics* **2009**, *28*, 1369–1376. [CrossRef]
29. Arteaga, O.; Baldrís, M.; Antó, J.; Canillas, A.; Pascual, E.; Bertran, E. Mueller matrix microscope with a dual continuous rotating compensator setup and digital demodulation. *Appl. Opt.* **2014**, *53*, 2236–2245. [CrossRef]
30. Arteaga, O.; Nichols, S.M.; Antó, J. Back-focal plane Mueller matrix microscopy: Mueller conoscopy and Mueller diffractometry. *Appl. Surf. Sci.* **2017**, *421*, 702–706. [CrossRef]
31. Bueno, J.M. Liquid-crystal variable retarders for aerospace polarimetry applications. *J. Opt. A Pure Appl. Opt.* **2000**, *46*, 689–698. [CrossRef]
32. Stokes, G.G. On the composition and resolution of streams of polarized light from different sources. *Trans. Cambridge Philos. Soc.* **1852**, *9*, 399. [CrossRef]
33. Wu, S.T.; Efron, U.; Hess, L.D. Birefringence measurements of liquid crystals. *Appl. Opt.* **1984**, *23*, 3911–3915. [CrossRef] [PubMed]
34. Bass, M.; DeCusatis, C.; Enoch, J.; Lakshminarayanan, V.; Li, G.; Macdonald, C.; Mahajan, V.; Stryland, E.V. Polarimetry. In *Handbook of Optics Volume II: Design, Fabrication and Testing, Sources and Detectors, Radiometry and Photometry*; McGraw-Hill: New York, NY, USA, 2010.

35. Azzam, R.M.A. Stokes-vector and Mueller-matrix polarimetry. *Opt. Soc. Am.* **2016**, *33*, 1396–1408. [CrossRef] [PubMed]
36. Arteaga, O.; Canillas, A. Analytic inversion of the Mueller–Jones polarization matrices for homogeneous media. *Opt. Lett.* **2010**, *35*, 559–561. [CrossRef]

© 2020 by the authors. Licensee MDPI, Basel, Switzerland. This article is an open access article distributed under the terms and conditions of the Creative Commons Attribution (CC BY) license (http://creativecommons.org/licenses/by/4.0/).

Article

Zn-Doped Calcium Copper Titanate Synthesized via Rapid Laser Sintering of Sol-Gel Derived Precursors

Yanwei Huang [1,2,*], Yu Qiao [1], Yangyang Li [3], Jiayang He [1] and Heping Zeng [1,2,4,*]

1. State Key Laboratory of Precision Spectroscopy, East China Normal University, Shanghai 200062, China; 5119092722@stu.ecnu.edu.cn (Y.Q.); 51170920010@stu.ecnu.edu.cn (J.H.)
2. Chongqing Institute of East China Normal University, Chongqing 401120, China
3. School of Materials and Environment, Hangzhou Dianzi University, Hangzhou 310018, Zhejiang, China; liyy@hdu.edu.cn
4. Jinan Institute of Quantum Technology, Jinan 250101, Shandong, China
* Correspondence: ywhuang@lps.ecnu.edu.cn (Y.H.); hpzeng@phy.ecnu.edu.cn (H.Z.)

Received: 12 May 2020; Accepted: 12 June 2020; Published: 13 June 2020

Abstract: Zn-doped calcium copper titanate (CCTO) was successfully synthesized by rapid laser sintering of sol-gel derived precursors without the conventional long-time heat treatment. The structural, morphological, and crystalline properties were characterized, and the performances of dielectrics and impedance were measured and discussed. The X-ray diffractometer results show that Zn-doped CCTO is polycrystalline in a cubic structure, according to the doping ratio of $Ca(Cu_2Zn)Ti_4O_{12}$. Electron microscopy showed that Zn-doped CCTO has a denser microstructure with better uniformness with shrunken interplanar spacing of 2.598 nm for the plane (220). Comparing with undoped CCTO, the permittivity almost remains unchanged in the range of 10^2–10^6 Hz, demonstrating good stability on frequency. The electrical mechanism was investigated and is discussed through the impedance spectroscopy analysis. The resistance of grain and grain boundary decreases with rising temperature. Activation energies for the grain boundaries for Zn- doped CCTO were calculated from the slope for the relationship of $\ln\sigma$ versus $1/T$ and were found to be 0.605 eV, smaller than undoped CCTO. This synthesis route may be an efficient and convenient approach to limit excessive waste of resources.

Keywords: permittivity; impedance; sol-gel; laser sintering

1. Introduction

Today, with the development of the electronics industry, the active demand for better performances, higher capacity and lower cost in various electron devices have become the focus of attention. One promising solution is developing high density capacitors and related materials which possess better dielectric properties, as well as thermal and frequency stability. Calcium copper titanate ($CaCu_3Ti_4O_{12}$, CCTO) has been paid much attention due to its novel performances, especially the huge dielectric permittivity (above 10^4) and wide frequency stability. The value of permittivity is almost frequency-independent up to 10^6 Hz which shows wide frequency stability. These performances are significant for high density energy storage devices [1–5]. CCTO belongs to the Im3 space group, with a perovskite-like structure with a lattice parameter of 7.391 Å, presenting an extraordinarily high dielectric permittivity (ε_r) and moderate dielectric loss (tanδ) [6]. The ε_r is almost independent of frequency in a broad range and it shows excellent phase stability from 20 K to 600 K. This special characteristic could be ascribed to the polarizability mechanism related to the peculiar CCTO crystal structure according to reports [7–9]. To date, although different theoretic models have been proposed to explain the electrical properties, consensus has not yet been reached. However, the generally accepted view is the internal barrier layer capacitor (IBLC) model relying on the Maxwell–Wagner polarization [10]. The IBLC model considers that the huge ε_r value could be ascribed to the grains and grain boundaries [2,10,11]. Polycrystalline CCTO usually consists of grains and grain boundaries, which work as semiconductor and insulator, respectively. The grains work as electrodes of connected micro-capacitors and the grain boundaries contribute to the dielectric properties [12]. Researchers have investigated the electrical mechanism of CCTO ceramic using impedance spectroscopy (IS) and demonstrated that it was strongly dependent on the inhomogeneous microstructure, containing conducting grains and insulating boundaries [8]. In addition, comparing with conventional ceramics like $PbTiO_3$, $BaTiO_3$ or PZT, CCTO demonstrates several disadvantages, such as high dielectric losses at low frequency and low breakdown voltage [13,14]. The disadvantages of CCTO ceramics limit its application in capacitors because higher dielectric loss cannot maintain stored charges for a long time. There are a lot of methods to minimize the dielectric loss and enhance dielectric performance of CCTO materials. Doping of a metallic element at Ca^{2+} or Cu^{2+} sites can depress effectively the oxygen vacancies and alleviate the inner stresses to improve the dielectric properties [15–18], but the performance of CCTO is much more easily mutable during the synthesis process [4,19].

Various routes to synthesize CCTO have been employed, which include high temperature solid- state sintering, chemical methods, the sol-gel method, co-precipitation, microwave heating, and mechanical mixing methods [15–21]. The conventional solid-state sintering reaction is a method welcomed by most researchers for preparation of CCTO using powders of oxides or salts as precursors, however it involves higher reaction temperature with long sintering time, even several days. Not only that, it also requires at least two steps to react in the furnace: one is calcination close to 1000 °C and 8–12 h, and the other is sintering at 1150 °C for more than 10 h. Even so, the high temperature solid-state reaction suffers disadvantages of inhomogeneity, and requirement of complex repetitive ball milling with prolonged reaction time at higher temperature. These processes cause excessive consumption of energy which is not suitable for mass production. Other chemical methods, such as the solvothermal method, the sol-gel method, and coprecipitation, allow synthesis of CCTO at lower temperature with a reduced synthesizing time. Although the resultant product is homogeneous in stoichiometric ratio at atomic scale, the final acquisition of pure CCTO is inseparable from the processes of calcination and sintering in the furnace at high temperature [22,23].

Here, we adopted a novel route to synthesize CCTO: rapid laser sintering of sol-gel derived precursors, substituting the conventional furnace. By combining the dual advantages of sol-gel and laser sintering, we first synthesized CCTO precursors by the sol-gel method, which can easily prompt nanostructured and nanosized micro-grain formation. With a focused laser beam through a convex lens directed on the precursors from the sol-gel process, the reactive sintering time is significantly reduced down to several seconds or minutes, while guaranteeing that the CCTO has good crystal structure and high dielectric permittivity. Laser is directed on the precursors to remove the organic ionic group through irradiation at low power first of all and then it induces the solid-state reaction at high power, finally sintered CCTO can be obtained. Meanwhile, the sol-gel route to derive CCTO precursor is beneficial for good stoichiometry, uniform composition, and nanosized particles which can enhance dielectric performance. In this experiment, the precursor for CCTO is first synthesized by the sol-gel process, and then the rapid laser sintering technique is used to substitute the conventional muffle furnace. The obtained Zn-doped CCTO ceramic exhibits high dielectric permittivity and moderate dielectric loss at a wide frequency range, with good frequency-stability. The microstructure, grain and grain boundary resistance were analyzed by studying the impedance spectroscopy, and the conduction mechanism is discussed based on the equivalent circuit constituted of resistors and capacitors for semiconducting grains and insulating boundaries. According to the relationship of $\ln\sigma$ versus $1/T$, the activation energies for the grain boundaries were deduced.

2. Experimental Details

CCTO and Zn-doped CCTO precursors were obtained first by the sol-gel method. The original materials included Ti(OC$_4$H$_9$)$_4$(99%), Ca(CH$_3$COO)$_2$·H$_2$O(99%), Cu(NO$_3$)$_2$·3H$_2$O(99%), and Zn(CH$_3$COO)$_2$·2H$_2$O(99%), which are bought from the Sinopharm Chemical Reagent Co., Ltd. (SCR, China). Glacial acetic acid (bought from SCR) was used as a complexing agent. Ti^{4+}, Cu^{2+}, Ca^{2+}, and Zn^{2+}-containing sol were, respectively, made from the raw materials according to the molar ratio 4:3:1 and 4:2:1:1. A weighed amount of Ti(OC$_4$H$_9$)$_4$ was added in ethanol by stirring and with glacial acetic acid through magnetic stirring formed solution A. A clear and blue solution was formed by mixing and stirring a mixture of Cu(NO$_3$)$_2$·3H$_2$O with ethanol and then added into solution A to form solution B through vigorous stirring with a magnetic stirrer. Similarly, an appropriate amount of Ca(CH$_3$COO)$_2$·H$_2$O with solvent deionized water was stirred to form solution C by stirring and then was mixed with solution B. Zn^{2+}-containing sol was prepared similarly from Zn(CH$_3$COO)$_2$·2H$_2$O with deionized water. Finally, a blue-green sol was obtained after stirring for 1h and aging for 12 h. The sol was then baked at 80 °C for 12 h to form a dry gel and then it was ground into fine powder. After drying at 150 °C for 8 h, the fine powders were again ground in an agate mortar for 1 h ready for laser treatment.

A multimode diode laser (Optotools, 980 nm, CW, 1200 W, DILAS, Mainz-Galileo-Galilei, Germany) was well adjusted and then directed on samples through a quartz convex lens. In order to make the precursors react uniformly with laser radiation, we adjusted the laser in defocus so that the spot could cover the samples. The fine dry gel powders were thermally treated by laser irradiation at a relatively low power to remove the organic matter and the CCTO or Zn-doped CCTO precursor powders were thus obtained. As-prepared precursors were pressed in an oil press into pellets with thickness of 2 mm, and diameter of 10 mm under a pressure of 100 MPa. The pellets were put into a copper crucible for laser sintering and the solid- state reaction occurred for the precursors in the pellet when irradiated under high laser power. To guarantee the structural homogeneity of samples, we adjusted the laser spot diameter in the defocusing state to cover the pressed pellet. Then the pellet was cooled down to room temperature naturally and the finally obtained product was CCTO-based ceramic.

X-ray diffraction (XRD) patterns for samples were collected on a Rigaku Ultima VI X-ray diffractometer (Cu-Kα radiation, λ = 1.5418Å, Rigaku, Tokyo, Japan). The morphologies and microstructures of CCTO were measured by scanning electron microscope (SEM, ZEISS MERLIN Compact, Oberkochen, Baden-Württemberg, Germany) and high resolution transmission electron microscopy (HRTEM, FEI Tecnai G20, Hillsboro, OR, USA). Raman experiments were performed using a Renishaw in Via Raman spectrometer (London, UK) with 532 nm excitation. For dielectric measurements, electrodes of silver paste were painted on both sides of the thin ceramic cylindrical piece. The dielectric constants, dielectric loss, and complex impedance of CCTO-based ceramic samples were measured with a high-precision LCR meter (HP 4990A; Agilent, Palo Alto, CA, USA).

3. Results and Discussion

The precursors were thermally treated by laser irradiation at 30 W for 10 min to remove the organic group. The precursor pellets were sintered under laser at 120 W for 5 s to form the CCTO- based ceramics. Figure 1 shows XRD spectra of the prepared ceramics. The diffraction peaks identified were found to be consistent with the diffraction planes of body-centered cubic structure of CCTO, agreeing with the PDF file No. 75-1149. The major peaks include (211), (220), (310), (222), (321), (400), (411), (422), and (440) for CCTO and Zn-doped CCTO samples. We can see that the samples demonstrate better crystalline quality. The Zn-doped CCTO also has a pseudo-cubic structure according to the doping ratio of Ca(Cu$_2$Zn)Ti$_4$O$_{12}$. The main diffraction peaks almost show no splitting peak and only a small peak appears at 36.1 which can be attributed to peak (101) of CuO. Meanwhile, the Zn-doped CCTO shows that the diffraction peak moved to a high diffraction angle, signifying d-spacing reduction which could easily result in smaller grain size to enhance the compactness of the ceramics. Figure 2 shows SEM and HRTEM images of prepared CCTO and Zn- doped CCTO. It shows that the doped CCTO has better uniformity in particle size and better densification. Figure 2c,d present the HRTEM images which clearly demonstrate a well-defined crystalline lattice structure. The determined interplanar spacing of the plane (220) shows it was shrunken slightly from 2.618 nm to 2.598 nm with Zn-doping which is consistent with analysis of the results of XRD and SEM.

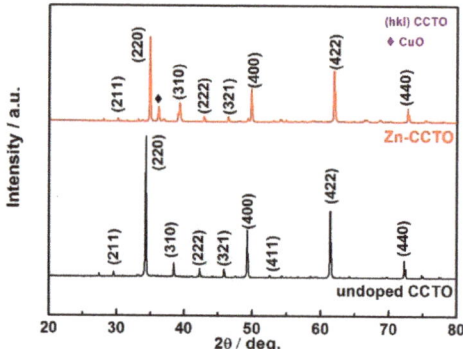

Figure 1. X-ray diffraction (XRD) patterns of undoped and Zn-doped calcium copper titanate (CCTO) prepared by rapid laser sintering of sol-gel derived precursors.

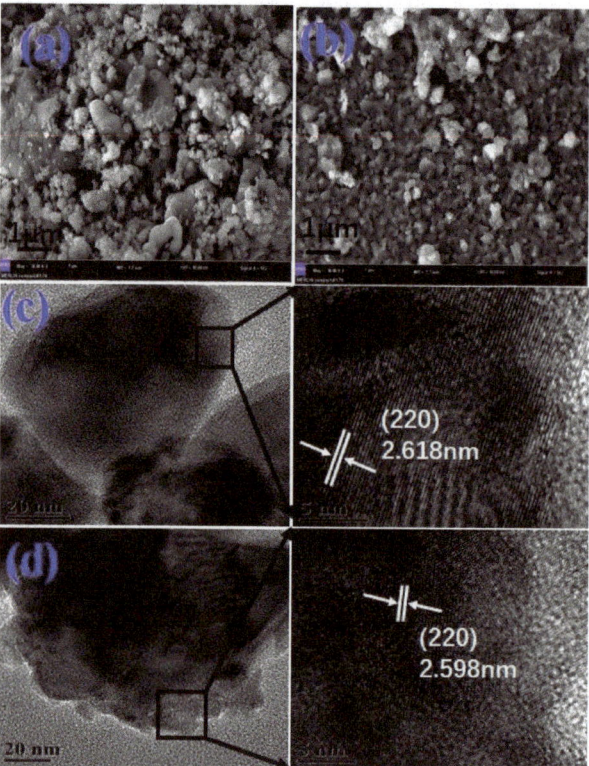

Figure 2. The SEM and HRTEM images for prepared undoped (**a**,**c**) and Zn-doped CCTO (**b**,**d**).

Raman spectra measurements were performed on samples after laser sintering of the precursors, shown in Figure 3. As the modes of Raman show, there are peaks located at 446 cm^{-1}, 513 cm^{-1}, and 576 cm^{-1} for undoped CCTO which is in accord with previous reports [24–26]. The Raman shift of 446 cm^{-1} corresponds to the modes of A$_g$(1) and F$_g$(2), and the peaks at 513 and 576 cm^{-1} correspond to A$_g$(2) and F$_g$(3), respectively. The modes of A$_g$(1), A$_g$(2) and F$_g$(2) are ascribed to lattice vibrations in TiO$_6$, and F$_g$(3) reflects the vibration mode of adverse stretching for the O–Ti–O bond [24,27]. The Raman vibration modes of Zn-doped CCTO show a shift to lower frequencies at 442 cm^{-1}, 506 cm^{-1}, and 570 cm^{-1}, assigned to A$_g$(1)/F$_g$(2), A$_g$(2), and F$_g$(3), respectively. This could be attributed to Zn ion incorporation affecting the structure of CCTO. The weakness of the relative intensity for the Raman peaks in Zn-doped CCTO, especially for the A$_g$(2) mode, demonstrates that Zn doping changes slightly the lattice vibrations of the titanium–oxygen octahedron.

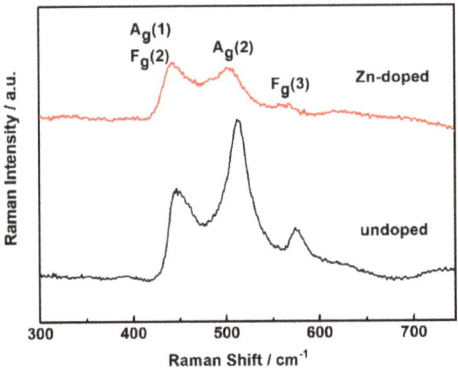

Figure 3. Raman measurement for undoped and Zn-doped CCTO prepared by laser treatment of sol-gel derived precursors.

Figure 4a,b shows the frequency relationship of permittivity and tan δ for undoped and Zn-doped CCTO in the range of 10–10^7 Hz. The samples show permittivity of above 10^3 at the broad frequency region, and the dielectric permittivity decreases gently over the entire frequency range shown in Figure 4a. It shows that the permittivity almost remains unchanged in the range of 10^2–10^6 Hz, demonstrating good stability on frequency. The frequency-stability of the permittivity should be closely associated with size of grain, density of grain boundaries, and the compactness of ceramics. The Zn-doped CCTO has higher permittivity than undoped CCTO. This could be ascribed to the denser structure and finer particles which can be seen from the SEM images and HRTEM. Above 10^6 Hz, the permittivity descends linearly, which could be attributed to Debye-like relaxation [28,29]. In Figure 4b we can see that tan δ of the specimen declines first in low-frequency to a platform in middle-frequency and then gently increases in high-frequency. The Zn-doped CCTO demonstrates lower dielectric loss than the undoped sample over the entire frequency region, especially in the low frequency range. Therefore, Zn doping effectively reduces the dielectric loss and enhances the dielectric properties. The mechanism of giant dielectric permittivity of CCTO is still a hotspot of controversy, but people generally believe that the behavior of CCTO could be related to the microstructure which may play a vital role for the huge permittivity. Several techniques have been adopted to improve the dielectric properties of CCTO, including nanosizing, recombination, modifying etc. Among them element doping in based materials is a simple and efficient method. A lot of elements have been doped into CCTO, but unanticipated results often occur, with accompanying dielectric loss. Most literatures reported element doping raised the permittivity as well as loss arising [9,18,30,31]. Said Senda et al. [32] reported Ni-doped CCTO prepared by the routine solid state reaction method. They studied on grain growth, morphological evolution with Ni doping content and found the dielectric properties including the permittivity, loss, and resistivity were improved. It was interpreted that dopants could easily segregate between the grain boundaries, and further depress the grain growth which changes the Cu/Ti stoichiometry. Finally, it greatly influences the microstructure of materials. This is in agreement with results of this work. Sonia De Almeida-Didry et al. [33] studied alumina doped CCTO prepared by the core-shell approach and they found that the grain boundaries of CCTO materials play a key role in the huge permittivity by controlling the density of grain boundaries. They demonstrated that core-shell design is an efficient technique to obtain a high dielectric constant, which shows obvious performance improvement. Furthermore, the various synthesized methods for doped CCTO have been very important to improve the dielectric performance. M.A. dela Rubia et al. [34] reported the dielectric properties for Hf doping CCTO ceramics prepared by comparing these two synthesizing routes. They concluded that the reactive sintering method is better than the conventional synthesis technique. The reactive sintering method offers the convenience for Hf incorporation in the CCTO lattice. Dong Xu et al. [35] reported CCTO with Zn element doping by the sol-gel method and found Zn-doping does not improve the dielectric properties because of the mesoporous structure of the synthesized materials. Therefore, the improved sol-gel method here with laser treatment can effectively enhance the dielectric properties. It can be concluded that a compact structure composed of a number of smaller grains can benefit the improvement of overall dielectric performance.

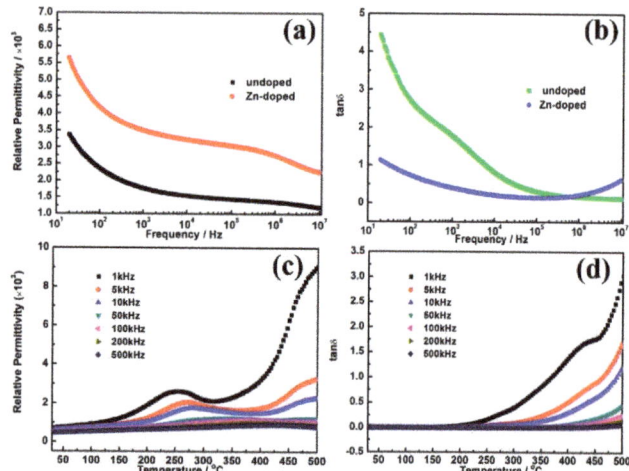

Figure 4. The relationships of frequency and permittivity, loss for undoped (a) and Zn-doped CCTO (b). Permittivity and loss for Zn-doped CCTO measured from 30 to 500 °C (c,d).

Figure 4c,d shows the permittivity and loss of Zn-doped CCTO with temperature at fixed frequencies. The values of dielectric permittivity obviously increase with rising temperature and decrease with incremental frequency. At higher temperature region (>400 °C), the dielectric permittivity increases with a large amplitude, which is attributed to enhanced conductivity by high temperature. The decrease in dielectric permittivity at higher frequencies could be explained in respect of the space charge polarization occurring at the interface [36], which is due to heterogeneous microstructures. This polarization mechanism is enhanced at higher temperature which could be ascribed to the direct current conductivity rising. Thus, the polarization arises quickly in permittivity at high temperature and low frequencies. Figure 4d shows the changes in the dielectric loss (tan δ) of the Zn-doped CCTO ceramic with temperature at some fixed frequencies. It can be seen that the tan δ demonstrates a low platform from 30 to 250 °C, then an obvious increase above the temperature of 250 °C for all given frequencies. However, tan δ remains at lower values at higher frequency. The dielectric loss of Zn-doped CCTO materials is strongly temperature dependent.

Figure 5a,b shows that the real part of the impedance changes at some fixed temperatures with frequency for undoped and Zn-doped CCTO, respectively. Below 10^3 Hz when the temperature increases, the values of Z' decreases and then the Z' values almost merge above 10^4 Hz. The different Z' values indicate the various electrical properties related to the microstructures. For the undoped CCTO, the Z' values are more divergent than the Zn-doped sample. Researchers attributed this dispersion at lower frequency to reduction of space charge polarization and discharging, which may be attributed to charge carrier hopping between para-electric phases [37–39]. Figure 5c,d shows the Nyquist plot for undoped and Zn-doped CCTO with a measured temperature of 25–300 °C. Impedance spectroscopy is an efficient tool to figure out the contribution mechanism of grains, boundaries, and interfaces to the total electrical conduction performance of materials by the equivalent RC circuit model. The curves commonly consist of two semicircular arcs standing for the resistances of the grain (R_g) and the grain boundary (R_{gb}).

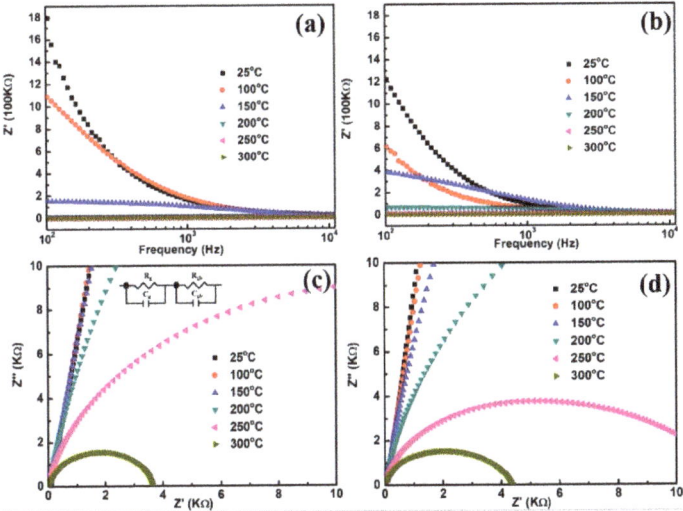

Figure 5. The real part (Z') versus frequency (a,b) and complex impedance plots (c,d) for undoped CCTO (a,c) and Zn-doped CCTO (c,d), respectively, with a measured temperature of 25–300 °C. Inset in (c) shows the two parallel RC equivalent circuits in series.

Respectively, it can be seen that the data in this work only show arcs partly due to the measuring range limit. According to the approximate conduction mechanisms of the IBLC model, the samples are formed by conducting grains and insulating boundaries. The inset of Figure 5c shows the RC equivalent circuit. It is apparent from Figure 5c,d that the resistance values of the grain and grain boundaries decrease with rising temperature, which is consistent with the results of enhanced conductivity with high temperature from Figure 4c,d. The Zn-doping affected the resistances of the CCTO bulk (R_g) and boundary (R_{gb}) which can be seen as the specimen demonstrates a smaller diameter of the semicircle than the undoped one around 200 °C. Moreover, the decreases in diameter of the semicircle with rising temperature indicated that the resistance of the samples demonstrates a negative temperature coefficient [40]. Based on the equivalent circuit, the fitted values with Z-View software, are shown in Table 1.

Table 1. Grain and grain boundary resistance for undoped and Zn-doped calcium copper titanate (CCTO) at temperature range of 25–300 °C.

Temperature (°C)	Undoped		Zn-Doped	
	R_g (Ω)	R_{gb} (Ω)	R_g (Ω)	R_{gb} (Ω)
25	906.7	8,998,000,000	837.4	312,700,000
100	422.7	23,223,000	406.9	3,288,100
150	223.8	2,225,400	276.2	512,790
200	105.9	188,050	102.2	70,520
250	90.5	22,282	50.6	10,454
300	50.0	3660	40.2	4076

To reveal the effectiveness of grain boundaries on Zn-doped CCTO, we studied the dependence of temperature on the conductivity of grain boundaries, and the grain boundary resistance (R_{gb}) was deduced from the impedance spectra as described in Table 1. According to the Arrhenius equation, the resistance has a close relationship with temperature, which could be summarized in this equation:

$$\sigma(T) = \sigma \exp\left(-\frac{E}{KT}\right) \quad (1)$$

where σ the pre-exponential factor, E presents the activation energy of the grain boundary, K is the Boltzmann constant and T is the absolute temperature, respectively. For the plot of lnσ vs. 1/T as shown in Figure 6, the solid lines show the results fitted using the Arrhenius equation. The experimental data approximately obeys the equation. The calculated activation energy for grain boundaries E_{gb} is 0.712 eV and 0.605 eV for undoped and doped CCTO, respectively, which is close to the reported values. The decrease in activation energy of E for the Zn-doped sample suggests a change in electrical conductivity of the grain boundaries, which is due to the intensive motion of the charge carriers accumulated at the boundaries.

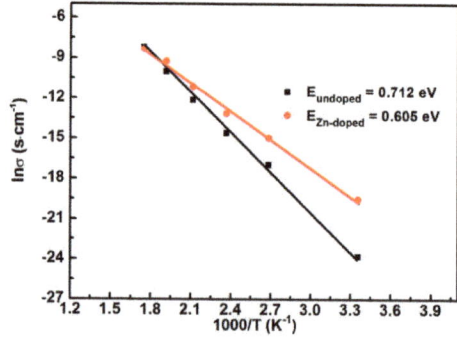

Figure 6. lnσ vs. temperature for undoped and Zn doping CCTO in the form of Arrhenius plots.

4. Conclusions

Zn-doped CCTO was prepared using a novel method of rapid laser sintering of sol-gel derived precursors. By comparing with undoped CCTO, the microstructure, dielectric properties, and impedance response were investigated in detail. HRTEM images demonstrate a well-defined crystalline lattice structure. The determined interplanar spacing of the plane (220) shows that it was shrunken slightly from 2.618 nm to 2.598 nm with Zn-doping. Raman spectra analysis demonstrated that the Zn doping changes slightly the lattice vibrations of the titanium–oxygen octahedron. The permittivity almost remains unchanged in the range of 10^2–10^6 Hz, demonstrating good stability on frequency. The impedance mechanisms were discussed according to the IBLC model by impedance spectroscopy analysis, and the results show that Zn-doping obviously affected the grain and grain boundary resistance. The activation energy for the grain boundaries was deduced from the slope of $\ln\sigma$ versus $1/T$. The activation energy of E_A for the Zn-doped sample is 0.605 eV, smaller than the undoped sample, which suggests increased charge carrier motion at the grain boundaries. This technique overcomes the shortcomings of long-time thermal energy supply by a furnace and presents references to synthesize the ceramic materials through a combination of soft-chemical methods.

Author Contributions: Data curation, Y.L.; Formal analysis, J.H.; Resources, H.Z.; Supervision, Y.H. and H.Z.; Writing—original draft, Y.H.; Writing—review & editing, Y.H. and Y.Q. All authors have read and agreed to the published version of the manuscript.

Funding: This work was supported by the National Natural Science Fundation of China (11974112, 11704127, 61875243, 11804100). This research received no external funding.

Conflicts of Interest: The authors declare no conflict of interest.

References

1. Ahmadipour, M.; Ain, M.F.; Ahmad, Z.A. A Short Review on Copper Calcium Titanate (CCTO) Electroceramic: Synthesis, Dielectric Properties, Film Deposition, and Sensing Application. *Nano-Micro Lett.* **2016**, *8*, 291–311. [CrossRef] [PubMed]
2. Irvine, J.T.S.; Sinclair, D.C.; West, A.R. Electroceramics: Characterization by Impedance Spectroscopy. *Adv. Mater.* **1990**, *2*, 132–138. [CrossRef]
3. Wang, C.-C.; Zhang, M.-N.; Xu, K.-B.; Wang, G.-J. Origin of High-Temperature Relaxor-Like Behavior in $CaCu_3Ti_4O_{12}$. *J. Appl. Phys.* **2012**, *112*, 034109. [CrossRef]
4. Ashokbabu, A.; Thomas, P. Structural, thermal and dielectric behavior of Polyaryletherketone (PAEK)/$CaCu_3Ti_4O_{12}$(CCTO). *Ceram. Int.* **2019**, *45*, 25052–25059. [CrossRef]
5. Zhang, J.; Zheng, J.; Liu, Y.; Zhang, C.; Hao, W.; Lei, Z.; Tian, M. The dielectric properties of CCTO ceramics prepared via different quick quenching methods. *Mater. Res. Bull.* **2019**, *115*, 49–54. [CrossRef]
6. Yang, W.; Yu, S.; Sun, R.; Du, R. Nano- and microsize effect of CCTO fillers on the dielectric behavior of CCTO/PVDF composites. *Acta Mater.* **2011**, *59*, 5593–5602. [CrossRef]
7. Subramanian, M.A.; Li, D.; Duan, N.; Reisner, B.A.; Sleight, A. High Dielectric Constant in $ACu_3Ti_4O_{12}$ and $ACu_3Ti_3FeO_{12}$ Phases. *J. Solid State Chem.* **2000**, *151*, 323–325. [CrossRef]
8. Adams, T.B.; Sinclair, D.C.; West, A.R. Giant Barrier Layer Capacitance Effects in $CaCu_3Ti_4O_{12}$ Ceramics. *Adv. Mater.* **2002**, *14*, 1321–1323. [CrossRef]
9. Lee, J.-W.; Koh, J.-H. Enhanced dielectric properties of Ag-doped CCTO ceramics for energy storage devices. *Ceram. Int.* **2017**, *43*, 9493–9497. [CrossRef]
10. Sinclair, D.C.; Adams, T.B.; Morrison, F.D.; West, A.R. $CaCu_3Ti_4O_{12}$: One-step internal barrier layer capacitor. *Appl. Phys. Lett.* **2002**, *80*, 2153–2155. [CrossRef]
11. Nazari, T.; Alizadeh, A.P. Microstructure and dielectric properties of CCTO glass-ceramic prepared by the melt-quenching method. *Ceram. Int.* **2019**, *45*, 19316–19322. [CrossRef]
12. Adams, T.B.; Sinclair, D.C.; West, A.R. Characterization of grain boundary impedances in fine- and coarse-grained $CaCu_3Ti_4O_{12}$ ceramics. *J. Am. Ceram. Soc.* **2006**, *89*, 3129–3135. [CrossRef]
13. Chen, G.; Lin, X.; Li, J.; Fisher, J.G.; Zhang, Y.; Huang, S.; Cheng, X. Enhanced dielectric properties and discharged energy density of composite films using submicron PZT particles. *Ceram. Int.* **2018**, *44*, 15331–15337. [CrossRef]
14. Maldonado, F.; Rivera, R.; Villamagua, L.; Maldonado, J. DFT modelling of ethanol on $BaTiO_3$(001) surface. *Appl. Surf. Sci.* **2018**, *456*, 276–289. [CrossRef]
15. Herle, J.V.; Vasquez, R. Conductivity of Mn and Ni-doped stabilized zirconia electrolyte. *J. Eur. Ceram. Soc.* **2004**, *24*, 1177–1180. [CrossRef]
16. Srivastava, A.; Kumar, D.; Singh, R.K.; Venkataraman, H.; Eisenstadt, W.R. Improvement in electrical and dielectric behavior of (Ba, Sr) TiO_3 thin films by Ag doping. *Phys. Rev. B* **2000**, *61*, 7305–7307. [CrossRef]
17. Amhil, S.; Choukri, E.; Moumen, S.B.; Bourial, A.; Essaleh, L. Evidence of large hopping polaron conduction process in Strontium doped Calcium Copper Titanate ceramics. *Phys. B: Condens. Matter* **2019**, *556*, 36–41. [CrossRef]

18. Kwon, S.; Huang, C.; Patterson, E.A.; Cann, D.P.; Alberta, E.F.; Kwon, S.; Hackenberger, W.S.; Cann, D.P. The effect of Cr_2O_3, Nb_2O_5 and ZrO_2 doping on the dielectric properties of $CaCu_3Ti_4O_{12}$. *Mater. Lett.* **2008**, *62*, 633–636. [CrossRef]
19. Xiao, M.; Meng, J.; Li, L. Non-Ohmic behavior of copper-rich CCTO thin film prepared through magnetron sputtering method. *J. Mater. Sci. Mater. Electron.* **2019**, *30*, 9266–9272. [CrossRef]
20. Wang, B.; Pu, Y.P.; Wu, H.D.; Chen, K.; Xu, N. Influence of sintering atmosphere on dielectric properties and microstructure of $CaCu_3Ti_4O_{12}$ ceramics. *Ceram. Int.* **2013**, *39*, S525–S528. [CrossRef]
21. Sung, Y.L.; Young, H.K.; Kyoung, J.C.; Sung, M.J.; Sang, I.Y. Effect of copper-oxide segregation on the dielectric properties of CaCu3Ti4O12 thin films fabricated by pulsed-laser deposition. *Thin Solid Films* **2010**, *518*, 5711–5714.
22. Fan, H.Q.; Kim, H.E. Microstructure and electrical properties of sol–gel derived $Pb(Mg_{1/3}Nb_{2/3})_{0.7}Ti_{0.3}O_3$ thin films with single perovskite phase. *Jpn. J. Appl. Phys.* **2002**, *41*, 6768–6772. [CrossRef]
23. Sun, D.L.; Wu, A.Y.; Yin, S.T. Structure, Properties, and Impedance Spectroscopy of $CaCu_3Ti_4O_{12}$ Ceramics Prepared by Sol–Gel Process. *J. Am. Ceram. Soc.* **2008**, *91*, 169–173. [CrossRef]
24. Schmidt, R.; Stennett, M.C.; Hyatt, N.C.; Pokorny, J.; Prado-Gonjal, J.; Li, M.; Sinclair, D.C. Effects of sintering temperature on the internal barrier layer capacitor (IBLC) structure in $CaCu_3Ti_4O_{12}$ (CCTO) ceramics. *J. Eur. Ceram. Soc.* **2012**, *32*, 3313–3323. [CrossRef]
25. Valim, D.; Souza Filho, A.G.; Freire, P.T.C.; Fagan, S.B.; Ayala, A.P.; Mendes Filho, J.; Almeida, A.F.L.; Fechine, P.B.A.; Sombra, A.S.B.; Staun Olsen, J.; et al. Raman scattering and x-ray diffraction studies of polycrystalline $CaCu_3Ti_4O_{12}$ under high-pressure. *Phys. Rev. B* **2004**, *70*, 132103. [CrossRef]
26. Rhouma, S.; Saîd, S.; Autret, C.; de Almeida-Didry, S.; Amrani, M.E.I.; Megriche, A. Comparative studies of pure, Sr-doped, Ni-doped and co-doped $CaCu_3Ti_4O_{12}$ ceramics: Enhancement of dielectric properties. *J. Alloy. Compd.* **2017**, *717*, 121–126. [CrossRef]
27. Kolev, N.; Bontchev, R.P.; Jacobson, A.J.; Popov, V.N.; Hadjiev, V.G.; Litvinchuk, A.P.; Iliev, M.N. Raman spectroscopy of $CaCu_3Ti_4O_{12}$. *Phys. Rev. B* **2002**, *66*, 132102. [CrossRef]
28. Bueno, P.R.; Tararan, R.; Parra, R.; Joanni, E.; Ramirez, M.A.; Ribeiro, W.C.; Longo, E.; Varela, J. A polaronic stacking fault defect model for $CaCu_3Ti_4O_{12}$ material: An approach for the origin of the huge dielectric constant and semiconducting coexistent features. *J. Phys. D Appl. Phys.* **2009**, *42*, 055404. [CrossRef]
29. Liu, P.; Lai, Y.; Zeng, Y.; Wu, S.; Huang, Z.; Han, J. Influence of sintering conditions on microstructure and electrical properties of $CaCu_3Ti_4O_{12}$ (CCTO) ceramics. *J. Alloy. Compd.* **2015**, *650*, 59–64. [CrossRef]
30. Tang, L.; Xue, F.; Guo, P.; Luo, Z.; Xin, Z.; Li, W. Multiferroic properties of $BiFeO_3$–$Pb(Zr_{0.52}Ti_{0.48})O_3$ solid solution. *J. Mater. Sci. Mater. Electron.* **2018**, *29*, 9435–9441. [CrossRef]
31. Thongbai, P.; Meeporn, K.; Yamwong, T.; Maensiri, S. Extreme Effects of Na Doping on Microstructure, Giant Dielectric Response and Dielectric Relaxation Behavior in $CaCu_3Ti_4O_{12}$ Ceramics. *Mater. Lett.* **2013**, *106*, 129–132. [CrossRef]
32. Senda, S.; Rhouma, S.; Torkani, E.; Mefriche, A.; Autret, C. Effect of nickel substitution on electrical and microstructural properties of $CaCu_3Ti_4O_{12}$ ceramic. *J. Alloys Compd.* **2017**, *698*, 152–158. [CrossRef]
33. De Almeida-Didry, S.; Nomel, M.M.; Autret, C.; Honstettre, C.; Lucas, A.; Pacreau, F.; Gervais, F. Control of grain boundary in alumina doped CCTO showing colossal permittivity by core-shell approach. *J. Eur. Ceram. Soc.* **2018**, *38*, 3182–3187. [CrossRef]
34. Dela Rubia, M.A.; Leret, P.; del Campo, A.; Alonso, R.E.; López-Garcia, A.R.; Fernández, J.F.; de Frutos, J. Dielectric behaviour of Hf-doped $CaCu_3Ti_4O_{12}$ ceramics obtained by conventional synthesis and reactive sintering. *J. Eur. Ceram. Soc.* **2012**, *32*, 1691–1699. [CrossRef]
35. Xu, D.; Zhang, C.; Cheng, X.; Fan, Y.; Yang, T.; Yuan, H. Dielectric Properties of Zn-Doped CCTO Ceramics by Sol-Gel Method. *Adv. Mater. Res. Vols.* **2011**, *197*, 302–305. [CrossRef]
36. Singh, L.; Rai, U.S.; Mandal, K.D.; Sin, B.C.; Lee, S.I.; Lee, Y. Dielectric, AC-impedance, modulus studies on $0.5BaTiO_3$ $0.5CaCu_3Ti_4O_{12}$ nano-composite ceramic synthesized by one-pot, glycine-assisted nitrate-gel route. *Ceram. Int.* **2014**, *40*, 10073–10083. [CrossRef]
37. Ahmadipour, M.; Cheah, W.K.; Ain, M.F.; Rao, K.V.; Ahmad, Z.A. Effects of deposition temperatures and substrates on microstructure and optical properties of sputtered CCTO thin film. *Mater. Lett.* **2018**, *210*, 4–7. [CrossRef]
38. Wieczorek, W.; Płocharski, J.; Przyłuski, J.; Głowinkowski, S.; Pajak, Z. Impedance spectroscopy and phase structure of PEO NaI complexes. *Solid State Ion.* **1998**, *28*, 1014–1017. [CrossRef]

39. Varghese, O.K.; Malhotra, L.K. Studies of ambient dependent electrical behavior of nanocrystalline SnO_2 thin films using impedance spectroscopy. *J. Appl. Phys.* **2000**, *87*, 7457–7465. [CrossRef]
40. Xiao-Yu, G.; Ji, Z.; Rui-Xue, W.; Xian-Zhu, D.; Lei, S.; Zheng-bin, G.; Shan-Tao, Z. ZnO-enhanced electrical properties of $Bi_{0.5}Na_{0.5}TiO_3$-based incipient ferroelectrics. *J. Am. Ceram.* **2017**, *100*, 5659–5667.

© 2020 by the authors. Licensee MDPI, Basel, Switzerland. This article is an open access article distributed under the terms and conditions of the Creative Commons Attribution (CC BY) license (http://creativecommons.org/licenses/by/4.0/).

Article

Fabrication and Deposition of Copper and Copper Oxide Nanoparticles by Laser Ablation in Open Air

Mónica Fernández-Arias [1,*], Mohamed Boutinguiza [1], Jesús del Val [1], Antonio Riveiro [1], Daniel Rodríguez [2], Felipe Arias-González [3], Javier Gil [3] and Juan Pou [1]

1. Applied Physics Department, University of Vigo, 36310 Vigo, Spain; mohamed@uvigo.es (M.B.); jesusdv@uvigo.es (J.d.V.); ariveiro@uvigo.es (A.R.); jpou@uvigo.es (J.P.)
2. Biomaterials, Biomechanics and Tissue Engineering Group, Materials Science and Metallurgical Engineering Department, UPC-Barcelona TECH, 08930 Barcelona, Spain; Daniel.rodriguez.rius@upc.edu
3. School of Dentistry, Universitat Internacional de Catalunya, 08017 Barcelona, Spain; farias@uic.es (F.A.-G.); xavier.gil@uic.cat (J.G.)
* Correspondence: monfernandez@uvigo.es

Received: 1 January 2020; Accepted: 6 February 2020; Published: 10 February 2020

Abstract: The proximity of the "post-antibiotic era", where infections and minor injuries could be a cause of death, there are urges to seek an alternative for the cure of infectious diseases. Copper nanoparticles and their huge potential as a bactericidal agent could be a solution. In this work, Cu and Cu oxide nanoparticles were synthesized by laser ablation in open air and in argon atmosphere using 532 and 1064 nm radiation generated by nanosecond and picosecond Nd:YVO$_4$ lasers, respectively, to be directly deposited onto Ti substrates. Size, morphology, composition and the crystalline structure of the produced nanoparticles have been studied by the means of field emission scanning electron microscopy (FESEM), high resolution transmission electron microscopy (HRTEM), the energy dispersive spectroscopy of X-rays (EDS), selected area electron diffraction (SAED) and X-ray diffraction (XRD). The UV-VIS absorbance of the thin layer of nanoparticles was also measured, and the antibacterial capacity of the obtained deposits tested against *Staphylococcus aureus*. The obtained deposits consisted of porous coatings composed of copper and copper oxide nanoparticles interconnected to form chain-like aggregates. The use of the argon atmosphere contributed to reduce significantly the formation of Cu oxide species. The synthesized and deposited nanoparticles exhibited an inhibitory effect upon *S. aureus*.

Keywords: copper nanoparticles; laser ablation; antibacterial effects

---

1. Introduction

The decreasing effectiveness of antibiotics and other antimicrobial agents is a global concern. According to the World Health Organization (WHO), "the post-antibiotic era", where no treatment for infections and minor injuries exists, is near [1]. Currently, antibiotic resistance kills an estimated 700,000 people each year, and some experts predict that the number could rise to 10 million by 2050, if efforts to curtail resistance or develop new antibiotics are not made [2].

Staphylococcus aureus, in addition to being related to a large number of infectious diseases, is one of the bacteria that presents a greater resistance to current commercial antibiotics [3]. Furthermore, several researchers have demonstrated the important role of this *S. aureus* in some oral infections such as peri-implantitis, which is considered the main cause of dental implant failure [4]. To promote an antimicrobial response from implants, some metallic antibacterial elements have been incorporated into implants' surfaces and matrices [5,6]. Among these elements, noble metal and transition metal nanoparticles are attracting great interest due to their remarkable antibacterial properties.

Copper, besides the fact that it is potentially effective against different bacterial pathogens [7], is a trace element. That is, an element necessary for the proper functioning of the human body in amounts less than 100 mg per day, there being recommended a daily copper intake of around 1.4 mg for an adult weighing about 70 kg [8]. It works as an agent to help integrate iron, zinc and vitamin C, besides being essential for the brain and its neurotransmissions, energy production, and to regulate several hormonal processes. In addition, copper is part of a good number of enzymes, and is involved in tissue respiration [9,10].

In particular, the antibacterial activity of copper and its oxides when the size is reduced to nanoscale is of great interest because of a high surface to volume ratio, which allows doctors to kill the pathogens without affecting the healthy tissue that surrounds it [11]. But not only is the size behind these particular properties, but also shape of the particles, which depends upon the fabrication method. The technique of laser ablation in gaseous media used in the present study allows researchers to obtain nanoparticles with reduced average sizes. On the other hand, the absence of potentially toxic chemicals makes this technique a preferred choice for the preparation of nanoparticles. In other techniques, the use of surfactants or the use of chemical precursors produces the contamination of the nanoparticles with agents potentially toxic for human cells [12–14].

In previous works, silver nanoparticles were obtained by laser ablation in open air [15], and their antibacterial properties against *Lactobacillus salivarius* were then probed [16]. More recently, copper nanoparticles with different degrees of oxidation were also obtained by laser ablation in water and methanol, in order to demonstrate their cytocompatibility and bacteriological activity against *Aggregatibacter actinomycetemcomitans* [17]. In the present work, we report the synthesis and deposition of copper and copper oxide nanoparticles on cp Ti substrates in a one-step process by laser ablation. The process is carried out in the open air and in an argon atmosphere using two different laser sources. Results and formation process, including the influence of laser parameters, are discussed. Antibacterial activity against *Staphylococcus aureus* (a gram-positive aerobic bacteria normally associated with surgical wounds in orthopedic and dental patients) is also studied for being one of the most dangerous antibiotic-resistant bacteria.

2. Materials and Methods

2.1. Laser Ablation

A Copper foil with 99.99% of purity (Alfa Aesar) previously cleaned, was used as a laser ablation target. Titanium discs Grade 2 (Goodfellow Cambridge Limited, Huntingdon, UK) with 10 mm diameter and about 200 nm of average surface roughness were used as substrates to collect the ablated material. The target was set at a 30 degree angle with the horizontal plane, and the laser beam was focused on its upper surface. In each experiment, one Ti substrate was tilted to be placed almost parallel to the copper target, 10 mm away from the incident point of the laser beam, as detailed in Figure 1.

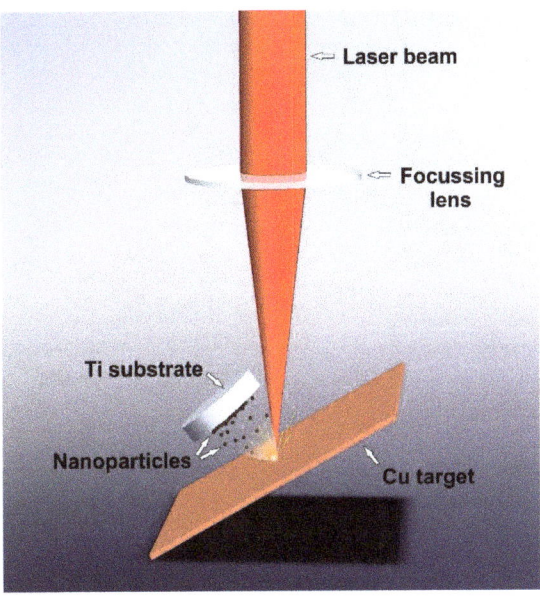

Figure 1. Laser ablation process.

Two different laser sources were used in the process. The first system was a diode-pumped Nd:YVO4 laser, providing pulses of 14 ns at a wavelength of 532 nm with 0.26 mJ of pulse energy. The second laser source was a Nd:YVO4 laser providing pulses of 800 ps at 1064 nm of wavelength and 0.03 mJ of pulse energy. The laser beam spot diameter on the target surface was estimated to be 132 µm, giving a fluence of 1.90 J/cm^2 in the case of the Green-Nanosecond laser and 196 µm, giving a fluence of 0.09 J/cm^2, for the IR–Picosecond laser. In all cases, the laser beam was kept in relative movement with respect to the target at 50 mm/s of scanning speed. The processing parameters used are listed in Table 1.

Table 1. Processing parameters.

Laser Source	Wavelength (nm)	Pulse Length (ns)	Pulse Frequency (kHz)	Pulse Energy (mJ)	Scanning Speed (mm/s)
Green-Nanosecond	532	14	20	0.26	50
IR-Picosecond	1064	0.8	200	0.03	50

To analyze in depth the formation process of nanoparticles and the influence of oxygen on this process, two different atmospheres were used with each laser source: In open air and in the argon environment by using an airtight chamber which kept the oxygen content below 50 ppm during the process. Note that all the assays were performed at atmospheric pressure (1 atm). Sample nomenclature with the corresponding assay conditions are listed in Table 2.

Table 2. Samples produced and analyzed.

Sample	Laser Source	Atmosphere	Time (min)
a	Green-Nanosecond	air	5
b	Green-Nanosecond	argon	7
c	IR-Picosecond	air	2
d	IR-Picosecond	argon	3

In order to compare the obtained samples, the same ablated mass (2 mg) for each condition was deposited on a titanium plate. Since the process parameters (shown in Table 1) and atmosphere changed, different processing time was required to obtain the same ablated mass as reported in Table 2. In this sense, it is noteworthy to mention that several previous assays were performed with each condition before the final preparation, in order to adjust carefully the required time to obtain the same mass. Additionally, in each final assay, the ablated mass was confirmed by weighing the targets before and after the ablation process ($\Delta m \pm 0.001$ g) for a higher accuracy.

2.2. Sample Preparation and Characterization Technics

The morphology and composition of the nanoparticles on the titanium surface were characterized by using a Ga ion beam in the FEI Helios NanoLab 600 (FEI, Hillsboro, OR, USA) dual beam microscope. Scanning electron microscopy (SEM) micrographs and energy dispersive spectroscopy X-rays (EDS) on the sectioning layer of the coating were obtained by Focused Ion Beam (FIB).

The deposited nanoparticles on the titanium discs were observed by Field Emission Scanning Electron Microscopy (FESEM) with a JEOL JSM 6700F microscope (JEOL, Akishima, Japan). Carbon-coated copper microgrids were also used as substrates to collect the nanoparticles for a size and morphology analysis. Transmission Electron Microscopy (TEM) images were acquired with a JEOL JEM 1010 (JEOL, Akishima, Japan) microscope and the nanoparticle size distribution was derived from a histogram obtained by measuring the diameter of about 300 particles.

Nanoparticles were also deposited on carbon-free copper microgrids in order to accomplish a crystallographic characterization. High-Resolution Transmission Electron Microscopy (HRTEM) and Selected Area Electron Diffraction (SAED) images were acquired with a JEOL JEM 2010F (JEOL, Akishima, Japan) high-resolution transmission electron microscope, equipped with a slow digital camera scan, using a 200 kV accelerating voltage. Identification of phases was achieved by comparing the measured distances with the diffraction patterns from the ICDD (JCPDS) database.

In order to corroborate the composition of the obtained nanoparticles, X-ray diffraction (XRD) analysis was carried out by means of a PANanalytical X'Pert Pro X-ray diffractometer using monochromated Cu-Kα radiation (wavelength 1.54 Å) over the 20–100° 2θ range with a step size of 0.02°. To facilitate the phase identification, the nanoparticles were deposited on a zero-background holder. The diffraction peaks of each sample were compared with the reference pattern of pure copper and different copper oxides from the ICDD (JCPDS) database.

The same ablation process was repeated by using a piece of glass to collect the nanoparticles, with the purpose of studying the optical properties. The ultraviolet to visible (UV/VIS) absorption spectrum of the copper nanoparticles' layer was measured in the range from 280 to 800 nm, using a Hewlett Packard HP 8452 diode array spectrophotometer.

Finally, three replicas (titanium discs with NPs) of each condition were submerged in 25 mL of ultrapure, deionized water in order to study the influence of copper ions in the bactericidal process. From each replica, 1.5 mL were laid away regularly during the first 21 days to be centrifuged. The extracted volume (1.5 mL) was gently replaced with fresh, deionized water. Afterwards, possible nanoparticles were separated from the solutions by using an Eppendorf miniSpin centrifuge at 13,400 rpm for 30 min at room temperature. After centrifuging, only 1 mL from the upper surface was taken, leaving the heaviest matter (NPs) in the bottom. In this regard, in order to ensure that only the ions in suspension are measured, UV-VIS spectroscopy was also used before analyzing. Finally, the ions content in the solutions was measured by Inductively Coupled Plasma Optical Emission Spectrometry (ICP-OES) with an Optima 4300 DV (Perkin Elmer, Waltham, MA, USA).

2.3. Antimicrobial Activity

The antibacterial activity of the treated samples was studied with *Staphylococcus aureus* (CECT (Colección Española de Cultivos Tipo) 435, Valencia, Spain) cultured in BHI broth (Scharlab SL, Sentmenat, Spain). A bacteria inoculum was incubated for 24 h at 37 °C before the assay.

The optical density for the bacterial suspension was adjusted to 0.2 ± 0.01 at 600 nm, equivalent to 1×10^8 colony-forming units (CFU)/mL.

The control and treated samples were immersed in ethanol and distilled water for 15 min each, and put into a 24-multiwell plate (Nunc, Rochester, NY, USA) with 1 mL bacterial suspension at 37 °C for 2 h. Afterwards, the samples were washed thrice with phosphate buffered solution (PBS) to wash off the nonadherent bacteria. The adhered bacteria were collected, sonicating the samples in 1 mL sterile PBS for 5 min. The PBS was serially diluted, and the diluted bacterial suspensions were seeded onto agar plates supplemented with BHI medium. The agar plates were then incubated at 37 °C for 24 h, and the CFUs were counted.

2.4. Statistical Analysis

Data collected from the antibacterial assay were statistically analyzed. All quantitative values were presented as mean ± standard deviation (SD). All experiments were performed using three replicates. Tukey's test was applied for comparing the mean of each group with the mean of the control group, and the means of all groups in pairs with a level of significance p-value < 0.05.

3. Results and Discussion

The characteristics and properties of the obtained nanoparticles were investigated, and the results analyzed concerning the influence of different parameters in the ablation process.

3.1. Ablation Rate

In order to ensure that the same quantity of ablated material was produced for each condition, the ablation time was carefully adjusted, being required to be at a different duration for each processing condition (see Table 2). This information gives us an idea about the efficiency of the process depending on the laser source and the atmosphere used, and also makes it possible to analyze the influence of time on the resultant nanoparticles.

As detailed in Table 2, the Green-Nanosecond laser requires more than twice as long as the IR–Picosecond to ablate the same amount of material from the copper plate surface. These results evidence the influence of laser parameters on the process. According to previous works, the ablation rate of pure metals is higher at low fluence due to the plasma shielding effect [18]. During the laser–matter interaction, material from the surface of the target is ejected and the plasma plume appears. This plasma reaches temperatures of tens of thousands Kelvin in the ablation process, but it is even higher in case of high fluences. When this occurs, as a result of the temperature increase, the plasma expands, and a protective shield on the upper surface of the material appears [19]. In addition, the absorption coefficient of the copper is higher for the 532 nm (green) than for the 1064 nm (invisible) wavelength, leading to a self-absorption process, where part of the incoming laser radiation is absorbed by the ablated material [20]. In the aforementioned processes (plasma shielding and self-absorption), the ablated material absorbs part of the incoming laser radiation, and consequently, the energy which reaches the surface is attenuated. Both phenomena have a great influence on the plasma development, and consequently on the amount of the ablated material.

Although laser parameters such as laser fluence, wavelength and pulse duration are important for controlling the process, the ablation environment determines to a large extent the laser–material interaction [21,22]. In this sense, the ablation process seems to be more productive in open air than in argon. This is likely due to the presence of oxygen and the high reactivity of copper. With about 20% of the background, air seems to contribute to increase the ablation efficiency by means of an exothermal reaction. As it was addressed by other authors, oxygen molecules dissociated in the process due to the high temperatures, and oxidized the surface of the target favoring the reaction and the heating of the material [18]. In addition, the atomic weight of the gaseous media has a great influence upon the plasma development. In this sense, Ar with a higher atomic weight than air conduces to more confinement of the plasma plume and a slow expansion.

3.2. Physicochemical Characterization of the Film

FIB-prepared cross-section shows the arrangement of the deposited nanoparticles on the titanium surface (Figure 2A,B). SEM micrographs show micrometer nanoparticles immersed inside a spongy coating of small nanoparticles. The adhesion between titanium and nanoparticles seems to be weak due to the small contact surface between them.

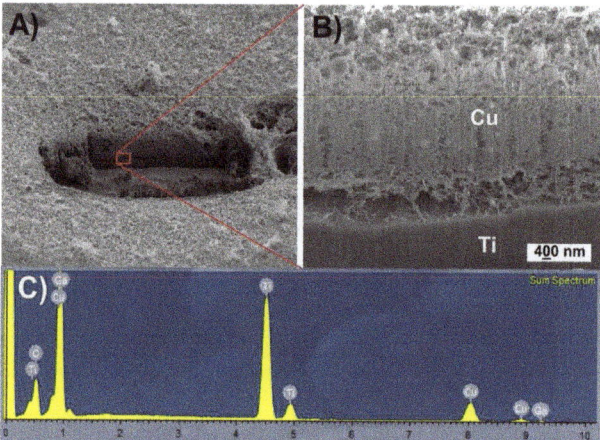

Figure 2. (**A**) Scanning electron microscopy (SEM) micrograph of the nanoparticles coating on the titanium disc, (**B**) SEM micrograph of the Focused Ion Beam (FIB)-prepared cross-section, (**C**) Sum spectrum obtained from the cross-section.

EDS performed confirms the composition of the coating.

3.3. Characterization of the Obtained Nanoparticles

3.3.1. Size and Morphology

Several images were recorded and used to perform an analysis of morphology and size distribution. As reference, a FESEM image of the Titanium substrate surface was also acquired (see Figure 3).

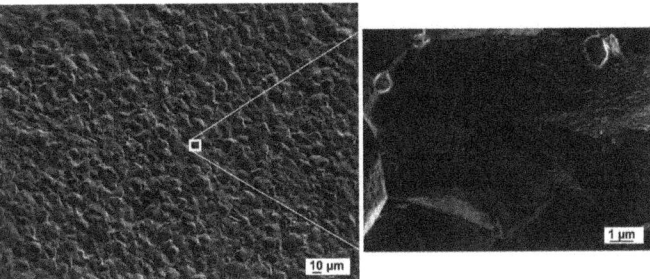

Figure 3. Field emission scanning electron microscopy (FESEM) micrographs of titanium disc surface used as substrate.

Figure 4 shows the aspect of the obtained copper nanoparticles anchored on the titanium surface at low magnification.

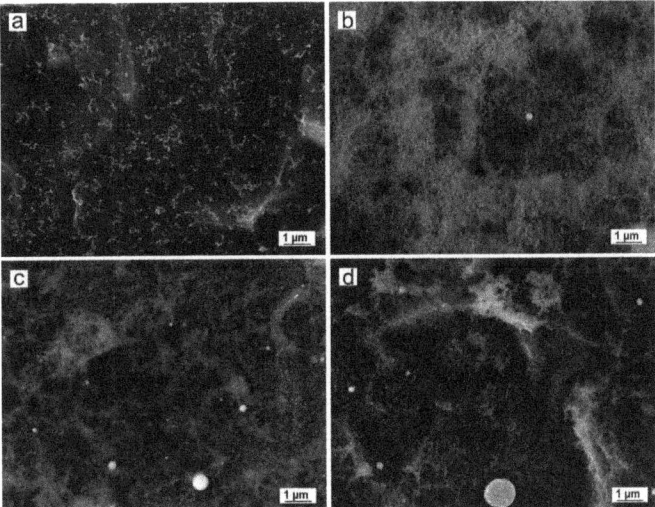

Figure 4. FESEM micrographs of Cu nanoparticles from samples a, b, c and d obtained by laser ablation using (**a**) Green-Nanosecond laser in air, (**b**) Green-Nanosecond laser in argon, (**c**) IR-Picosecond laser in air, (**d**) IR-Picosecond laser in Argon, as reported in Table 2.

Note that large particles can be observed on the titanium surface. This result is a common consequence of laser ablation due to the formation mechanism of nanoparticles. When laser pulses of high intensity strike on the copper target, the temperature of the interaction zone is increased up to its boiling point, leading to an explosive boiling, where nanodroplets together with ionized matter (ions, clusters, free atoms) are ejected from the target to the substrate surface. These nanodroplets, can form spherical micrometric and submicrometric particles and solidify on the substrate, while the ionized matter nucleates and grows. The EDS analysis confirms that even the micrometric particles are copper or copper oxide.

TEM micrographs of the obtained copper nanoparticles are shown in Figure 5.

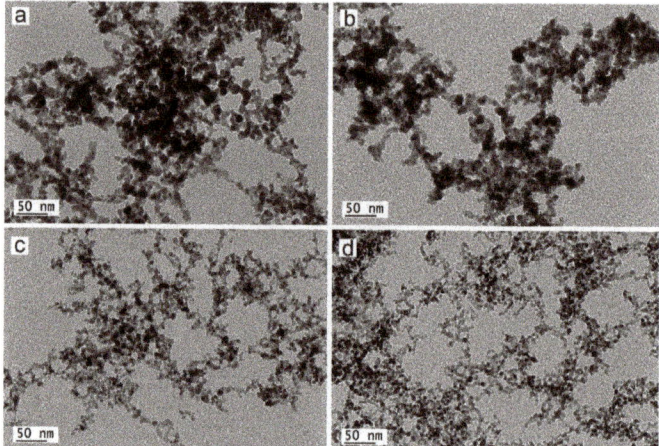

Figure 5. Transmission electron microscopy (TEM) micrographs of Cu nanoparticles from samples a, b, c and d obtained by laser ablation using (**a**) Green-Nanosecond laser in air, (**b**) Green-Nanosecond laser in argon, (**c**) IR-Picosecond laser in air, (**d**) IR-Picosecond laser in argon, as reported in Table 2.

As can be observed at higher magnification, copper nanoparticles obtained by laser ablation exhibit a rounded shape with a high tendency to agglomerate, forming chain-like structures. This is because of the metallic nature of the starting material and the formation mechanism of the nanoparticles. Concurrently with the nucleation and growth process of nanoparticles, the absorbed radiation increases the temperature of the already formed NPs above the melting point, melting and joining with others by coalescence, leading to the formation of chain-like structures.

The represented histograms (see Figure 6) were obtained from several representative TEM images by measuring the diameter of about 300 particles of each sample.

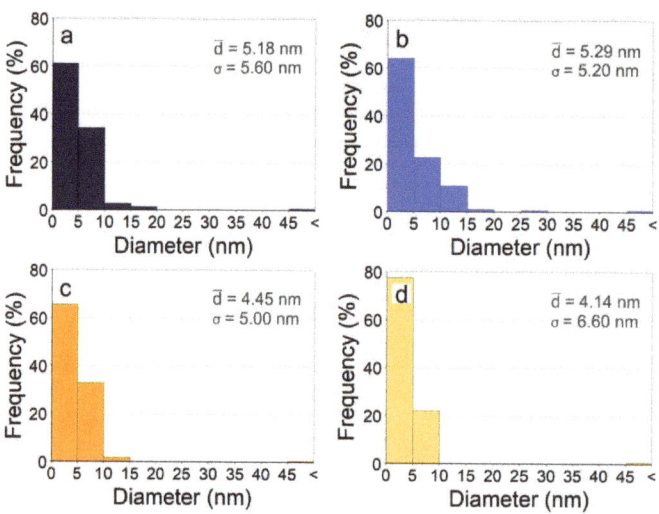

Figure 6. Histograms of size distribution from Cu nanoparticles obtained by laser ablation with (**a**) Green-Nanosecond laser in air, (**b**) Green-Nanosecond laser in argon, (**c**) IR-Picosecond laser in air, (**d**) IR-Picosecond laser in argon.

Close inspection of Figure 6 reveals that the sizes of the obtained nanoparticles are deeply determined by the nucleation time and the growth time of nuclei. In this growth kinetics, parameters such as fluence, wavelength and the repetition rate of the laser source, but also the environmental conditions, are critical factors [21,23].

As can be seen, copper nanoparticles obtained by the IR–Picosecond laser in argon present the lowest average size and the highest dispersion. On one hand, the fraction of particles of small size (between 0 and 5 nm) is greater when the laser ablation takes place in argon. These results bring to light the great influence of the atmosphere on the ablation process. In argon, the absorption of the laser beam by the plasma is higher than in air, and consequently, less of the sample is vaporized [21]. On the other hand, the broad size dispersion is due to the presence of large particles from the material ejection during the laser ablation process. These micrometric particles occur more frequently with the IR-Picosecond laser, because, although the power is similar, the pulse energy is concentrated in shorter pulses.

3.3.2. Composition and Crystallography

HRTEM, FFT and SAED

A high resolution transmission electron microscope was used to reveal the crystalline structure of the obtained nanoparticles. All of the particles obtained, even the smallest ones, are crystalline. This aspect can be observed in Figure 7, showing HRTEM of polycrystalline nanoparticles with clear lattice fringes and their corresponding Fast Fourier Transform (FFT) of the characteristic crystal as insets.

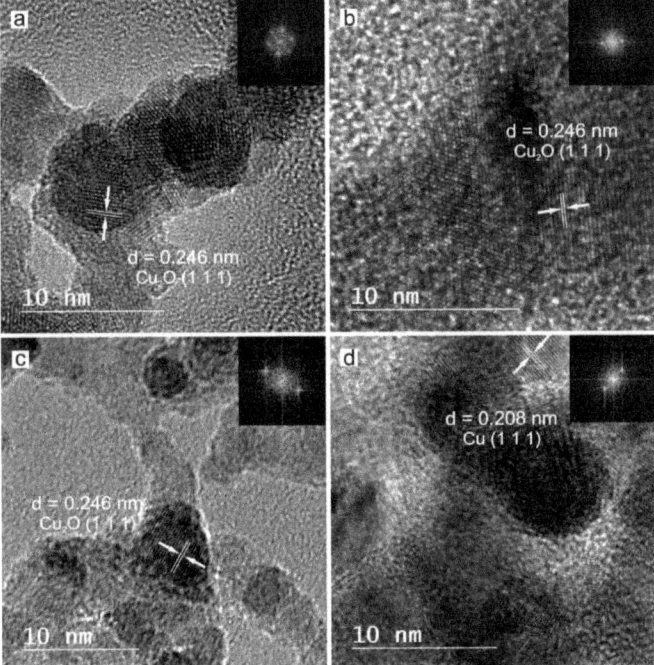

Figure 7. This is high resolution transmission electron microscopy (HRTEM) images of copper crystalline nanoparticles obtained by laser ablation with (**a**) Green-Nanosecond laser in air, (**b**) Green-Nanosecond laser in argon, (**c**) IR-Picosecond laser in air, (**d**) IR-Picosecond laser in argon. Measured interplanar distances and the corresponding Fast Fourier Transform (FFT) of single nanoparticles inset.

In order to elucidate the crystalline phases of each sample, SAED was performed on several groups of particles, as shown in Figure 8. Identification of phases was achieved by comparing the measured distances from SAED with the diffraction patterns from the ICDD (JCPDS) database.

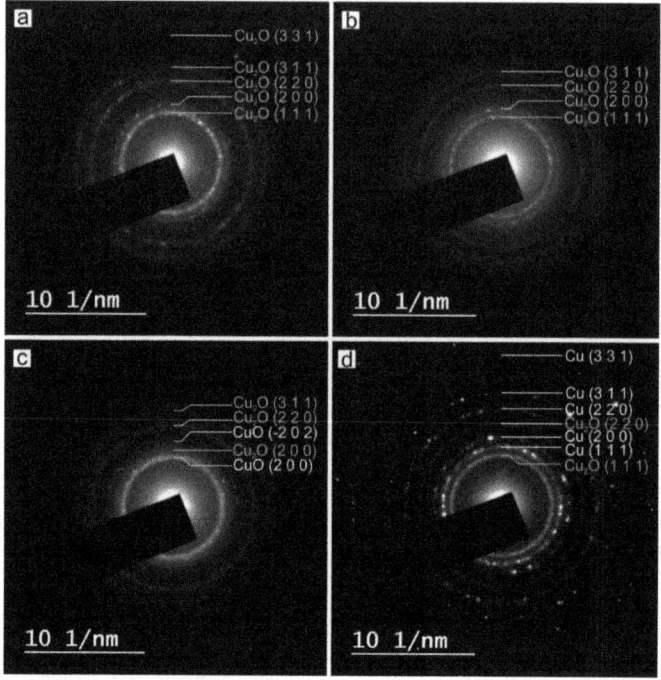

Figure 8. Selected area electron diffraction (SAED) pattern obtained over a group of copper nanoparticles obtained by laser ablation with (**a**) Green-Nanosecond laser in air, (**b**) Green-Nanosecond laser in argon, (**c**) IR Picosecond laser in air, (**d**) IR-Picosecond laser in argon.

The measured interplanar distances from Figure 8 are listed in Table 3.

Table 3. Lattice spacing measured in nm from SAED and the corresponding diffraction patterns from the ICDD database of metallic Copper, CuO and Cu_2O.

a	b	c	d	Cu (hkl)	CuO (hkl)	Cu_2O (hkl)
0.247	0.246		0.248			0.247 (111)
		0.234			0.231 (200)	
			0.208	0.209 (111)		
0.214	0.216	0.213				0.214 (200)
			0.183		0.181 (200)	
		0.187			0.187 (−202)	
0.152	0.151	0.151	0.151		0.150 (−113)	0.151 (220)
0.129	0.132	0.129	0.128		0.128 (220)	0.129 (311)
			0.109	0.109 (311)		
0.098			0.083		0.083 (331)	0.098 (331)

According to the data collected in Table 3, the main reflections of each sample correspond with diffraction patterns of metallic copper and different oxidation states of copper from the ICDD (JCPDS) database. The main reflections hkl of Cu-NPs obtained by laser ablation using the Green-Nanosecond laser in air (sample a) and argon (sample b), can be indexed on the cubic crystal lattice of Cu_2O (JCPDS-ICDD ref. 005-0667), while the Cu-NPs obtained by laser ablation using the IR-Picosecond laser in air (sample c) corresponds with a combination of Cu_2O and CuO (JCPDS-ICDD ref. 041-0254), and a combination of Cu_2O and Cu (JCPDS-ICDD ref. 004-0836) if the IR-Picosecond laser is used in argon (sample d).

XRD

To elucidate the crystalline phases, X-ray diffractometry (XRD) was performed on the obtained nanoparticles and the precursor copper plate. The corresponding patterns of the samples are depicted in Figure 9.

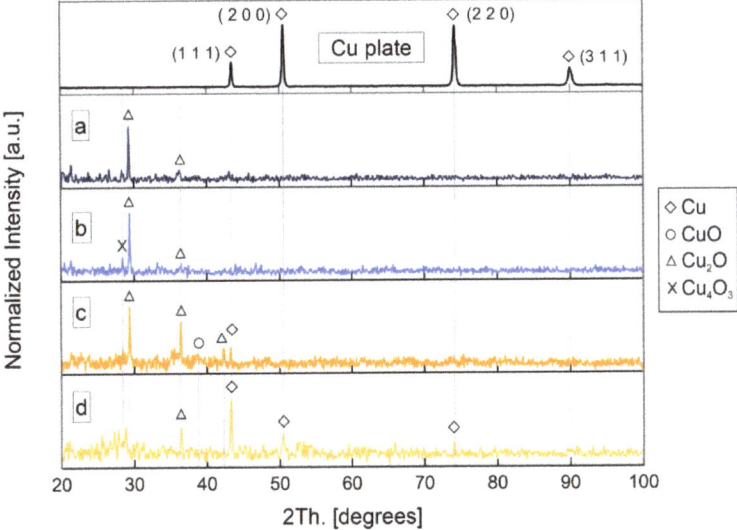

Figure 9. X-ray diffractometry (XRD) of Cu nanoparticles obtained with a wavelength of 532 nm by Los Alamos National Laboratory (LASL) with (a) the Green-Nanosecond laser in air, (b) Green-Nanosecond laser in argon, (c) an IR-Picosecond laser in air, (d) and the IR-Picosecond laser in argon. Gray lines represent the position of the representative diffraction peaks.

As can be clearly seen, the diffraction pattern of the target corresponds to a cubic crystal system with characteristic diffraction peaks (111), (200), (220) and (311) at 2θ values of 43.3°, 50.4°, 74.1° and 89.9°, respectively, according to JCPDS-ICDD ref. 004-0836. The elemental nature of the obtained Cu NPs was confirmed through XRD analysis. Samples obtained by laser ablation with the Green-Nanosecond laser (samples a and b) are mainly composed of Cu_2O. Although there is the presence of an intensity peak at 28.5° corresponding to Cu_4O_3 (JCPDS-ICDD ref. 003-0879) in sample b, it suggests that copper nanoparticles obtained in argon are less oxidized. On its behalf, Cu NPs obtained by laser ablation with the IR–Picosecond laser in air (sample c) corresponds with Cu_2O and a low presence of Cu and CuO, while in argon (sample d) NPs correspond mainly with metallic copper.

It is noteworthy to mention, that although the samples obtained in an inert atmosphere present a lower oxidation degree, the insignificant difference between samples a and b compared to the samples obtained when the IR–Picosecond laser, is probably due to the different mechanisms of NPs formation. When the Picosecond laser is used, ablated material is directly ejected from the surface target as a result of the direct sublimation. On the contrary, in the Nanosecond regime, the ablated material is melted and vaporized, growing in the plasma as a consequence of the interaction with the ambient gas, and subsequently by coalescence on the substrate. This thermal process favors the oxidation of the final NPs.

Note that although the results from XRD are in good agreement with those from SAED, the first one gives us more significant information. This is because, while in SAED the diffraction is performed on a group of particles, the scanned area in XRD is much larger. Furthermore, the corresponding peak positions of Cu and Cu_2O among samples match, but the intensities are quite different.

As reported by Ingham et al. [24], these results are common in thin films, and reveal a preferential orientation of each sample.

3.3.3. UV-VIS Absorption

In order to study the optical properties of the obtained nanoparticles, the absorbance of each sample was measured in the range of 280–800 nm by UV-VIS spectroscopy (Figure 10).

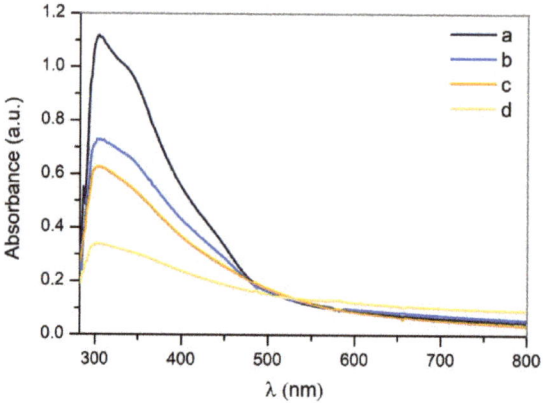

Figure 10. UV-VIS spectrum of Cu nanoparticles obtained by laser ablation using the Green-Nanosecond laser in (**a**) air, (**b**) argon and using the IR-Picosecond laser in (**c**) air, (**d**) Argon.

The mean peak, present in all samples at approximately 300 nm, is characteristic of surface plasmon resonance (SPR), a feature of CuO nanoparticles [25,26], although curves corresponding to samples a and b seem to be a convolution of two peaks. The mean peak corresponds to CuO and there is another at 350 nm, which corresponds with Cu_2O [25]. This hypothesis is supported by the results previously obtained by means of the crystallographic analysis. It is noteworthy to mention that SPR is a resonance condition, and occurs when the frequency of the incident light matches with the surface electron frequency of NPs [27]. As it was addressed by previous authors, this effect may not be observed in all nanoparticles, since only nanoparticles larger than 20 nm present SPR [28]. Furthermore, the reactive nature of copper extensively used in catalysis may be responsible for the subsequent oxidation by contact with atmospheric oxygen [29]. On the other hand, the SPR is associated to the nanoparticles' shape, size and surrounding medium. In this sense, the presence of a single surface plasmon peak implies that they are spherical [30], and the broadening of the absorbance peak is characteristic of a wide size distribution [31], which is in agreement with the TEM analysis.

3.3.4. Copper Ion Release

In order to study the ion release kinetics from the deposited nanoparticles, the copper ions content in water was measured during the first 21 days after ablation.

As can be seen in Figure 11, nanoparticles deposited by laser ablation in open air (samples a and c) show a higher rate of ion release than Cu-NPs deposited in argon (samples b and d), being almost four times higher in the first 8 h. This suggests that although the mass of copper nanoparticles deposited in each sample is the same (2 mg), the composition of the environment determines to a large extend the laser–matter interaction. Thus, when argon is used as a gaseous environment, nanoparticles could be deposited onto the surface of the substrate being embedded. This would reduce the nanoparticle surface in contact with the release medium. On the contrary, Cu nanoparticles obtained in open air would lead to a more superficial coating.

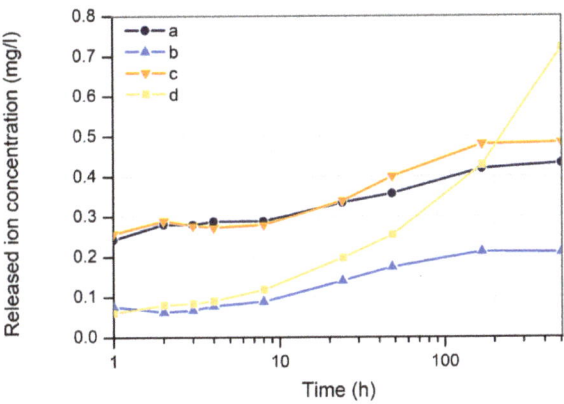

Figure 11. Kinetics of copper ion release from the immobilized copper nanoparticles on Ti plates.

Note that after the first 8 h, Cu nanoparticles obtained in Argon with the IR–Picosecond increase the ions' release ratio. After 7 days (168 h), samples a, b and c seem to become stable.

3.4. Analysis of Antimicrobial Activity

In order to assess the bactericidal activity of the obtained NPs, the analysis was performed with a Gram-positive bacteria, *Staphylococcus aureus*. The obtained relative absorbance of each sample with the corresponding error is shown in Figure 12. Taking into account that the higher the absorbance, the more the bacterial growth, these values were compared with a positive control (titanium discs with bacteria without NPs). As is shown, after 24 h, the bacterial growth values are noteworthily reduced by all samples. Copper nanoparticles obtained by laser ablation in argon by using the IR–Picosecond laser (sample d), exhibit the highest bactericidal capacity.

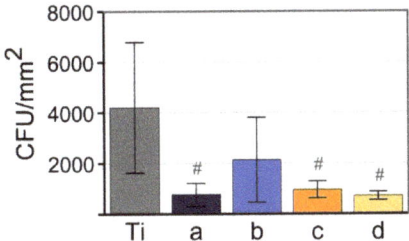

Figure 12. Bacterial adhesion on Ti with and without copper nanoparticles after 24 h of culture. Significant difference of each group with the control group (# p-value < 0.05).

The statistical analysis performed shows differences between control and samples a, c and d. On the contrary, no statistically significant differences between samples a, c and d was observed.

The notable dispersion in the results is consistent with the wide range of surface roughness on the titanium discs. In this sense, several authors have addressed the high impact of titanium surface topography on the bacterial adhesion [32,33].

Note that copper nanoparticles with the highest bactericidal effect (sample d) correspond to those that underwent a lower oxidation degree and smaller mean size with a high size dispersion. On one hand, previous work showed that a high oxidation state promotes the bactericidal activity but in conditions of similar oxidation states, particle size seems to be a decisive parameter.

The smaller the size, the greater the antibacterial capacity. This is consistent with previous works, and demonstrates that NPs with a large surface-to volume ratio provide more antibacterial activity [7]. On the other hand, although previous works state a direct relationship between the bactericidal activity of copper nanoparticles and their ions' release [6], our results provide compelling evidence that ions' release is not crucial compared to other parameters, such as particle size or oxidation state. In general terms, our results are consistent with previous studies [7,34], where copper exhibits excellent antibacterial properties when tested on *S. aureus*.

4. Conclusions

Crystalline copper and copper oxide nanoparticles have been obtained by means of a laser ablation technique in gaseous media, using two different nanosecond Nd:YVO4 lasers working at 532 nm and 1064 nm of wavelength, and without any chemical reagent or contamination.

The laser ablation ratio is higher with the IR–Picosecond laser than with the Green–Nanosecond. The presence of oxygen contributes to the process efficiency, obtaining more ablated material when laser ablation is carried out in open air than in argon. Nanoparticles obtained in argon do present in general terms a lower degree of oxidation than those obtained in air.

The antibacterial assay proved the strong antibacterial activity of the obtained copper nanoparticles against *S. aureus*. The best inhibitory effects are provided by copper nanoparticles obtained by laser ablation in argon with 1064 nm of wavelength. These results confirm the influence of size, crystallographic structure and oxidation state in the bactericidal effects of copper nanoparticles.

Author Contributions: Conceptualization, M.B. and J.P.; Methodology, M.F.-A. and D.R.; Validation, J.d.V. and A.R.; Formal analysis, J.d.V. and F.A.-G.; Investigation, M.F.-A., M.B. and D.R.; Data curation, J.d.V., J.G. and A.R.; Writing—original draft preparation, M.F.-A.; Writing—review and editing, M.B. and J.P.; Supervision, M.B., J.G. and J.P.; All authors have read and agreed to the published version of the manuscript.

Funding: This work was partially supported by the EU research project CVmar+i (INTERREG V A España-Portugal (POCTEP). By the Government of Spain [RTI2018-095490-J-I00 (MCIU/AEI/FEDER, UE)] and by Xunta de Galicia ED431C 2019/23, ED481D 2017/010, ED481B 2016/047-0).

Acknowledgments: The technical staff from CACTI (University of Vigo) is gratefully acknowledged.

Conflicts of Interest: The authors declare no conflict of interest. The funders had no role in the design of the study; in the collection, analyses, or interpretation of data; in the writing of the manuscript, or in the decision to publish the results.

References

1. Reardon, S. WHO warns against 'post-antibiotic' era. *Nature* **2014**. [CrossRef]
2. Willyard, C. Drug-resistant bacteria ranked. *Nature* **2017**, *543*, 7643. [CrossRef]
3. Ondusko, D.S.; Nolt, D. Staphylococcus aureus. *Pediatr. Rev.* **2018**, *39*, 287–298. [CrossRef] [PubMed]
4. Goudouri, O.; Kontonasaki, E.; Lohbauer, U.; Boccaccini, A.R. Antibacterial properties of metal and metalloid ions in chronic periodontitis and peri-implantitis therapy. *Acta Biomater.* **2014**, *10*, 3795–3810. [CrossRef] [PubMed]
5. Massa, M.A.; Covarrubias, C.; Bittner, M.; Fuentevilla, I.A.; Capetillo, P.; von Marttens, A.; Carvajal, J.C. Synthesis of new antibacterial composite coating for titanium based on highly ordered nanoporous silica and silver nanoparticles. *Mater. Sci. Eng. C* **2014**, *45*, 146–153. [CrossRef] [PubMed]
6. Palza, H.; Quijada, R.; Delgado, K. Antimicrobial polymer composites with copper micro- and nanoparticles: Effect of particle size and polymer matrix. *J. Bioact. Compat. Polym.* **2015**, *30*, 1–15. [CrossRef]

7. Khalid, H.; Shamaila, S.; Zafar, N.; Sharif, R.; Nazir, J.; Rafique, M.; Ghani, S.; Saba, H. Antibacterial Behavior of Laser-Ablated Copper Nanoparticles. *Acta Metall. Sin. Engl. Lett.* **2016**, *29*, 748–754. [CrossRef]
8. World Health Organization. *Trace Elements in Human Nutrition and Health*; World Health Organization: Geneva, Switzerland, 1996; pp. 1–360.
9. Karlin, K.D. Metalloenzymes, Structural Motifs, and Inorganic Models. *Science* **1993**, *261*, 701–708. [CrossRef]
10. Skalnaya, M.G.; Skalny, A.V. *Essential Trace Elements in Human Health: A Physician'S View*; Publishing House of Tomsk State University: Tomsk, Russian, 2018.
11. Vimbela, G.V.; Ngo, S.M.; Fraze, C.; Yang, L.; Stout, D.A. Antibacterial properties and toxicity from metallic nanomaterials. *Int. J. Nanomed.* **2017**, *12*, 3941–3965. [CrossRef]
12. Wagener, P.; Jakobi, J.; Rehbock, C.; Chakravadhanula, V.S.K.; Thede, C.; Wiedwald, U.; Bartsch, M.; Kienle, L.; Barcikowski, S. Solvent-surface interactions control the phase structure in laser-generated iron-gold core-shell nanoparticles. *Sci. Rep.* **2016**, *6*, 1–12. [CrossRef]
13. Boutinguiza, M.; Pou, J.; Lusquiños, F.; Comesaña, R.; Riveiro, A. Laser-assisted production of tricalcium phosphate nanoparticles from biological and synthetic hydroxyapatite in aqueous medium. *Appl. Surf. Sci.* **2011**, *257*, 5195–5199. [CrossRef]
14. Boutinguiza, M.; Lusquiños, F.; Riveiro, A.; Comesaña, R.; Pou, J. Hydroxylapatite nanoparticles obtained by fiber laser-induced fracture. *Appl. Surf. Sci.* **2009**, *255*, 5382–5385. [CrossRef]
15. Boutinguiza, M.; Comesaña, R.; Lusquiños, F.; Riveiro, A.; del Val, J.; Pou, J. Production of silver nanoparticles by laser ablation in open air. *Appl. Surf. Sci.* **2014**, *336*, 108–111. [CrossRef]
16. Boutinguiza, M.; Fernández-Arias, M.; del Val, J.; Buxadera-Palomero, J.; Rodríguez, D.; Lusquiños, F.; Gil, F.J.; Pou, J. Synthesis and deposition of silver nanoparticles on cp Ti by laser ablation in open air for antibacterial effect in dental implants. *Mater. Lett.* **2018**, *231*, 126–129. [CrossRef]
17. Fernández-Arias, M.; Boutinguiza, M.; del Val, J.; Covarrubias, C.; Bastias, F.; Gómez, L.; Maureira, M.; Arias-González, F.; Riveiro, A.; Pou, J. Copper nanoparticles obtained by laser ablation in liquids as bactericidal agent for dental applications. *Appl. Surf. Sci.* **2019**, *507*, 145032. [CrossRef]
18. Vadillo, J.M.; Fernández, J.M.; Rodríguez, C.; Laserna, J.J. Effect of Plasma Shielding on Laser Ablation Rate of Pure Metals at Reduced Pressure. *Surf. Interface Anal.* **1999**, *27*, 1009–1015. [CrossRef]
19. Gottfried, J.L. Influence of exothermic chemical reactions on laser-induced shock waves. *Phys. Chem. Chem. Phys.* **2014**, *16*, 21452–21466. [CrossRef]
20. Maina, M.; Okamoto, Y.; Inoue, R.; Nakashiba, S.; Okada, A.; Sakagawa, T. Influence of Surface State in Micro-Welding of Copper by Nd:YAG Laser. *Appl. Sci.* **2018**, *8*, 2364. [CrossRef]
21. Hermann, J.; Gerhard, C.; Axente, E.; Dutouquet, C. Comparative investigation of laser ablation plumes in air and argon by analysis of spectral line shapes: Insights on calibration-free laser-induced breakdown spectroscopy. *Spectrochim. Acta Part B At. Spectrosc.* **2014**, *100*, 189–196. [CrossRef]
22. Hamad, A.H. Laser Ablation in Different Environments and Generation of Nanoparticles Generation. *Appl. Laser Ablation Thin Film Depos. Nanomater. Synth. Surf. Modif.* **2016**, 177–196. [CrossRef]
23. Yang, G.W. Laser ablation in liquids: Applications in the synthesis of nanocrystals. *Prog. Mater. Sci.* **2007**, *52*, 648–698. [CrossRef]
24. Ingham, B. X-Ray Diffraction for Characterizing Metallic Films. In *Metallic Films for Electronic, Optical and Magnetic Applications*; Woodhead Publishing: Sawston, UK, 2014; pp. 3–38. [CrossRef]
25. Raghav, R.; Aggarwal, P.; Srivastava, S. Tailoring oxides of copper-Cu_2O and CuO nanoparticles and evaluation of organic dyes degradation. *AIP Conf. Proc.* **2016**, *1724*. [CrossRef]
26. Zhu, J.; Li, D.; Chen, H.; Yang, X.; Lu, L.; Wang, X. Highly dispersed CuO nanoparticles prepared by a novel quick-precipitation method. *Mater. Lett.* **2004**, *58*, 3324–3327. [CrossRef]
27. Khashan, K.S.; Jabir, M.S.; Abdulameer, F.A. Carbon Nanoparticles decorated with cupric oxide Nanoparticles prepared by laser ablation in liquid as an antibacterial therapeutic agent. *Mater. Res. Express* **2018**, *5*, 035003. [CrossRef]
28. Díaz-Visurraga, J.; Daza, C.; Pozo, C.; Becerra, A.; von Plessing, C.; García, A. Study on antibacterial alginate-stabilized copper nanoparticles by FT-IR and 2D-IR correlation spectroscopy. *Int. J. Nanomed.* **2012**, *7*, 3597–3612. [CrossRef]
29. Goncharova, D.; Lapin, I.; Svetlichnyi, V. Structure and optical properties of nanoparticles obtained by pulsed laser ablation of copper in gases. *J. Phys. Conf. Ser.* **2019**, *1145*, 1–9. [CrossRef]

30. Mafune, F.; Kohno, J.; Takeda, Y.; Kondow, T. Formation and Size Control of Silver Nanoparticles by Laser Ablation in Aqueous Solution. *J. Phys. Chem. B* **2000**, *104*, 9111–9117. [CrossRef]
31. Dadras, S.; Torkamany, M.J.; Jafarkhani, P. Analysis and optimization of silver nanoparticles laser synthesis with emission spectroscopy of induced plasma. *J. Nanosci. Nanotechnol.* **2012**, *12*, 3115–3122. [CrossRef]
32. Truong, V.K.; Lapovok, R.; Estrin, Y.S.; Rundell, S.; Wang, J.Y.; Fluke, C.J.; Crawford, R.J.; Ivanova, E.P. The influence of nano-scale surface roughness on bacterial adhesion to ultrafine-grained titanium. *Biomaterials* **2010**, *31*, 3674–3683. [CrossRef]
33. Schubert, A.; Wassmann, T.; Holtappels, M.; Kurbad, O.; Krohn, S.; Bürgers, R. Predictability of Microbial Adhesion to Dental Materials by Roughness Parameters. *Coatings* **2019**, *9*, 456. [CrossRef]
34. Drelich, J.; Li, B.; Bowen, P.; Hwang, J.; Mills, O.; Hoffman, D. Vermiculite decorated with copper nanoparticles: Novel antibacterial hybrid material. *Appl. Surf. Sci.* **2011**, *257*, 9435–9443. [CrossRef]

© 2020 by the authors. Licensee MDPI, Basel, Switzerland. This article is an open access article distributed under the terms and conditions of the Creative Commons Attribution (CC BY) license (http://creativecommons.org/licenses/by/4.0/).

Article

Comparative Study of the Structure, Composition, and Electrocatalytic Performance of Hydrogen Evolution in MoS$_{x\sim2+\delta}$/Mo and MoS$_{x\sim3+\delta}$ Films Obtained by Pulsed Laser Deposition

V. Fominski [1,*], M. Demin [2], D. Fominski [1], R. Romanov [1], A. Goikhman [2] and K. Maksimova [2]

[1] Moscow Engineering Physics Institute, National Research Nuclear University MEPhI (Moscow Engineering Physics Institute), 115409 Moscow, Russia; dmitryfominski@gmail.com (D.F.); limpo2003@mail.ru (R.R.)
[2] Russia Immanuel Kant Baltic Federal University, 236016 Kaliningrad, Russia; sterlad@mail.ru (M.D.); aygoikhman@gmail.com (A.G.); xmaksimova@gmail.com (K.M.)
* Correspondence: vyfominskij@mephi.ru

Received: 2 December 2019; Accepted: 22 January 2020; Published: 24 January 2020

Abstract: Systematic and in-depth studies of the structure, composition, and efficiency of hydrogen evolution reactions (HERs) in MoS$_x$ films, obtained by means of on- and off-axis pulsed laser deposition (PLD) from a MoS$_2$ target, have been performed. The use of on-axis PLD (a standard configuration of PLD) in a buffer of Ar gas, with an optimal pressure, has allowed for the formation of porous hybrid films that consist of Mo particles which support a thin MoS$_{x\sim2+\delta}$ (δ of ~0.7) film. The HER performance of MoS$_{x\sim2+\delta}$/Mo films increases with increased loading and reaches the highest value at a loading of ~240 µg/cm^2. For off-axis PLD, the substrate was located along the axis of expansion of the laser plume and the film was formed via the deposition of the atomic component of the plume, which was scattered in Ar molecules. This made it possible to obtain homogeneous MoS$_{x\sim3+\delta}$ (δ~0.8–1.1) films. The HER performances of these films reached saturation at a loading value of ~163 µg/cm^2. The MoS$_{x\sim3+\delta}$ films possessed higher catalytic activities in terms of the turnover frequency of their HERs. However, to achieve the current density of 10 mA/cm^2, the lowest over voltages were −162 mV and −150 mV for the films obtained by off- and on-axis PLD, respectively. Measurements of electrochemical characteristics indicated that the differences in the achievable HER performances of these films could be caused by their unique morphological properties.

Keywords: pulsed laser deposition; nanocatalysts; buffer gas; transition metal chalcogenides; hydrogen evolution reaction

1. Introduction

Chalcogenides of transition metals, specifically, molybdenum sulfides (MoS$_x$), form compounds with different local packings of metal (Mo) and chalcogen (S) atoms, which are organized into several crystalline and amorphous phases. If structures are crystalline, such as the layered 2H-MoS$_2$ and 1T-MoS$_2$ phases, which are nanometer sized, they can initiate a strong catalytic effect that activates a hydrogen evolution reaction (HER) during water electrolysis [1–5]. For this reason, thin-film molybdenum sulfide coatings are among the most used nonprecious nanoelectrocatalysts for hydrogen evolution. However, for widespread electrocatalytic applications, the HER activities and long-term stabilities of such nanomaterials are paramount. In the thermodynamically stable crystalline 2H-MoS$_2$ phase, basal plane activation and edge site exposure/orientation on the cathode surface are the most frequently employed strategies [6,7]. The edge sites of the crystalline 2H-MoS$_2$ phase exhibit the highest catalytic activity; however, in order to activate the basal planes, they should be modified by defect formation [8–11].

The catalytic activity of the metastable 1T-MoS$_2$ phase does not depend so much on the orientation of the nanocrystals relative to the surface of the cathode; however, the synthesis of this phase is a difficult task [5].

Thin films of amorphous MoS$_x$ also have catalytic properties for HER activation. However, they require other approaches to enhance their activities [12–15]. Amorphous thin-film MoS$_x$ materials can greatly vary in their concentration of sulfur ($1 \leq x \leq 10$), which affects the local packing (chemical state) of atoms in the films. Currently, the factors that influence the catalytic activities of amorphous MoS$_x$ catalysts are being actively experimentally and theoretically studied [16–20]. An increased concentration of sulfur ($x \geq 2$) and the formation of specific S ligands, and particularly the bridging and terminal S ligands in Mo$_3$-S clusters, have been established to contribute to an increase in the catalytic activities of amorphous forms of MoS$_x$. An adequate synergistic interaction of the amorphous MoS$_x$ film with the substrate/supporting material can also enhance the electrocatalytic HER activated by MoS$_x$ [21,22].

The most common methods for producing amorphous catalytic MoS$_x$ films are chemical and electrochemical synthesis/deposition [12–14,16,17,19]. The selection of precursors and the variation of the synthesis conditions makes it possible to obtain films with controllable morphologies, compositions, and chemical states. Advances in chemical synthesis have not yet resolved the urgency of the problem of obtaining analogical or more efficient amorphous electrocatalysts by the method of physical vapor deposition (PVD). The interest in this method is due to its environmental friendliness and the possibility of implementing other original conditions for the formation of new forms of amorphous MoS$_x$.

Amorphous MoS$_x$ films began to be created intensively by PVD methods (i.e., ion sputter deposition and pulsed laser deposition (PLD)) more than 30 years ago, and their main use was related to the preparation of solid lubricating (low friction) coatings for the complicated working conditions of friction pairs [23–26]. Only very recently was the attention given to the potential of the use of PVD/PLD methods to obtain electrocatalytic films for HER activation [15,17,27–29]. In published articles, it has been reported that MoS$_x$ films obtained by the PLD method, under the standard configuration, possess sufficiently good electrocatalytic properties in terms of their reaction of hydrogen evolution. However, the task of these reports was not to determine the maximum attainable performance of MoS$_x$-based electrocatalysts and to find the optimal conditions for the preparation of the most effective electrocatalyst by PLD. In particular, the problem of the possibility of variation of S content in a wide range, as well as the effect of the MoS$_x$-based catalyst nanostructure and loading on the HER characteristics, have not been clarified.

However, changing the functionality of amorphous MoS$_x$ films requires significant modification in the conditions/regimes of their formation when using the PVD/PLD methods. For tribological application, MoS$_x$ films of a substoichiometric composition ($x \leq 2$) were deemed most suitable. For application in the catalysis of HER, it is desirable to increase the concentration of sulfur significantly, and to implement such chemical states of amorphous MoS$_x$ films that will enhance the electrocatalytic performance of the created films. The use of reactive H$_2$S gas with reactive PVD/PLD allows for variation of the S concentration in the formed MoS$_x$ films over a wide range [17,27]. However, this gas is dangerous, and the use of this gas causes technical and environmental problems. It is desirable to perform the PVD/PLD of MoS$_x$ films with designed characteristics using only MoS$_2$ targets and nonreactive gas. These targets may be easily made from abundant earth material.

Also, for effective catalysis, it is important to obtain porous films that are not applicable in tribology. The peculiarity of pulsed laser ablation of MoS$_2$ targets, which causes the formation of porous MoS$_x$ films, was revealed in the early studies of PLD concerning such films [30,31]. The originality of PLD from targets of transition metal dichalcogenides also lies in the formation of "core-in-shell" particles [32–34]. There is reasonable interest in investigating the influences of these factors on the electrocatalytic properties of the MoS$_x$ films obtained by PLD more deeply.

The use of a buffer gas during PLD alters the transport processes of the expansion of a laser plume from the MoS$_2$ target and makes it possible to increase the S concentration in MoS$_x$ films above a stoichiometric value [29,35,36]. The actual composition of the deposited MoS$_x$ films depends on both the buffer gas pressure and the location of the substrate relative to the direction of the laser plume expansion. In cases where the substrate location is on the axis of the laser plume expansion and for locations perpendicular to it (i.e., using the on-axis PLD mode), all laser-ablated products, including the atomic flux and particles of submicron and nanometer dimensions, are deposited on the substrate. In cases where the substrate location is at a distance from the axis and parallel to it (i.e., using the off-axis PLD mode), the depositing flux consists of atoms that have been taken off from the target during ablation and scattered at large angles after collision with the buffer gas molecules.

Obviously, the on- and off-axis PLD methods have specific features in terms of their influences on the structures and chemical states of the deposited MoS_x films. The objective of this work was to perform a rather in-depth study of the main characteristics of MoS_x films prepared by on- and off-axis PLD, and to determine the highest performance of HER that can be achieved with MoS_x films deposited on glassy carbon substrates by these methods. Within this aim, it was necessary to solve the problem of determining the optimal pressure of the buffer gas for the PLD of high performance MoS_x films and to identify the criteria/features that allow for the appropriate selection of this parameter.

2. Materials and Methods

2.1. Experimental Methods for the on- and off-Axis PLD of MoS_x Films

Figure 1 shows time-integrated pictures of laser plumes measured during the pulsed laser ablation of MoS_2 targets under vacuum conditions and at different pressures of buffer gas (Ar). In the case of on-axis PLD in different conditions, the substrates were located at a normal angle to the axis of laser plume expansion and were 3.5 cm from the target. Off-axis PLD was performed at an Ar pressure of 8 Pa, and the substrates were located 2 cm from the target. The substrate surface was parallel to the axis of the laser plume expansion, and the shift of the substrate from this axis was ~0.3 cm. This setup precluded the influence of the shadow effect from the edge of the substrate on the distribution of the MoS_x film over the surface of the substrate. The dimension of the substrate in the plane was approximately 0.7×0.7 cm^2.

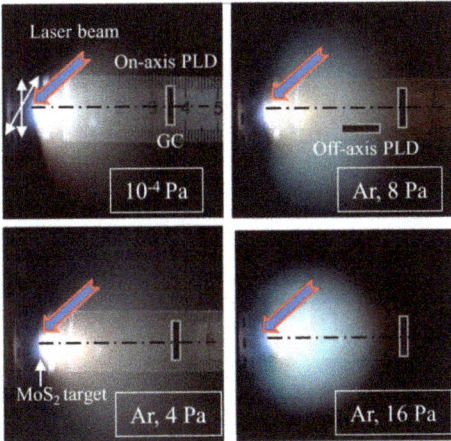

Figure 1. Time integrated optical images of the laser plume formed during the pulsed laser deposition of MoS_x films from a MoS_2 target under vacuum, considering various pressures of the Ar buffer gas. The arrangement of glassy carbon (GC) substrates, with respect to the axis of laser plume expansion, is shown for the modes of on- and off-axis pulsed laser deposition (PLD).

The MoS_2 target was ablated by pulses from a Solar LQ529 laser (Solar, Minsk, Belarus). The laser beam fell at an angle of 45° to the surface of the target. The laser generation parameters were as follows: Radiation wavelength of 1064 nm, pulse duration of 15 ns, pulse energy of 14 mJ, laser fluence of 5 J/cm^2, and pulse repetition rate of 50 Hz. The selected parameters provided a weak erosion of the target surface when exposed to a single laser pulse.

Preliminary studies from the authors have found that the use of laser radiation with a shorter wavelength of 266 nm from the ultraviolet region of the spectrum (the 4th harmonic of laser radiation) does not cause noticeable changes in the ablation future of the MoS_2 target, as with the structure/composition of the deposited MoS_x films.

A significant increase of energy in the laser pulse made it possible to increase the film growth rate; however, a strongly inhomogeneous distribution of the Mo and S elements over the substrate surface could be formed. In this case, in the center of the substrate, an area of metallization could arise [24]. An increase in the laser fluence, leading to strong ionization of vapors during the ablation of MoS_2 target, results in a more isotropic expansion of the laser plasma. This could cause the formation of the most uniform distribution of the thickness/composition of the MoS_x film over the surface of the substrate during on-axis PLD. However, the film growth rate decreases, due to the self-sputtering of the MoS_x film by high-energy laser plasma ions. The preferential sputtering of S atoms by ions could result in the formation of a MoS_x film with a very low S content ($x \sim 1$).

As illustrated in Figure 1, the target was moved in two directions, allowing for the maintenance of a relatively smooth surface of the target in the ablation zone under the repeatable laser ablation. The chamber for PLD was pumped with a turbomolecular pump to a pressure of $\sim 10^{-4}$ Pa. The regime for the preparation of the MoS_x films under this condition is indicated in the text as vacuum PLD. To conduct PLD in a buffer gas, after vacuum pumping, Ar gas was introduced into the chamber. The Ar pressure during the PLD ranged from 4 to 16 Pa. The deposition was performed in a small chamber with a volume of ~ 500 cm^3, which made it possible to increase the partial pressure of the sulfur vapor during the prolonged ablation of the MoS_2 target.

To analyse the character of the laser plume expansion under different vacuum conditions, plasma plume images were collected through the viewport (orthogonal to the axis of the plume motion) using a digital camera with an exposure time of 0.5 s. This time corresponded to an average value of more than 20 pulses. Different plumes were assumed to be equivalent, and therefore, the times, shapes, and sizes of the integrated plumes (i.e., the visible parts) were quite adequately recorded. In addition to the photoregistration of the pulsed laser plumes, measurements of the time-of-flight signal of a pulsed laser plasma were performed. The ion probe was placed 3.5 cm from the target. To reflect the electrons of the laser plasma, a negative bias of 40 V was applied to the probe.

2.2. Structural and Electrochemical Characterization of the Prepared MoSx Films

The MoS_x films with different thicknesses were deposited at room temperature on substrates made of polished glassy carbon (GC), polished silicon plates (including those with a layer of oxide (Si/SiO$_2$)), and NaCl crystals. The time of PLD varied within the range of 1 min to 2 h. To measure the catalyst loading and to determine the S/Mo atom content ratio, Rutherford backscattering spectroscopy (RBS) was used (He ion energy of 1.3 MeV; scattering angle of 160°; ion beam diameter of 100 µm). Mathematical modelling of the RBS spectra was performed using the SIMNRA program.

The structures of the thin films deposited on the NaCl were studied using transmission electron microscopy and selected-area electron diffraction (TEM and SAED (selected-area electron diffraction), respectively, JEM-2100, JEOL, Tokyo, Japan). For this, the NaCl crystals with a deposited film were dipped in distilled water. After separation of the films from the substrate, the films were placed on fine-grained metal grids and transferred to an electron microscope. The structures, compositions and surface morphologies of the MoS_x films prepared on the GC and Si substrates were studied using scanning electron microscopy with energy dispersive X-ray spectroscopy (SEM and EDS, Tescan LYRA 3), as well as micro-Raman spectroscopy (MRS), using a 632.8-nm (He-Ne) laser. The cross section of the laser beam was <1 µm. The chemical states of the MoS_x films were studied using X-ray photoelectron spectroscopy (XPS, K-Alpha apparatus, Thermo Scientific, Madison, WI53711, USA) with Al Kα radiation (1486.6 eV). The Si/SiO$_2$ substrates with deposited MoS_x films were split, and the fracture of the samples was studied with SEM. This procedure allowed us to examine the cross-sectional morphologies of the prepared films.

It should be noted that SEM and EDS studies of the MoS_2 target have shown that the polished surface of the target was noticeably modified after irradiation with one laser pulse. However, the composition and morphology of the target surface formed after the first laser pulse were not modified substantially after prolonged laser ablation. This indicated that the main characteristics of the laser plume would not be changed during the deposition of the electrocatalytic MoS_x films with growing loading.

The electrochemical studies of the MoS_x films deposited on the GC substrates were performed in an H_2-sparged 0.5 M H_2SO_4 aqueous solution using an Elins Instruments electrochemical analyzer (Model P-5X, Chernogolovka, Russia). A saturated silver chloride electrode (Ag/AgCl) was used as the reference electrode, and the GC with the pulsed laser-deposited MoS_x film was the working electrode. For the used modes of on- and off-axis PLD, quite uniform distributions of the MoS_x catalyst over the entire surface areas of the samples were achieved. All potentials reported in this work have been measured versus the reversible hydrogen electrode (RHE), and they were calculated according to the following formula: U(RHE) = U(Ag/AgCl) + (0.205 + 0.059 pH) (pH ~0.3). A bare GC plate was used as a counter electrode. For the electrochemical testing of a desired area on the GC substrate, the sample was placed in a special holder that was made of Teflon.

The main electrochemical characteristics of the MoS_x films that were accepted for use in determining the performances of the HER catalysts included the measurements/evaluations of the cathodic polarization curves, turnover frequencies (TOFs) of the HER, double-layer capacitances (C_{dl}), electrochemical impedances, and long-term stabilities. Explanations of the techniques of the electrochemical measurements that were used to study the MoS_x-based catalysts can be found in some published works, e.g., References [14,37,38].

The cathodic polarization curves were measured using linear voltammetry (LV), with a sweep of the applied potential from 0 to −350 mV and a scan rate of 2 mV/s. Before the LV measurements, all samples were subjected to cathodic pre-treatment at −350 mV. The optimal catalyst loading was determined in accordance with the minimum overvoltage (U_{10}) that was needed to achieve a current density of j_{10} = 10 mA/cm^2.

The procedure for calculating the TOF of HER has previously been proposed [16,38]. Cyclic voltammograms (CV) were measured after applying the cathodic polarization of the sample in the same electrolyte in the potential range from 0 to 1200 mV. Integration of the current peaks which resulted from the oxidation of the active sites allowed for the measurement of the number of active sites on the catalyst surface. The relation of the current density (j), measured by LV, to the number of catalytically active sites allowed for the evaluation of the TOF of the catalyst using the expression TOF = $j/2Q$, where Q is the total charge. This charge was determined by integrating the current (i), as measured by CV; specifically, $Q = \frac{1}{U_{sr}} \int_{E_1}^{E_2} i dE$, where E is the potential and U_{sr} is the potential scan rate during the CV measurements (50 mV/s).

For the estimation of C_{dl}, the current densities versus the potential data were acquired by CV, while sweeping the applied potential at various scan rates (40–200 mV/s) in a potential range (30–140 mV), within which no Faradaic electron-transfer processes were observed. The double-layer capacitances of the MoS_x films were calculated as the slopes of plots of the scan rates versus the current densities at 100 mV. Electrochemical impedance spectroscopy (EIS) was performed at an overpotential of −180 mV in the frequency range of 10^5 to 10^{-1} Hz, with a perturbation voltage amplitude of 20 mV. All LV curves measured in this work were iR_s-corrected. The equivalent series resistance (R_s) was extracted from the EIS data. The stabilities of the prepared films were tested by continuously cycling the voltage between 100 and −350 mV at a scan rate of 50 mV/s.

3. Results

3.1. Selection of the Conditions (Buffer Gas Pressure) for the PLD of MoS$_x$ Films

The dynamics of a laser plume in a vacuum are characterized by a practically free and collision-less propagation regime. The spatial distribution is strongly forward directed, and the observed light emission is weak. If a background gas is present in the deposition chamber, the light emission of the plume increases, due to particle collisions. These collisions produce radiative de-excitation of the ablated species, both in the body of the plume and particularly in the expansion front. The plume edge can be better defined, due to the presence of a shock wave front. The plume is slowed down and spatially confined. The time-integrated visible plume length might relate to the maximum distance reached by the shock wave front (i.e., the stopping distance) from the ablated target. Obviously, the composition, structure, and functional properties of the deposited films can significantly depend on both the buffer gas pressure and the location of the substrate, relative to the stopping distance of the laser plume [39,40].

Figure 1 shows that, in a vacuum and at an Ar gas pressure of 4 Pa, the laser plume expanded over the large volume of the chamber and enveloped the substrate for the deposition of the film in the on-axis PLD mode. At a pressure of 8 Pa, a noticeable limitation of the plume volume was observed, but the front of its expansion reached the substrate located for the on-axis PLD. At a pressure of 16 Pa, the stopping distance of the laser plume was less than the distance from the target to this substrate. Measurements of the ion signals of the pulsed laser plasma revealed that, for the ion flux bombarding the substrate, an increase in Ar pressure caused both a decrease in the intensity and a change in the ion energy (Figure 2). At a pressure of 8 Pa, along with a peak from the high-speed ions (time of ion flight up to 5 µs), a peak with a time of ion flight of more than 10 µs appeared. Moreover, at the pressure of 16 Pa, the time of flight for the main ion flux reaching the substrate increased to 15 µs.

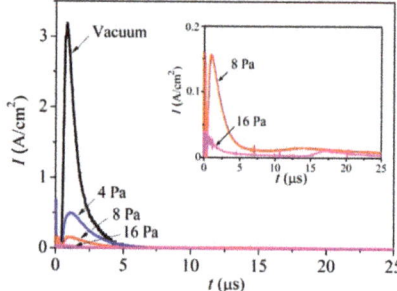

Figure 2. Pulsed ion signals that were detected by an ion probe during the pulsed laser ablation of the MoS$_2$ target, both in a vacuum and at different pressures of Ar buffer gas.

Wood et al. [41] proposed that during ablation in a buffer gas, the laser plume is broken into orders corresponding to the numbers of collisions made with the background. The first order reaches the detector without any scattering, the second order undergoes one scattering event, and so forth. Scattered atoms move mainly by the mechanism of collective motion with the buffer gas molecules captured by the laser plume, i.e., those in the shock wave. Analyses of the results of the laser plume optical recording and the ion pulse measurement suggest that, at 8 Pa, a shock wave was indeed formed, but some of the atoms in the laser plume reached the substrate without scattering. This finding indicated that the formation of the MoS$_x$ films proceeded, due to the deposition of a scattered atomic flux. However, the flux-containing atoms from the second scattering order could retain the energetic motion toward the substrate, and therefore, could ensure good adhesion of the films to the substrate. For the higher pressure of 16 Pa, the formation of the MoS$_x$ film was proceeded by the deposition of atoms that had experienced multiple scattering events on the gas molecules and had lost the energy of directed motion to the substrate.

It is shown below that a pressure of 8 Pa was optimal for obtaining a highly efficient HER catalyst using the on-axis PLD mode. We considered this fact and also analyzed the optical images of the laser plume (Figure 1). The images indicated that, at the pressure of 8 Pa, there was no noticeable gradient of brightness along the axis of the expansion of the plume in the range of 2 to 3 cm from the target. This made it possible to obtain rather uniform MoS_x films by off-axis PLD on substrates of up to 1 cm in size.

3.2. Deposition of the $MoS_{x\sim2+\delta}$/Mo Films by on-Axis PLD

Figure 3a shows the experimental and model RBS spectra for a thin MoS_x film obtained on a Si substrate over 1 min of on-axis PLD at an Ar pressure of 8 Pa. According to the SIMNRA simulation, the composition of this film was $MoS_{2.7\pm0.2}$. The loadings of the model thin-film catalyst were $\sim6.6 \times 10^{15}$ Mo atom/cm² (1.1 µg/cm²) and $\sim17.8 \times 10^{15}$ S atom/cm² (0.95 µg/cm²). At a density of 5.06 g/cm³, the film thickness was ~4 nm.

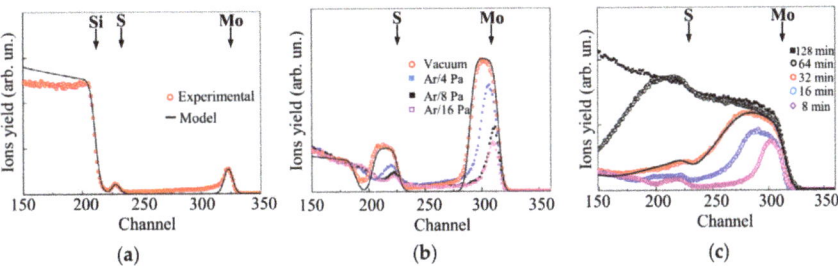

Figure 3. Rutherford backscattering spectroscopy (RBS) spectra for $MoS_{x\sim2+\delta}$/Mo films obtained by on-axis pulsed laser deposition (PLD) on the Si and Si/SiO₂ substrates. (**a**) A thin film deposited for 1 min on the Si substrate. (**b**) Films of middle thickness deposited on the Si/SiO₂ substrates for 4 min in a vacuum and at different Ar pressures. (**c**) Thicker films deposited on the Si/SiO₂ substrate in Ar at a pressure of 8 Pa for various deposition times. The solid lines indicate the model/fitted RBS spectra for the corresponding experimental spectra. The arrows indicate the channels of accumulation of He ions which were back scattered by Mo, S, and Si atoms located on the surface of the samples.

Notably, the actual load was greater, due to the Mo nanoparticles, which caused a long "tail" in the RBS spectrum (channels 240–315). Calculation of the "Mo peak/Mo tail" intensity ratio revealed that the Mo content in the thin MoS_x film was approximately equal to that of the Mo nanoparticles. The total $MoS_{2.7}$/Mo catalyst load was estimated to be ~3.4 µg/cm², corresponding to a deposition rate of ~3.4 µg/cm²/min. For the used energy of the analyzed ion beam, the imposition of the RBS peak from the Mo atoms on the RBS peak from the S atoms occurred when the He ions were scattered by Mo particles larger than 230 nm. The use of the buffer gas reduced the deposition rate of the atomic component of the laser plume; however, this allowed for an increase in the S concentration in that component of the hybrid $MoS_{x\sim2+\delta}$/Mo films, which was formed, due to the deposition of the atomic flux.

Figure 3b shows the RBS spectra for the $MoS_{x\sim2+\delta}$/Mo films deposited on the Si/SiO₂ substrate under different conditions for 4 min. These spectra indicated that the relative contribution of the Mo peak can be attributed to "Mo atoms in the $MoS_{x\sim2+\delta}$ film", which reduced with increasing Ar pressure, where the yield of He ions in the range of channels 275 to 240 did not depend on the Ar pressure, which implies that the deposition flux of Mo particles persisted unchanged under various conditions of the on-axis PLD.

Due to the greater deposition rate of atomic flux during vacuum on-axis PLD, the RBS spectrum for a corresponding $MoS_{x\sim2+\delta}$ film could be adequately processed by SIMNRA if the model object consisted of a homogeneous $MoS_{x=2.1-0.2}$ film on a SiO₂ substrate. The catalyst loading included

~1.7×10^{17} Mo atom/cm^2 (27.3 µg/cm^2) and ~3.6×10^{17} S atom/cm^2 (19.4 µg/cm^2). The deposition rate for the vacuum on-axis PLD of the MoS$_x$/Mo catalyst was 11.7 µg/cm^2/min.

Figure 3c shows the dependence of the loading of the MoS$_{x\sim2+\delta}$/Mo catalyst on the time of its on-axis PLD in Ar at a pressure of 8 Pa. A monotonic growth of the catalyst loading was observed with increases in deposition time. The shapes of the RBS spectra are strongly distorted, due to the accumulation of Mo particles. An example of the mathematical fitting of the RBS spectra is shown in Figure 3c for a sample obtained after 32 min of on-axis PLD deposition. A fairly good match was obtained for a model that contained a thin surface layer with the composition MoS$_{x\sim1.8}$ and a thicker underlayer with the composition MoS$_{x\sim1.4}$. A satisfactory fit was achieved, due to the assumption of large surface roughness for the model film. The model film contained ~4.8×10^{17} Mo atom/cm^2 (76 µg/cm^2) and ~7.5×10^{17} S atom/cm^2 (44 µg/cm^2). The calculated deposition rate of the MoS$_{x\sim2+\delta}$/Mo film was 3.75 µg/cm^2/min, which correlated well with the data regarding the deposition rate at the initial stage (i.e., for 1 min).

The spectra of the MoS$_{x\sim2+\delta}$/Mo films obtained with longer deposition times were difficult to process with SIMNRA (Figure 3c). Below, we show that the optimum catalyst loading was formed with an on-axis PLD time of 64 min. We assumed that, for this time of on-axis PLD, the catalyst loading increased to 240 µg/cm^2 and included ~9.6×10^{17} Mo atom/cm^2 (152 µg/cm^2) and ~1.5×10^{18} S atom/cm^2 (88 µg/cm^2). After an increase in deposition time of up to 128 min, the RBS spectrum of the MoS$_{x\sim2+\delta}$/Mo film corresponded to a thick layer of a homogeneous mixture of Mo and S, and the atomic ratio of these components was S/Mo ≤ 1.

The formation of Mo nanoparticles during the pulsed laser ablation of MoS$_2$ targets and their transfer to the surfaces of the MoS$_{x\sim2+\delta}$/Mo films during the on-axis PLD was confirmed by the TEM, and MD results from a thin film deposited on NaCl (Figure 4). At low magnification, round-shaped dark particles were detected. The sizes of these particles varied in the range of 10–200 nm. The MD pattern of these particles corresponded to the cubic lattice of Mo. In high-resolution TEM imaging of a separate Mo nanoparticle, atomic planes with an interplanar distance of 0.22 nm, characteristic of Mo (110), were observed. The Mo particles were surrounded by an amorphous ~5 nm-thick shell. The MoS$_{x\sim2+\delta}$ matrix of the MoS$_{x\sim2+\delta}$/Mo film was amorphous. The TEM image of this film contained rounded nanosized areas with lighter contrast. These areas probably corresponded to the sites at which the Mo nanoparticles were localized. However, when manipulating the film for the TEM studies, these particles were removed, due to weak adhesion to the thin MoS$_{x\sim2+\delta}$ matrix.

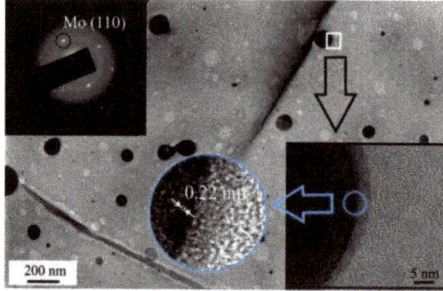

Figure 4. TEM image and selected-area electron diffraction (SAED) pattern of the thin MoS$_{x\sim2+\delta}$/Mo thin film obtained by on-axis PLD for 1 min in Ar at a pressure of 8 Pa. The bottom inserts show high resolution TEM images of the film.

Figure 5a shows SEM images of a relatively thin MoS$_{x\sim2+\delta}$/Mo film deposited by on-axis PLD onto the smooth surface of the Si/SiO$_2$ substrate. Here, the deposition time was 4 min, and at this stage of film growth, the film had a relatively dense structure. This structure was formed from sufficiently small Mo nanoparticles and a scattered flux of Mo and S atoms. The sizes of most of the Mo nanoparticles did not exceed 200 nm, and the scattered flux of Mo and S atoms provided a conformal deposition of a MoS$_{x\sim2+\delta}$ shell on these particles and enveloped them with approximately the same efficiency over the entire rounded surface.

Figure 5. SEM images of MoS$_{x\sim2+\delta}$/Mo films obtained on the (**a**) Si, (**b**) SiO$_2$, and (**c**) glassy carbon (GC) substrates by on-axis PLD in Ar at a pressure of 8 Pa. The deposition times were (**a**) 4 min, (**b**) 64 min, and (**c**) 64 min. (**b**) The top inset shows a cross-sectional image of the film.

Figure 5a shows that larger particles occasionally appeared in the surface of the MoS$_{x\sim2+\delta}$/Mo film. The sizes of these particles reached several fractions of a micrometer. The deposition of Mo particles of a submicron size had a significant effect on the formation of the morphologies of the thicker films. Figure 5b shows SEM images of the MoS$_{x\sim2+\delta}$/Mo film deposited on the Si/SiO$_2$ substrate for 64 min. The large conglomerates are formed from submicro- and nano-sized particles, and the structure of the film became porous here. The film thickness, as estimated based on the SEM images of the cleaved Si/SiO$_2$ substrate covered with this film, was ~1.5 µm.

An SEM image of the MoS$_{x\sim2+\delta}$/Mo catalyst obtained under the same conditions by on-axis PLD on the glassy carbon is shown in Figure 5c. The image indicates that the catalyst consisted of loose packaging of nano- and sub-micro-particles. The catalytic film covered the surface of the glassy carbon with a continuous layer. The GC substrate had noticeable roughness, and the lateral sizes of the surface cavities on this substrate were approximately several µm, and their depths varied in the range of ~0.1–1 µm. Due to the roughness of the glassy carbon substrate, the local loading of the MoS$_{x\sim2+\delta}$/Mo catalyst was somewhat less than 240 µg/cm^2. EDS analysis of the sample, shown in Figure 5c, indicated that the concentrations of the Mo, S, O, and C atoms were 10.6, 10.2, 5.2, and 74%, respectively. A surface area of 10 × 10 µm^2 was analyzed by EDS. The content of O atoms was practically unchanged when the EDS analysis was performed for a pure polished GC substrate. This result indicated that the O atom concentration in the films did not exceed several percent. The result of the EDS measurement of the ratio S/Mo in the thick MoS$_{x\sim2+\delta}$/Mo catalytic film coincided well with the result of its measurement by the RBS.

Figure 6a shows the micro-Raman spectrum for a MoS$_{x\sim2+\delta}$/Mo catalyst obtained on glassy carbon by on-axis PLD for 4 min in Ar at 8 Pa. For this spectrum, there were no peaks characteristic of the MoS$_2$ compound. Measurements of the Raman spectrum for the MoS$_2$ target showed that the characteristic and most intense E$_{2g}^1$ and A$_{1g}$ peaks were located at frequencies of 383 cm^{-1} and 408 cm^{-1}, respectively. This spectrum confirms the amorphous structure of the catalytic layer, which is characterized by the presence of several broad vibrational bands near the frequencies of 200, 330, 450, and 540 cm^{-1}. The same Raman data, with four very weak and broad bands, were obtained from MoS$_x$ films grown by traditional PLD elsewhere [42]. McDevitt et al. [42] proposed that this Raman spectrum indicates that the laser-deposited films represent a mixture of small domains of MoS$_2$ and amorphous sulfur. It is more reasonable to use a more recent model of a cluster-based polymeric structure that consists of Mo$_3$-S clusters with some different configurations of S ligands [4,16,18,27]. In the frame of this model, the characteristic Raman spectrum of amorphous MoS$_x$ contains the following vibration modes: ν(Mo-Mo) at ~200 cm^{-1}, ν(Mo-S)$_{coupled}$ at ~320 cm^{-1}, ν(Mo-S$_{apical}$) at ~450 cm^{-1}, ν(S-S)$_{terminal}$ at ~520 cm^{-1}, and ν(S-S)$_{bridging}$ at 540 cm^{-1}. The Raman spectra of the MoS$_{x\sim2+\delta}$/Mo catalyst contained bands that could be attributed to all these modes. However, the vibration peaks of the bridging S$_2^{2-}$ and terminal S$_2^{2}$ moieties overlapped, which caused the appearance of a broadened band in the 500–560 cm^{-1} range. This finding indicated a weak order of atom packing in the local regions, comparable to the sizes of the Mo$_3$-S clusters.

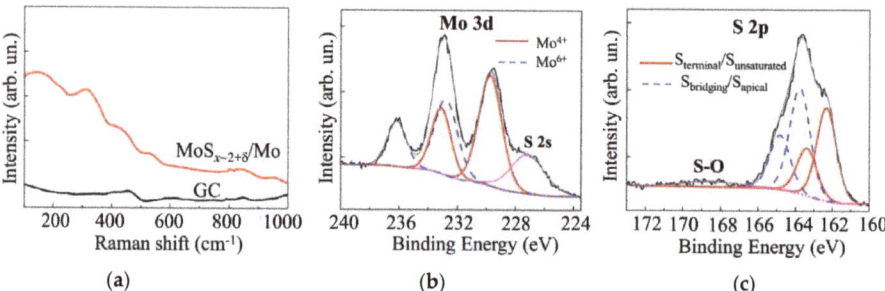

Figure 6. (**a**) Raman spectra for GC substrate, with and without the MoS$_{x\sim 2+\delta}$/Mo thin film catalyst obtained by on-axis PLD. (**b**,**c**) XPS spectra of Mo 3d and S 2p, measured on the surface of MoS$_{x\sim 2+\delta}$/Mo catalyst after prolonged exposure in the air.

The spectrum of the film obtained by on-axis PLD contained no peaks that are characteristic of nanocrystalline molybdenum oxides. In the case of the formation of MoO$_3$ nanocrystals, the distinctive peaks at ~820 and ~990 cm^{-1} were observed [18]. For the MoO$_2$ nanocrystals, vibrations at ~205, 229, 345, 365, 498, 572, and ~745 cm^{-1} are characteristic [43]. However, the appearance in the spectrum in Figure 6a, with wide vibration bands at ~820 and ~950 cm^{-1}, indicates the formation of disordered MoO$_{3-y}$ clusters in the MoS$_{x\sim 2+\delta}$/Mo film.

This conclusion was confirmed by the results of the XPS studies of the MoS$_{x\sim 2+\delta}$/Mo film, which are shown in Figure 6b, c. The measurements of the XPS spectra were performed after the prolonged exposure of the film in the air (for approximately six months). The film was prepared by the on-axis PLD method in Ar at 8 Pa for 64 min. In the spectrum of Mo 3d, in addition to the doublet Mo 3d$_{5/2}$-Mo 3d$_{3/2}$, which corresponds to the chemical bonding of Mo with S (Mo^{4+}, the binding energy E$_B$ of Mo 3d$_{5/2}$ is 229.7 eV), there was a doublet found that was attributable to Mo oxide (Mo^{6+}, Mo 3d$_{5/2}$ E$_B$~232.8).

The XPS studies of the MoS$_2$ target indicated that, in the case of effective Mo-S bond formation, the surface of the compound had a higher resistance to oxidation in the air (results not shown). The formation of the molybdenum oxide nanophase could have resulted from the ineffective interaction of Mo and S atoms during the film deposition. Unsaturated Mo bonds in the local structure of the MoS$_{x\sim 2+\delta}$ film interacted with O atoms when the sample was exposed to air. Another mechanism of molybdenum oxide formation is the oxidation of the surface of the Mo particles that were uncoated with the MoS$_{x\sim 2+\delta}$ thin shell. The former mechanism seems to be more likely, because amorphous MoS$_x$ films obtained by electrochemical deposition (i.e., those without Mo nanoparticles) also undergo a slow transformation from Mo^{4+} to Mo^{6+} under atmospheric conditions [18]. This process could have partially occurred between the preparation and characterization of these MoS$_x$ thin films.

The XPS spectrum of S 2p for the same MoS$_{x\sim 2+\delta}$/Mo film is shown in Figure 6c. This figure reveals the presence of different S ligands that were considered in the Mo$_3$-S cluster-based model of amorphous MoS$_x$. The S ligands with an S 2p$_{3/2}$ peak at 162.3 eV were assigned to the S$_2{}^{2-}$ terminal or unsaturated S^{2-} entities in the amorphous MoS$_x$ and S^{2-} in the crystalline MoS$_2$. The S 2p$_{3/2}$ peak at 163.7 eV corresponded to the bridging S$_2{}^{2-}$ and apical S^{2-} ligands of the Mo$_3$-S cluster [17,18,37,44,45]. This result agrees well with the abovementioned Raman spectra of the MoS$_{x\sim 2+\delta}$/Mo films (Figure 6a). The oxidative process of the MoS$_{x\sim 2+\delta}$/Mo film in the air involved, to some extent, the S atoms. Consequently, a broad band at ~169 eV appeared on the XPS spectrum, and this binding energy was assigned to the S-O bonds [17,18].

Quantitative compositional analysis by XPS indicated that the S content in the surface layer of the MoS$_{x\sim 2+\delta}$/Mo films was slightly higher than in the bulk of the films. The x value measured by XPS was 5–10% larger than that measured by RBS. This could be due to the adsorption of sulfur atoms on the surface of the films from the residual atmosphere in the deposition chamber after the PLD process finished. A similar result was earlier revealed in Reference [35]. The O concentration in the surface layer of the films did not exceed 10 at % after prolonged exposure in the air. The O concentration was reduced to 3 at % after ion sputtering of the layer of surface contamination for 30 s.

3.3. Deposition of the $MoS_{x~3+\delta}$ Films by off-Axis PLD

Figure 7 shows the RBS spectrum of a thin MoS_x film deposited on a Si/SiO_2 substrate for 1 min using the off-axis mode of PLD in Ar at a pressure of 8 Pa. Fitting of the spectrum revealed that the experimental RBS spectrum coincided well with the model RBS spectrum that was calculated for a continuous/smooth thin film with a composition of $MoS_{x~3.9}$ (results not shown). The thin film contained ~2.6 × 10^{16} Mo atom/cm^2 (~4 µg/cm^2) and ~1 × 10^{17} S atom/cm^2 (6 µg/cm^2). Increases in deposition time caused increases in catalyst loading with a sublinear dependence. To fit the experimental RBS spectrum from a thicker film obtained by off-axis PLD for 20 min, a two-layer film model was necessary (Figure 7). A layer of $MoS_{4.1}$ was formed on the surface of this film, and a sublayer was composed of $MoS_{3.8}$. The film contained ~4.2 × 10^{17} Mo atom/cm^2 (65 µg/cm^2) and ~1.6 × 10^{18} S atom/cm^2 (98 µg/cm^2). Thus, the catalyst loading was increased to 163 µg/cm^2.

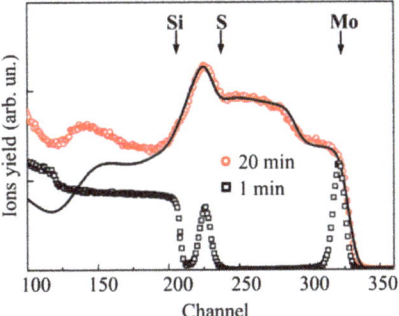

Figure 7. RBS spectra for $MoS_{x~3+\delta}$ films obtained by off-axis PLD on the Si/SiO_2 substrates in Ar at a pressure of 8 Pa for 1 and 20 min. The solid lines indicate the model/fitted RBS spectrum for the corresponding experimental spectrum.

RBS studies have shown that with increasing target ablation time, the composition of the laser plume can be altered to some extent. As a rule, the surface composition and roughness of the transition metal dichalcogenide target are significantly modified by prolonged pulsed laser irradiation [41]. However, an accumulation of sulfur vapor in the film production chamber is also possible with an increase in deposition time. The effects of these factors were also observed during the formation of a thicker MoS_x/Mo film by off-axis PLD.

Notably, upon calculating the model spectrum of a thicker film, the spectrum was found to be well-matched to the experimental spectrum in the channel range that corresponded to He ion scattering from Mo and S atoms. The carbon concentration in the film was set to no more than 20 at %. This value was determined by the X-ray energy dispersive spectroscopy of this film (results not shown). A disagreement between the model and experimental spectra was observed in the range of channels (less than 200), in which the ions yielded from the Si/SiO_2 substrate were accumulated. This result could be due to the structural features of a thicker film. Below, we demonstrate that the microstructure of the film (micro-crack formation) allowed for the "channeling" of ions through the film. It is difficult to achieve a good fit result for either the film or the substrate under these conditions.

The TEM and SAED studies revealed that the use of the off-axis PLD mode substantially decreased the Mo nanoparticle deposition on the surface of the $MoS_{x~3+\delta}$ film (Figure 8). The TEM image contrast and SAED pattern indicated the amorphous and quite homogeneous structure of the thin film. At low magnification, only individual Mo nanoparticles were observed on the TEM image. The high resolution TEM image of the $MoS_{x\sim3+\delta}$ film differed from that of the $MoS_{x\sim2+\delta}$/Mo film in terms of a pronounced contrast that contained nanosized threads of dark and light tones. This result could be due to the different natures of the local packings of atoms in the films obtained by on- and off-axis PLD.

Figure 8. TEM image and SAED pattern of thin $MoS_{x\sim3+\delta}$ thin film obtained by off-axis PLD in Ar at a pressure of 8 Pa. The bottom insert shows the high resolution TEM image of the film.

SEM studies of the $MoS_{x\sim3+\delta}$ films revealed the formation of a relatively dense thin film material with a quite smooth surface, and the morphology was slightly dependent on both the deposition time and the nature of the substrate (Figure 9). The main effect on the growth of the films during the off-axis PLD was the fragmentation of the films, which caused the formation of micro-blocks that were separated by grooves (micro-cracks). An SEM study of the cross section of the films revealed that the micro-cracks could have been formed, due to the development of a columnar structure in the films. The columnar units originated on the substrate-film interface and grew up to the surface of the film. This growth was characteristic of chemical compound films with a cauliflower structure in which bushes are formed, due to the deposition of the scattered flux of atoms and/or clusters of atoms of the laser plume [46].

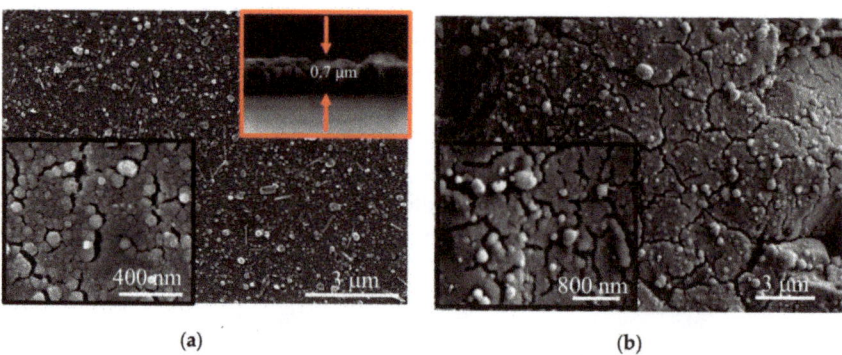

(a) (b)

Figure 9. SEM images (two magnifications) of $MoS_{x\sim3+\delta}$ films obtained on (a) Si/SiO_2 and (b) glassy carbon substrates by off-axis PLD in Ar at a pressure of 8 Pa for 20 min. The top inset in (a) shows a cross-sectional image of the film.

The separate rounded particles of submicron size present on the surfaces of these films (Figure 9) could have been formed by the deposition of Mo nanoparticles that subsequently grew in size, due to the deposition of a vapor. The needle-like submicroparticles that appeared on the film surface could have been formed, due to the spreading and solidification of larger liquid droplets that were ejected from the target at high speed and slid over the film surface.

The results of the MRS and XPS studies, shown in Figure 10, did not reveal essential differences in the local structural or chemical states of the catalysts formed by the on- and off-axis PLD. Indeed, the Raman spectra of the MoS$_{x\sim3+\delta}$ catalysts were wholly like those of the MoS$_{x\sim2+\delta}$/Mo catalysts (Figure 6a). The XPS Mo 3d spectrum for the catalyst obtained by off-axis PLD for 20 min consisted of two doublets that corresponded to Mo^{4+} and Mo^{6+}. Regarding the XPS S 2p spectrum of the MoS$_{x\sim3+\delta}$ catalyst, the intensity of the doublet with high binding energy was greater than that with low binding energy. Similar results were obtained for the MoS$_{x\sim2+\delta}$/Mo catalyst. However, in the S 2p spectrum of the MoS$_{x\sim3+\delta}$ film, the relative intensity of the band corresponding to the S-O bonds was noticeably larger than that in the spectrum of the MoS$_{x\sim2+\delta}$/Mo film. This finding suggests that, at higher S concentrations in the catalyst, not all S atoms formed perfect chemical bonds with Mo or other S atoms included in the Mo$_3$-S clusters. The S atoms possessing unsaturated bonds were subject to oxidation in the air environment.

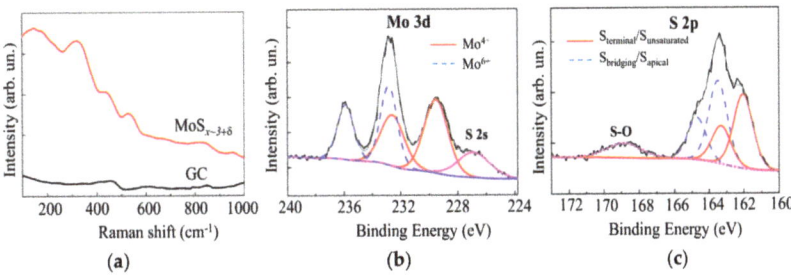

Figure 10. (a) Raman spectra for the GC substrate, with and without the MoS$_{x\sim3+\delta}$ thin film catalyst obtained by off-axis PLD. (b,c) XPS spectra of Mo 3d and S 2p, respectively, measured on the surface of the MoS$_{x\sim3+\delta}$ catalyst after prolonged exposure in the air.

3.4. Electrocatalytic Performances of the MoS$_{x\sim2+\delta}$/Mo and MoS$_{x\sim3+\delta}$ Films Prepared by on-Axis and off-Axis PLD

Figure 11 shows the results of an electrochemical study of the dependence of the electrocatalytic properties of the MoS$_{x\sim2+\delta}$/Mo films on the buffer Ar gas pressure. The results of LV measurements indicate that the smallest overpotential of HER was found for films deposited at pressures of 4 and 8 Pa (Figure 11a). A current density of 10 mA/cm^2 was achieved at a voltage of U_{10} ~206 mV. The Tafel slope was ~53.6 eV/dec. The MoS$_{x\sim2+\delta}$/Mo films deposited in a vacuum and at a higher Ar pressure (16 Pa) required much more overvoltage to achieve a current density of 10 mA/cm^2, and their Tafel slope was as large as 56.7 mV (Figure 11b). Additional anodic CV measurements and TOF calculations (Figure 11c,d) revealed that the relatively good performance of the catalyst, deposited at a pressure of 4 Pa, was caused by a larger loading compared to that for the catalyst obtained at 8 Pa. Indeed, for the film obtained at 8 Pa, the TOF value at the voltage of −200 mV was ~0.023 s^{-1}, and for the film deposited at 4 Pa, the TOF was ~0.014 s^{-1}. The films deposited in Ar at 16 Pa also had relatively large TOF (~0.023 s^{-1}). However, at the pressure of 16 Pa, the MoS$_{x\sim2+\delta}$/Mo film deposition rate was the lowest. These results were used to choose the Ar pressure of 8 Pa for the PLD of amorphous MoS$_x$ films in the present work.

Comparison of the shapes of the anodic CV curves (anodic stripping voltammograms) for the MoS$_{x\sim2+\delta}$/Mo films deposited at different Ar pressures indicated that an increase in Ar pressure of up to 8 Pa led to the formation of a curve in which a broad peak at ~700 mV was dominant (Figure 11c). Other peaks at higher voltages, which were present on the CV curves for the MoS$_{x\sim2+\delta}$/Mo films deposited in a vacuum and at Ar at 4 Pa, disappeared. This finding suggests that Ar pressure increases during the on-axis PLD, resulting in the formation of a homogeneous local structure of MoS$_{x\sim2+\delta}$ (0.7 ≤ δ ≤0.9) catalysts, in which all active sites possess an identical nature and participate in the HER. Notably, the oxidation of active sites in the amorphous MoS$_x$ catalysts obtained by electrodeposition was registered at ~900 mV [18,37]. This finding indicates a possible difference in the local atomic structure of the active sites formed during the PLD and electrochemical deposition of amorphous MoS_x films.

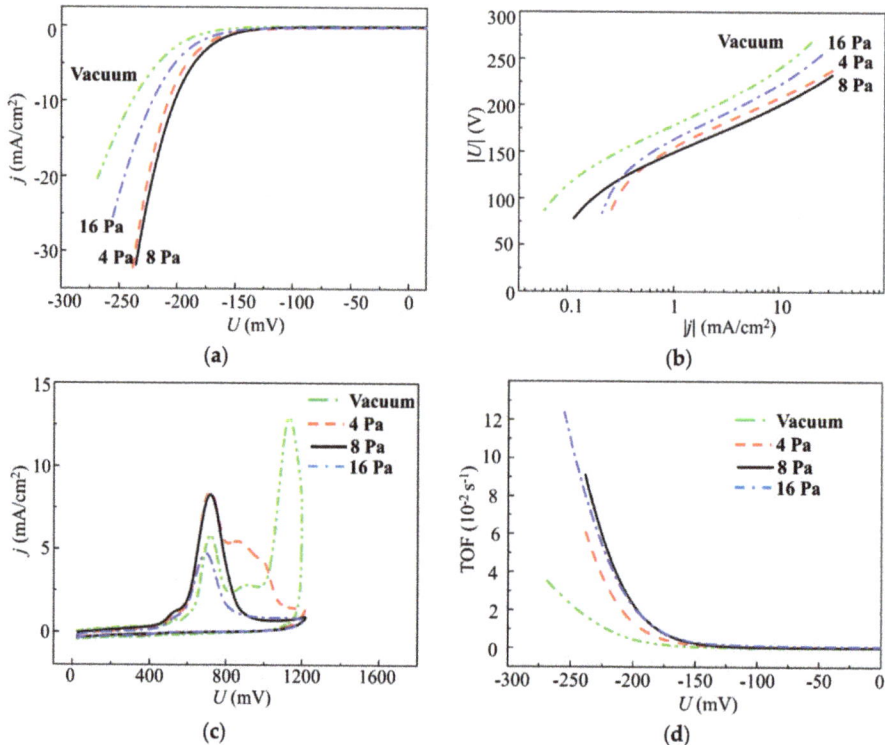

Figure 11. (a) Polarization curves. (b) Tafel plots. (c) Anodic cyclic voltammograms. (d) Turnover frequency (TOF) dependence on voltage for the $MoS_{x\sim2+\delta}$/Mo catalysts obtained on GC substrate by on-axis PLD in a vacuum and at different Ar pressures for 4 min.

Figure 12 shows the results of the studies of the electrocatalytic properties of the $MoS_{x\sim2+\delta}$/Mo films that were deposited by on-axis PLD onto the glassy carbon in Ar at 8 Pa for different times. The polarization curves (Figure 12a) show the benchmark activities of the $MoS_{x\sim2+\delta}$/Mo films, where both the apparent geometric area and the catalyst loading are known. The LV measurements indicated that a noticeable catalytic effect from a thin film was observed for very short deposition times (~2 min). With $MoS_{x\sim2+\delta}$/Mo catalyst loading of 6.8 µg/cm², an overpotential of −222 mV was required to achieve a current density of 10 mA/cm². With an increase in the deposition time to 64 min, the U_{10} value decreased (in absolute value) to −154.5 mV. A significant time increase up to 128 min caused only a slight decrease of the U_{10} to −150 mV. The Tafel slopes of the linear portions of the LV curves decreased from 53.7 to 50 mV/dec.

All anodic CV curves, shown in Figure 12b, had approximately identical shapes in which a broad peak dominated. This peak shifted from ~700 to ~800 mV under loading growth. The intensity of the peak revealed an outrunning growth with catalyst loading increase that caused a change of the TOFs for these films. The highest TOF value, −200 mV (equal to 0.026 s⁻¹), was found for the catalyst with minimal loading. As the loading increased, the TOFs decreased, and the TOF values were in the range of 0.01 ± 0.002 s⁻¹ for the films with higher loadings.

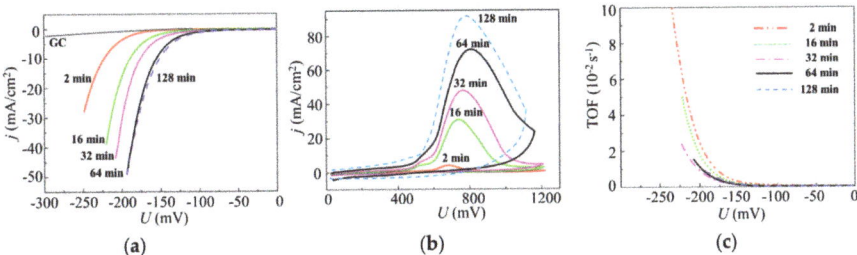

Figure 12. (a) Polarization curves. (b) Anodic cyclic voltammograms. (c) TOF dependence on voltage for the $MoS_{x\sim2+\delta}/Mo$ catalysts obtained on the GC substrate by on-axis PLD in Ar at 8 Pa for different deposition times.

The results of the study of the electrocatalytic properties of the $MoS_{x\sim3+\delta}$ films deposited on glassy carbon by off-axis PLD in Ar at 8 Pa for different times are shown in Figure 13. For the lowest deposition time of 1 min ($MoS_{x\sim3+\delta}$ catalyst loading 10 µg/cm^2), an overpotential of −209 mV was required to achieve a current density of 10 mA/cm^2. An increase in the deposition time, up to a certain point (20 min), caused a monotonic decrease of the U_{10} to −165.5 mV (Figure 13a). The Tafel slope decreased from 48.2 to 44.4 mV/dec. Further time increases caused only a slight decrease of U_{10} to −162 mV. The Tafel slopes of 50 and 44.4 mV/dec indicated that the HER was actually proceeded by a mechanism that was identical for the films obtained by on- and off-axis PLD.

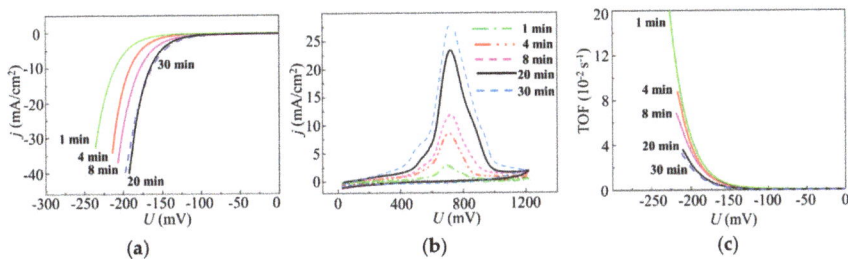

Figure 13. (a) Polarization curves. (b) Anodic cyclic voltammograms. (c) TOF dependence on voltage for the $MoS_{x\sim3+\delta}$ catalysts obtained on the GC substrate by off-axis PLD in Ar at 8 Pa for different deposition times.

The LV measurements indicated that the maximum achievable catalytic performance for the $MoS_{x\sim3+\delta}$ films was less than that for the $MoS_{x\sim2+\delta}/Mo$ films. This finding contradicted the results of the TOF measurements, which revealed that the TOFs were higher for the $MoS_{x\sim3+\delta}$ films (Figure 13c) than for the $MoS_{x\sim2+\delta}/Mo$ films (Figure 12c). Indeed, the highest TOF of ~0.05 s^{-1} (at −200 mV) was found for the thinnest $MoS_{x\sim3+\delta}$ film that was deposited for 1 min. An increase in loading caused a decrease in the TOF of the $MoS_{x\sim3+\delta}$ film, which was ~0.024 s^{-1} for the film deposited for ≥20 min. The larger TOFs for the $MoS_{x\sim3+\delta}$ films were due to both a lower current density during anodic striping CV and a narrower shape of the main peak, which is ascribable to the active site oxidation (Figure 13b). The narrower CV peak located at ~710 mV indicated a uniform chemical state of the atoms in the local areas of the catalytically active sites on the surfaces of the $MoS_{x\sim3+\delta}$ films.

4. Discussion

For both PLD deposition modes, an increase in catalyst loading caused a sublinear increase in the double-layer capacities of the obtained catalysts. However, at higher loadings, the C_{dl} for the $MoS_{x\sim3+\delta}$ films exceeded that of the $MoS_{x\sim2+\delta}/Mo$ films (Figure 14a). Thus, for the optimal loadings of 240 µg/cm² (MoS$_{x\sim2+\delta}$/Mo catalyst) and 163 µg/cm² (MoS$_{x\sim3+\delta}$ catalyst), the C_{dl} values were 20 and 22 mF/cm², respectively. These findings could be because there were significant contributions of heavy Mo particles to the loadings of the MoS$_{x\sim2+\delta}$/Mo films. For both films, the contents of S atoms that could form catalytically active sites (as S_2^{2-} units) in these types of catalysts were approximately equal to $1.5 \pm 0.1 \times 10^{18}$ cm^{-2}.

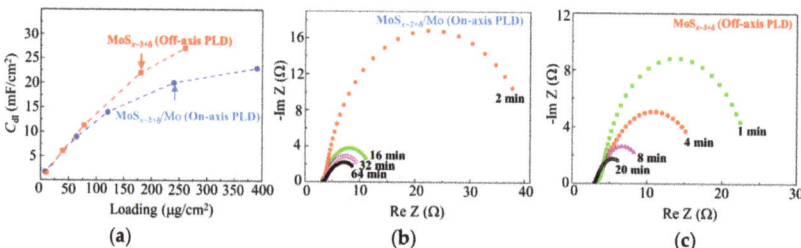

Figure 14. (a) Dependences of C_{dl} on the loading for the MoS$_x$-based catalysts obtained by on- and off-axis PLD. (b, c) Nyquist plots of the electrochemical impedance spectroscopy (EIS) measured for these catalysts obtained with different deposition times. The arrows indicate in (a) the optimal loadings of the catalysts.

Double layer capacitances are directly proportional to the effective electrochemical surface areas of the catalysts. However, the proportionality coefficient may depend on the nature of the catalyst-electrolyte interface. It is reasonable to assume that the specific capacities of two structures, i.e., semiconductor (MoS$_{x\sim3+\delta}$)-electrolyte and metal (Mo)-semiconductor (MoS$_{x\sim2+\delta}$)-electrolyte, might differ. Thus, it is difficult to correctly compare the effective electrochemical surface areas of the catalysts obtained by on- and off-axis PLD using the results of the C_{dl} measurements. Notably, the C_{dl} increased, even with an increase in loading past the optimal point. This allowed us to suggest that, under high loading, the entire electrochemical active surface area is not effectively involved in the HER. This supposition could be due to the fact that other mechanisms also influence the hydrogen evolution. Both the efficient transfer/diffusion of reagents to active sites and conditions for the free removal of molecular hydrogen are required [47–49]. The porous structures of the MoS$_{x\sim2+\delta}$/Mo films, with greater physical surface area, are expected to provide such conditions. Analysis of the anodic stripping voltammograms revealed that, on the physical surfaces of the MoS$_{x\sim2+\delta}$/Mo films, there were many more active centers than on the surfaces of the MoS$_{x\sim3+\delta}$ films (Figure 12b).

EIS revealed that the lower TOFs of the MoS$_{x\sim2+\delta}$/Mo films could have been caused not only by lower S concentrations in the catalytic MoS$_{x\sim2+\delta}$ films, but also by increased resistivities to the electric current transport through the hybrid layers. Indeed, the increase in load caused a monotonic decrease in the electrical resistances of the MoS$_{x\sim2+\delta}$/Mo and MoS$_{x\sim3+\delta}$ films (Figure 14b,c). However, with optimal loading, the resistances to current transport of the MoS$_{x\sim2+\delta}$/Mo and MoS$_{x\sim3+\delta}$ films were ~11 and ~7 Ω, respectively. As is known, the resistivity of a catalytic film has an important influence on the kinetics of the HER [50]. The increased resistivities of the MoS$_{x\sim2+\delta}$/Mo films could be caused by the hybrid structures of these films. Although metallic Mo is a good current conductor, in a hybrid structure, the electric current mainly passes through Mo-MoS$_{x\sim2+\delta}$ interfaces, which could produce a barrier effect (i.e., a Schottky barrier) for electron transport [51].

Long-term stability is an important aspect when evaluating the performance of an electrocatalyst. For the on- and off-axis PLD catalysts, nearly identical LV curves were detected after the first and 2000th cycles of CV testing. These results indicate no loss of catalytic performance and the remarkable stabilities of the electrocatalysts prepared on the GC substrate by the two PLD modes. Notably, the HER performances of both types of catalysts obtained with the two modes of PLD were quite high compared to other state-of-the-art nonprecious catalysts in acidic solutions. Comprehensive and updated information about the HER performances of amorphous MoS$_x$-based catalysts can be found in the literature [17,18,37].

To assess the potential feasibility of practical application of on-axis and off-axis PLD methods, the distribution of the thickness and composition of the MoS$_x$ films on the substrate should be considered. In the case of the vacuum on-axis PLD of the MoS$_{x\sim2+\delta}$/Mo film on the substrate installed at 3.5–5 cm from the MoS$_2$ target, almost all the laser ablated material is deposited on a substrate with a diameter of ~4–6 cm. The bell-shaped distribution of the film thickness is formed over the substrate surface [24]. The use of a buffer gas can improve the uniformity of film deposition, due to scattering of the laser plume (see, for example, Reference [52]), but the amount of the deposited material will be decreased. Analysis of Figure 3b revealed that at an optimal (for obtaining the highest HER efficiency of MoS$_{x\sim2+\delta}$/Mo film) Ar gas pressure, the amount of material deposited by on-axis PLD was decreased by ~6 times in comparison with the deposition by vacuum on-axis PLD.

The studies have shown that the atomic flux ablated by the laser from the MoS$_2$ target and scattered by Ar gas will not be wasted. The deposition of the scattered atomic flux also makes it possible to obtain rather effective MoS$_{x\sim3+\delta}$ electrocatalysts by off-axis PLD. For example, two substrates can be installed in parallel on opposite sides of the axis of expansion of the laser plume. Thus, most of the ablated material will be used to obtain the MoS$_{x\sim3+\delta}$ electrocatalyst. Studies of the film composition/thickness on the substrate at various distances from the target and the development of algorithms for moving/rotating substrates during the off-axis PLD processes will result in the preparation of rather uniform distributions of the catalytic MoS$_{x\sim3+\delta}$ films over the substrates with surface areas of ~10 cm^2.

A technical solution involving a combination of two methods of PLD may be promising. The on-axis PLD will cause the formation of a porous structure of the catalytic MoS$_{x\sim2+\delta}$/Mo layer, and off-axis PLD will result in the formation of a highly-effective MoS$_{x\sim3+\delta}$ thin-film electrocatalyst on the surface of the porous MoS$_{x\sim2+\delta}$/Mo layer.

5. Conclusions

The use of two modes of PLD from a MoS$_2$ target allowed us to obtain amorphous MoS$_x$-based catalysts with different compositions and morphologies. To obtain a high-quality MoS$_x$-based catalyst, it is important to choose the optimal buffer gas pressure during the PLD. During the deposition by traditional on-axis PLD, MoS$_{x\sim2+\delta}$/Mo films with hybrid porous structures and rather high concentrations of sulfur (δ ~0.7) in the catalytically active MoS$_{x\sim2+\delta}$ nanophase were obtained. Submicron and nanosized Mo-based particles were formed during the pulsed laser ablation of the target, and they were then incorporated into the bulk of the catalytic coatings deposited on the GC substrate. The hybrid structure somewhat slowed down the electric current transport through the deposited layer; however, this provided a greater physical surface area, and consequently, a greater number of active sites for HER. The porosity/morphology of the hybrid catalyst facilitated the effective growth of the exposed surface area with the increase in the MoS$_{x\sim2+\delta}$/Mo catalyst loading. This led to the possibility of achieving a higher HER performance at the optimal catalyst loading (~240 µg/cm^2) via the use of on-axis PLD compared to what could be achieved with off-axis PLD. To realize a current density of 10 mA/cm^2 on the GC substrate, the lowest overvoltages were −150 and −162 mV for the films obtained by on- and off-axis PLD, respectively.

The mode of off-axis PLD differs from the mode of on-axis PLD in terms of a higher deposition rate for catalytic MoS$_{x\sim3+\delta}$ films and a larger S concentration in the amorphous MoS$_{x\sim3+\delta}$ (δ ~0.8–1.1) phase. The depositions of Mo and S atomic fluxes scattered on Ar molecules facilitated the growth of the relatively dense structures of these films, which contributed to improved electric current transport through the deposited layer. This fact, combined with the greater S concentration, resulted in higher TOFs for the MoS$_{x\sim3+\delta}$ than the MoS$_{x\sim2+\delta}$/Mo films. However, the growth of the loading of these relatively dense catalytic films that possessed smoother surfaces did not effectively increase the numbers of active sites exposed in electrolyte for HER. The HER performance of the MoS$_{x\sim3+\delta}$ films obtained with off-axis PLD was saturated at a loading of ~163 µg/cm^2.

Author Contributions: Conceptualization and writing—original draft preparation, V.F.; Methodology, M.D.; Validation, A.G.; Formal analysis, R.R.; Investigation, D.F. and K.M. All authors have read and agreed to the published version of the manuscript.

Funding: This research was funded by the Russian Science Foundation, grant number 19-19-00081.

Conflicts of Interest: The authors declare no conflict of interest.

References

1. Vesborg, P.C.K.; Seger, B.; Chorkendorff, I. Recent development in hydrogen evolution reaction catalysts and their practical implementation. *J. Phys. Chem. Lett.* **2015**, *6*, 951–957. [CrossRef]
2. He, Z.; Que, W. Molybdenum disulfide nanomaterials: Structures, properties, synthesis and recent progress on hydrogen evolution reaction. *Appl. Mater. Today* **2016**, *3*, 23–56. [CrossRef]
3. Niyitanga, T.; Jeong, H.K. Thermally reduced graphite oxide/carbon nanotubes supported molybdenum disulfide as catalysts for hydrogen evolution reaction. *Int. J. Hydrog. Energy* **2019**, *44*, 977–987. [CrossRef]
4. Escalera-López, D.; Niu, Y.; Park, S.J.; Isaacs, M.; Wilson, K.; Palmer, R.E.; Rees, H.V. Hydrogen evolution enhancement of ultra-low loading, size-selected molybdenum sulfide nanoclusters by sulfur enrichment. *Appl. Catal. B* **2018**, *235*, 84–91. [CrossRef]
5. Liu, Z.; Gao, Z.; Liu, Y.; Xia, M.; Wang, R.; Li, N. Heterogeneous nanostructure based on 1T phase MoS2 for enhanced electrocatalytic hydrogen evolution. *ACS Appl. Mater. Interfaces* **2017**, *9*, 25291–25297. [CrossRef] [PubMed]
6. Kong, D.; Wang, H.; Cha, J.J.; Pasta, M.; Koski, K.J.; Yao, J.; Cui, Y. Synthesis of MoS$_2$ and MoSe$_2$ films with vertically aligned layers. *Nano Lett.* **2013**, *13*, 1341–1347. [CrossRef]
7. Li, S.; Wang, S.; Salamone, M.M.; Robertson, A.W.; Nayak, S.; Kim, H.; Tsang, S.C.E.; Pasta, M.; Warner, J.H. Edge-enriched 2D MoS$_2$ thin films grown by chemical vapor deposition for enhanced catalytic performance. *ACS Catal.* **2017**, *7*, 877–886. [CrossRef]
8. Hinnemann, B.; Moses, P.G.; Bonde, J.; Jørgensen, K.P.; Nielsen, J.H.; Horch, S.; Chorkendorff, I.; Nørskov, J.K. Biomimetic hydrogen evolution: MoS$_2$ nanoparticles as catalyst for hydrogen evolution. *J. Am. Chem. Soc.* **2005**, *127*, 5308–5309. [CrossRef]
9. Jaramillo, T.F.; Jorgensen, K.P.; Bonde, J.; Nielsen, J.H.; Horch, S.; Chorkendorff, I. Identification of active edge sites for electrochemical H$_2$ evolution from MoS$_2$ nanocatalysts. *Science* **2007**, *317*, 100–102. [CrossRef]
10. Li, H.; Tsai, C.; Koh, A.L.; Cai, L.; Contryman, A.W.; Fragapane, A.H.; Zhao, J.; Han, H.S.; Manoharan, H.C.; Pedersen, F.A.; et al. Activating and optimizing MoS$_2$ basal planes for hydrogen evolution through the formation of strained sulphur vacancies. *Nat. Mater.* **2016**, *15*, 48–53. [CrossRef]
11. Meng, C.; Lin, M.-C.; Du, X.-W.; Zhou, Y. Molybdenum disulfide modified by laser irradiation for catalyzing hydrogen evolution. *ACS Sustain. Chem. Eng.* **2019**, *7*, 6999–7003. [CrossRef]
12. Morales-Guio, G.G.; Hu, X. Amorphous molybdenum sulfides as hydrogen evolution catalysts. *Acc. Chem. Res.* **2014**, *47*, 2671–2681. [CrossRef] [PubMed]
13. Vrubel, H.; Hu, X. Growth and activation of an amorphous molybdenum sulfide hydrogen evolving catalyst. *ACS Catal.* **2013**, *3*, 2002–2011. [CrossRef]
14. Merki, D.; Fierro, S.; Vrubel, H.; Hu, X. Amorphous molybdenum sulfide films as catalysts for electrochemical hydrogen production in water. *Chem. Sci.* **2011**, *2*, 1262–1267. [CrossRef]
15. Fominski, V.Y.; Nevolin, V.N.; Romanov, R.I.; Fominski, D.V.; Dzhumaev, P.S. The influence of the local atomic packing of thin MoS$_x$ films on their electrocatalytic properties in hydrogen reduction. *Tech. Phys. Lett.* **2017**, *43*, 770–773. [CrossRef]
16. Deng, Y.; Ting, L.R.L.; Neo, P.H.L.; Zhang, Y.-J.; Peterson, A.A.; Yeo, B.S. Operando Raman spectroscopy of amorphous molybdenum sulfide (MoS$_x$) during the electrochemical hydrogen evolution reaction: Identification of sulfur atoms as catalytically active sites for H$^+$ reduction. *ACS Catal.* **2016**, *6*, 7790–7798. [CrossRef]
17. Xi, F.; Bogdanoff, P.; Harbauer, K.; Plate, P.; Höhn, C.; Rappich, J.; Wang, B.; Han, X.; van de Krol, R.; Fiechter, S. Structural transformation identification of sputtered amorphous MoS$_x$ as efficient hydrogen evolving catalyst during electrochemical activation. *ACS Catal.* **2019**, *9*, 2368–2380. [CrossRef]
18. Escalera-López, D.; Lou, Z.; Rees, N.V. Benchmarking the activity, stability, and inherent electrochemistry of amorphous molybdenum sulfide for hydrogen production. *Adv. Energy Mater.* **2019**, *9*, 1802614. [CrossRef]

19. Mabayoje, O.; Wygant, B.R.; Wang, M.; Liu, Y.; Mullins, C.B. Sulfur-rich MoS_6 as an electrocatalyst for the hydrogen evolution reaction. *ACS Appl. Energy Mater.* **2018**, *1*, 4453–4458. [CrossRef]
20. Afanasiev, P.; Jobic, H.; Lorentz, C.; Leverd, P.; Mastubayashi, N.; Piccolo, L.; Vrinat, M. Low-temperature hydrogen interaction with amorphous molybdenum sulfides MoS_x. *J. Phys. Chem. C* **2009**, *113*, 4139–4146. [CrossRef]
21. Song, X.; Chen, G.; Guan, L.; Zhang, H.; Tao, J. Interfacial engineering of MoS_2/TiO_2 hybrids for enhanced electrocatalytic hydrogen evolution reaction. *Appl. Phys. Express* **2016**, *9*, 095801. [CrossRef]
22. Liu, B.; Jin, Z.; Bai, L.; Liang, J.; Zhang, Q.; Wang, N.; Liu, C.; Wei, C.; Zhao, Y.; Zhang, X. Molybdenum-supported amorphous MoS_3 catalyst for efficient hydrogen evolution in solar-water splitting devices. *J. Mater. Chem. A* **2016**, *4*, 14204–14222. [CrossRef]
23. Fominskii, V.Y.; Markeev, A.M.; Nevolin, V.N. Pulsed ion beams for modification of metal surface properties. *Vacuum* **1991**, *42*, 73–74. [CrossRef]
24. Fominski, V.Y.; Nevolin, V.N.; Romanov, R.I.; Smurov, I. Ion-assisted deposition of MoS_x films from laser-generated plume under pulsed electric field. *J. Appl. Phys.* **2001**, *89*, 1449–1457. [CrossRef]
25. Muratore, C.; Voevodin, A.A. Control of molybdenum disulfide plane orientation during coating growth in pulsed magnetron sputtering discharges. *Thin Solid Films* **2009**, *517*, 5605–5610. [CrossRef]
26. Wang, J.; Lauwerens, W.; Wieers, E.; Stals, L.M.; He, J.; Celis, J.P. Structure and tribological properties of MoS_x coatings prepared by bipolar DC magnetron sputtering. *Surf. Coat. Technol.* **2001**, *139*, 143–152. [CrossRef]
27. Fominski, V.Y.; Romanov, R.I.; Fominski, D.V.; Shelyakov, A.V. Regulated growth of quasi-amorphous MoS_x thin-film hydrogen evolution catalysts by pulsed laser deposition of Mo in reactive H_2S gas. *Thin Solid Films* **2017**, *642*, 58–68. [CrossRef]
28. Wang, R.; Sun, P.; Wang, H.; Wang, X. Pulsed laser deposition of amorphous molybdenum disulfide films for efficient hydrogen evolution reaction. *Electrochim. Acta* **2017**, *258*, 876–882. [CrossRef]
29. Fominski, V.Y.; Romanov, R.I.; Fominski, D.V.; Dzhumaev, P.S.; Troyan, I.A. Normal and grazing incidence pulsed laser deposition of nanostructured MoS_x hydrogen evolution catalysts from a MoS_2 target. *Opt. Laser Technol.* **2018**, *102*, 74–84. [CrossRef]
30. Donley, M.S.; Murray, P.T.; Barber, S.A.; Haas, T.W. Deposition and properties of MoS_2 thin films grown by pulsed laser evaporation. *Surf. Coat. Technol.* **1988**, *36*, 329–340. [CrossRef]
31. Walck, S.D.; Zabinski, J.S.; Donley, M.S.; Bultman, J.E. Evolution of surface topography in pulsed-laser-deposited thin films of MoS_2. *Surf. Coat. Technol.* **1993**, *62*, 412–416. [CrossRef]
32. Grigoriev, S.N.; Fominski, V.Y.; Romanov, R.I.; Volosova, M.A.; Shelyakov, A.V. Pulsed laser deposition of nanocomposite $MoSe_x/Mo$ thin-film catalyst for hydrogen evolution reaction. *Thin Solid Films* **2015**, *592*, 175–181. [CrossRef]
33. Hu, J.J.; Zabinski, J.S.; Bultman, J.E.; Sanders, J.H.; Voevodin, A.A. Encapsulated nanoparticles produced by pulsed laser ablation of MoS_2-Te composite target. *Cryst. Growth Des.* **2008**, *8*, 2603–2605. [CrossRef]
34. Fominski, V.Y.; Grigoriev, S.N.; Gnedovets, A.G.; Romanov, R.I. On the mechanism of encapsulated particle formation during pulsed laser deposition of WSe_x thin-film coatings. *Technol. Phys. Lett.* **2013**, *39*, 312–315. [CrossRef]
35. Fominski, V.Y.; Markeev, A.M.; Nevolin, V.N.; Prokopenko, V.B.; Vrublevski, A.R. Pulsed laser deposition of MoS_x films in a buffer gas atmosphere. *Thin Solid Films* **1994**, *248*, 240–246. [CrossRef]
36. Fominski, V.; Gnedovets, A.; Fominski, D.; Romanov, R.; Kartsev, P.; Rubinkovskaya, O.; Novikov, S. Pulsed laser deposition of nanostructured MoS_3/np-Mo//WO_{3-y} hybrid catalyst for enhanced (photo) electrochemical hydrogen evolution. *Nanomaterials* **2019**, *9*, 1395. [CrossRef]
37. Deng, H.; Zhang, C.; Xie, Y.; Tumlin, T.; Giri, L.; Karna, S.P.; Lin, J. Laser induced MoS_2/carbon hybrids for hydrogen evolution reaction catalysts. *J. Mater. Chem. A* **2016**, *4*, 6824–6830. [CrossRef]
38. Ting, L.R.L.; Deng, Y.; Ma, L.; Zhang, Y.-J.; Peterson, A.A.; Yeo, B.S. Catalytic activities of sulfur atoms in amorphous molybdenum sulfide for the electrochemical hydrogen evolution reaction. *ACS Catal.* **2016**, *6*, 861–867. [CrossRef]
39. Bailini, A.; Di Fonzo, F.; Fusi, M.; Casari, C.S.; Li Bassi, A.; Russo, V.; Baserga, A.; Bottani, C.E. Pulsed laser deposition of tungsten and tungsten oxide thin films with tailored structure at the nano- and mesoscale. *Appl. Surf. Sci.* **2007**, *253*, 8130–8135. [CrossRef]

40. Fominski, V.Y.; Romanov, R.I.; Fominski, D.V.; Shelyakov, A.V. Preparation of MoSe$_{x>3}$/Mo-NPs catalytic films for enhanced hydrogen evolution by pulsed laser ablation of MoSe$_2$ target. *Nucl. Instrum. Methods Phys. Res. B* **2018**, *416*, 30–40. [CrossRef]
41. Wood, R.F.; Leboeuf, J.N.; Chen, K.R.; Geohegan, D.B.; Puretzky, A.A. Dynamics of plume propagation, splitting, and nanoparticle formation during pulsed-laser ablation. *Appl. Surf. Sci.* **1998**, *127–129*, 151–158. [CrossRef]
42. McDevitt, N.T.; Bultman, J.E.; Zabinski, J.S. Study of amorphous MoS$_2$ films grown by pulsed laser deposition. *Appl. Spectrosc.* **1998**, *52*, 1160–1164. [CrossRef]
43. Luo, J.; Xu, P.; Zhang, D.; Wei, L.; Zhou, D.; Xu, W.; Li, J.; Yuan, D. Synthesis of 3D-MoO$_2$ microsphere supported MoSe$_2$ as an efficient electrocatalyst for hydrogen evolution reaction. *Nanotechnology* **2017**, *28*, 465404. [CrossRef] [PubMed]
44. Kibsgaard, J.; Jaramillo, T.F.; Besenbacher, F. Building an appropriate active-site motif into a hydrogen-evolution catalyst with thiomolybdate [Mo$_3$S$_{13}$]$^{2-}$ clusters. *Nat. Chem.* **2014**, *6*, 248–253. [CrossRef] [PubMed]
45. Tran, P.D.; Tran, T.V.; Orio, M.; Torelli, S.; Truong, Q.D.; Nayuki, K.; Sasaki, Y.; Chiam, S.Y.; Yi, R.; Honma, I.; et al. Coordination polymer structure and revisited hydrogen evolution catalytic mechanism for amorphous molybdenum sulfide. *Nat. Mater.* **2016**, *15*, 640–646. [CrossRef]
46. Gnedovets, A.G.; Fominski, V.Y.; Nevolin, V.N.; Romanov, R.I.; Fominski, D.V.; Soloviev, A.A. Models of WO$_x$ films growth during pulsed laser deposition at elevated pressures of reactive gas. *J. Phys. Conf. Ser.* **2018**, *941*, 012064. [CrossRef]
47. Chaudhari, N.K.; Jin, H.; Kim, B.; Lee, K. Nanostructured materials on 3D nickel foam as electrocatalysts for water splitting. *Nanoscale* **2017**, *9*, 12231–12247. [CrossRef]
48. Kemppainen, E.; Halme, J.; Hansen, O.; Seger, B.; Lund, P.D. Two-phase model of hydrogen transport to optimize nanoparticle catalyst loading for hydrogen evolution reaction. *Int. J. Hydrog. Energy* **2016**, *41*, 7568–7581. [CrossRef]
49. Kemppainen, E.; Halme, J.; Lund, P.D. An analytical model of hydrogen evolution and oxidation reactions on electrodes partially covered with a catalyst. *Phys. Chem. Chem. Phys.* **2016**, *18*, 13616–13628. [CrossRef]
50. Huang, X.; Zhang, M.; Sun, R.; Long, G.; Liu, Y.; Zhao, W. Enhanced hydrogen evolution from CuO$_x$-C/TiO$_2$ with multiple electron transport pathways. *PLoS ONE* **2019**, *14*, e0215339. [CrossRef]
51. Bharathi, N.D.; Sivasankaran, K. Influence of metal contact on the performance enhancement of monolayer MoS$_2$ transistor. *Superlattices Microstruct.* **2018**, *120*, 479–486. [CrossRef]
52. Fominski, V.Y.; Romanov, R.I.; Gnedovets, A.G.; Nevolin, V.N. Formation of the chemical composition of transition metal dichalcogenide thin films at pulsed laser deposition. *Technical Phys.* **2010**, *55*, 1509–1516. [CrossRef]

© 2020 by the authors. Licensee MDPI, Basel, Switzerland. This article is an open access article distributed under the terms and conditions of the Creative Commons Attribution (CC BY) license (http://creativecommons.org/licenses/by/4.0/).

Article

Spectroscopic and Microscopic Analyses of Fe₃O₄/Au Nanoparticles Obtained by Laser Ablation in Water

Maurizio Muniz-Miranda [1,2,*], Francesco Muniz-Miranda [3] and Emilia Giorgetti [2]

1. Department of Chemistry "Ugo Schiff", University of Florence, Via Lastruccia 3, 50019 Sesto Fiorentino, Italy
2. Institute of Complex Systems (CNR), Via Madonna del Piano 10, 50019 Sesto Fiorentino, Italy; emilia.giorgetti@fi.isc.cnr.it
3. École Nationale Supérieure de Chimie de Paris and PSL Research University, CNRS, Institute of Chemistry for Life and Health Sciences (i-CLeHS), FRE 2027, 11, rue Pierre et Marie Curie, F-75005 Paris, France; f.muniz-miranda@chimieparistech.psl.eu
* Correspondence: maurizio.muniz@unifi.it

Received: 8 December 2019; Accepted: 8 January 2020; Published: 10 January 2020

Abstract: Magneto-plasmonic nanoparticles constituted of gold and iron oxide were obtained in an aqueous environment by laser ablation of iron and gold targets in two successive steps. Gold nanoparticles are embedded in a mucilaginous matrix of iron oxide, which was identified as magnetite by both microscopic and spectroscopic analyses. The plasmonic properties of the obtained colloids, as well as their adsorption capability, were tested by surface-enhanced Raman scattering (SERS) spectroscopy using 2,2′-bipyridine as a probe molecule. DFT calculations allowed for obtaining information on the adsorption of the ligand molecules that strongly interact with positively charged surface active sites of the gold nanoparticles, thus providing efficient SERS enhancement. The presence of iron oxide gives the bimetallic colloid new possibilities of adsorption in addition to those inherent to gold nanoparticles, especially regarding organic pollutants and heavy metals, allowing to remove them from the aqueous environment by applying a magnetic field. Moreover, these nanoparticles, thanks to their low toxicity, are potentially useful not only in the field of sensors, but also for biomedical applications.

Keywords: laser ablation; gold; magnetite; SERS; 2,2′-bipyridine

1. Introduction

Nanoparticles constituted of metals like silver, gold, or copper exhibit plasmonic properties and are widely employed as biosensors, drug vectors, and SERS (surface-enhanced Raman scattering) [1,2] and fluorescence markers, especially gold nanoparticles that are more biocompatible. Their applications can be realized by adding additional functionalities like magnetic properties. Hence, in nanomedicine they find diagnostic and/or therapeutic applications in magnetic resonance imaging or for generating hyperthermia by applying locally intense magnetic fields [3–6]. In this regard, Fe_3O_4 magnetic nanoparticles, presenting good biocompatibility and low toxicity [7–9], are widely used in these biomedical applications. Usually, different chemical procedures are employed to prepare these nanocomposites [10–17] to be used for sensoristic and biomedical applications, but they involve problems due to the presence of surfactants, stabilizers, residual reductants, and by-products, which could interfere in both the adsorption and the detection of ligands. In this regard, laser-assisted procedures have been recently employed to obtain metal nanoparticles with both plasmonic and magnetic properties [18–20]. In the past, some of us adopted the laser ablation procedure of metal targets in water to produce bimetallic colloidal nanoparticles [21,22]; in particular, two-step laser ablations of iron and silver [23] and of nickel and silver [24] were employed to obtain magneto-plasmonic colloidal nanoparticles.

Here, we propose the fabrication of bifunctional Fe_3O_4/Au nanoparticles obtained by the two-step laser ablation of iron and gold targets in water, along with microscopic and spectroscopic characterization. To this end, high-resolution transmission electron microscopy (HRTEM) and selected area electron diffraction (SAED) analyses have been performed, and visible absorption, XPS, Raman, and SERS spectra have been obtained. To obtain information on the type of ligand/metal adsorption provided by these nano-platforms, calculations based on density functional theory (DFT) have also been carried out using 2,2'-bipyridine as a molecular reporter.

The importance and novelty of the present investigation, in addition to producing "pure" colloidal suspensions—that is, without the aid of chemical reagents and surfactants—are due to the fact that gold nanoparticles are trapped in a ferromagnetic matrix, so they are preserved from colloidal collapse. In addition, it is possible to remove all the bimetallic material, including the possible load of adsorbed ligands, from the solvent and transport them thanks to the use of a magnetic field. In this regard, the presence of iron oxide gives the bimetallic colloid new possibilities for the adsorption of ligands, in addition to those inherent to gold nanoparticles, and also for removal of them from the aqueous environment, especially with regard to organic pollutants [25] and heavy metals [26]. Finally, our bimetallic colloids exhibit plasmonic properties, in addition to magnetic ones, due to the presence of gold nanoparticles, which allow application of the SERS technique for sensoristic purposes. In practice, SERS spectroscopy provides huge intensification of the Raman signal of molecules adsorbed on nanostructured gold or silver surfaces, usually up to 10^7 enhancement factors with respect to the normal Raman response of non-adsorbed molecules.

2. Materials and Methods

2.1. Laser Ablation

Iron (Sigma-Aldrich, St. Louis, Missouri (USA), 99.99% purity) and gold (Goodfellow, Huntingdon (UK), 99.95% purity) plates were used as targets for the laser ablation. Colloidal suspensions were prepared by laser ablations of iron in deionized water (18.2 MX cm @ 25 C), and then of gold, by using the fundamental wavelength (1064 nm) of a Q-switched Nd:YAG laser (Quanta System G90-10: rep. rate 10 Hz, pulse width at FWHM of 10 ns). The laser pulse energy was set at 20 mJ/pulse, corresponding to 200 mW average power, focusing the laser light into a laser spot of approximately 1 mm diameter and corresponding fluence of 2.5 J/cm^2. The target plate was fixed at the bottom of a glass vessel filled with 6 mL of liquid (height above the target: 2 cm). The irradiation time of the metal targets was about 20 min. To minimize effects due to crater formation in the metal targets, the glass vessel was manually rotated and translated, stopping the ablation process every three minutes. The laser pulse entered the vessel from above, thus impinging perpendicularly onto the target. These experimental procedures were chosen in order to obtain a valid colloidal stability, following the indications of our previous experiments [23].

2.2. UV–Visible Extinction Spectroscopy

UV–visible extinction spectra of the colloidal suspensions were obtained in the 200–800 nm region by using a Cary 5 Varian spectrophotometer (OPL (optical path length) = 2 mm). The observed bands were due to both absorption and scattering of the radiation.

2.3. Microscopic Techniques

TEM (transmission electron microscopy) and HRTEM (high-resolution TEM) images were obtained after dipping Ni grids in the colloidal suspensions. Microscopic measurements, EDX (energy-dispersive X-ray spectrometry) analysis, and SAED patterns were obtained using a Jeol 2010 instrument operating at 200 kV and equipped with an EDS Link ISIS EDX micro-analytic system.

2.4. Raman Spectroscopy

Raman spectra of the bimetallic nanoparticles deposited on aluminum plate were measured at different points of the dried sample by using a Renishaw RM2000 micro-Raman instrument equipped with a diode laser emitting at 785 nm. Sample irradiation was accomplished by using the 50× microscope objective of a Leica Microscope DMLM. The backscattered Raman signal was fed into the monochromator through 40 μm slits and detected by an air-cooled CCD (2.5 cm^{-1} per pixel) filtered by a double holographic Notch filters system. Spectra were calibrated with respect to a silicon wafer at 520 cm^{-1}.

SERS spectra of 10^{-4} M 2,2′-bipyridine (Sigma-Aldrich, St. Louis, Missouri (USA), 99% purity) in bimetallic colloid were obtained after addition of 10^{-2} M NaCl (Sigma-Aldrich, St. Louis, Missouri (USA), 99.999% purity) in order to increase the SERS enhancement without compromising the colloidal stability. The 647.1 nm line of a Kripton ion laser and a Jobin-Yvon HG2S monochromator equipped with a cooled RCA-C31034A photomultiplier were used. A defocused laser beam with 100 mW power was employed for impairing thermal effects. Power density measurements were made using a power meter instrument (model 362; Scientech, Boulder, CO, USA) giving ~5% accuracy in the 300–1000 nm spectral range.

2.5. X-ray Photoelectron Spectroscopy

XPS measurements were made using a non-monochromatic Mg Kα X-ray source (1253.6 eV) and a VSW HAC 5000 hemispherical electron energy analyzer operating in the constant pass energy mode at E_{pas} = 44 eV. The bimetallic colloidal samples were prepared just before the analysis by depositing a few drops of the colloidal suspensions on soda glass substrates and letting the solvent evaporate. In order to increase the amount of deposited nanoparticles, this procedure was repeated several times. Then, the glasses with bimetallic nanoparticles were introduced into the UHV system via a loadlock under inert gas (N$_2$) flux and kept in the introduction chamber overnight, allowing the removal of volatile substances as confirmed by the achieved pressure value (2×10^{-9} mbar), just above the instrument base pressure. The obtained spectra were referenced to the C 1s core peak at 284.8 eV assigned to the adventitious carbon. The spectra were fitted using CasaXPS software version 2.3.15.

2.6. Density Functional Theory Calculations

All DFT calculations were carried out using the GAUSSIAN 09 package [27]. Optimized geometries were obtained at the DFT level of theory, employing the widely adopted Becke 3-parameter hybrid exchange functional (B3) combined with the Lee–Yang–Parr correlation functional (LYP) [28,29], along with the Lanl2DZ basis set and pseudopotential [30–32]. All parameters were allowed to relax and all calculations converged toward optimized geometries corresponding to energy minima, as revealed by the lack of negative values in the frequency calculation. Dispersion interactions were taken into account using Grimme's D3 scheme along with Becke–Johnson damping [33]. A scaling factor of 0.98 for the calculated harmonic wavenumbers was employed, as usually performed in calculations at this level of theory [34–38]. The calculated Raman intensities were obtained by following the indications of reference [24].

3. Results and Discussion

3.1. Microscopic Investigation

The colloid obtained by laser ablation of an iron target in water has a zeta potential value of +20.0 mV, which is lowered to +13.9 mV when a gold target is also ablated. This lowering is due to the adsorption of negative ions deriving from the water environment on the (positive) surface of the gold nanoparticles. However, the zeta potential is sufficient to provide stability to the bimetallic colloid, with no precipitate visible a week after preparation. The zeta potential data are reported in Figure S1.

The bimetallic colloid presents a red color and exhibits magnetic properties, as shown in Figure S2. When approaching a magnet, the colloidal nanoparticles aggregate, until they appear as a dark red precipitate visible to the naked eye.

Observing the TEM images (see Figure 1), the colloid consists of spheroidal particles with dimensions ranging from a few nanometers to almost 20 nm in diameter. Based on the contrast, two kind of nanoparticles, with weaker contrast (low contrast, LC) and stronger contrast (high contrast, HC), can be distinguished. LC particles are mainly particles of a few nanometers, whereas the HC particles have two size classes: particles of a few nanometers and particles with a diameter of 10–20 nm. From the point of view of the metallic composition, the sample contains Fe and Au (in addition to O). The large HC particles are substantially composed of Au. Large LC particles are composed of Fe. The small particles, for which it is not possible to make EDX measurements on single individuals, show both Au and Fe. The EDX analyses of typical HC and LC nanoparticles are reported in Figure S3.

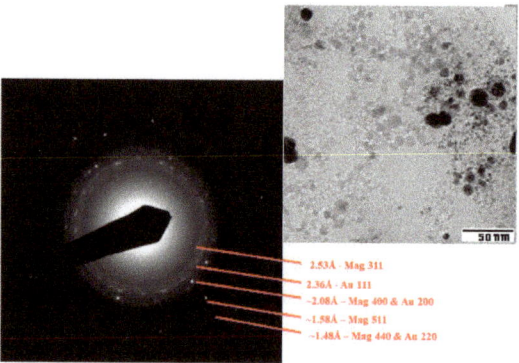

Figure 1. Low-magnification TEM micrograph of the Fe_3O_4/Au colloid (right), showing Au particles embedded in low-contrast matrix, along with SAED analysis (Mag: magnetite).

SAED on an enlarged field of the sample (Figure 1) shows interplanar distances consistent with magnetite (Fe_3O_4) and metallic gold. In particular, the ring around 2.36 Å is quite strong and must be attributed to the 111 reflection of gold [39]. Analysis of the high-resolution microscopic images (HRTEM) (Figure 2) shows interplanar distances typical of metallic Au for HC particles, both large and small. Crystalline growth in HC particles is observed as icosahedrons. LC particles show interplanar distances typical of magnetite [40], but also the presence of some small particles with amorphous characteristics. In conclusion, the gold nanoparticles appear to be embedded in a mucilaginous matrix consisting of magnetite in the form of nanoparticles of various sizes, with scarce tendency to aggregation, which can be separated from the aqueous environment under the action of a magnetic field.

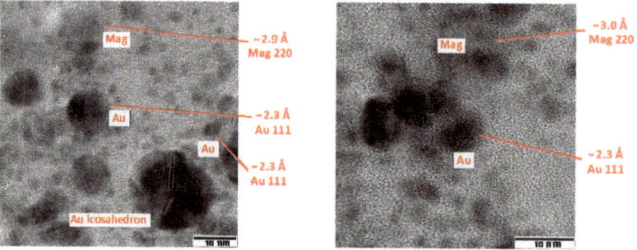

Figure 2. High-resolution TEM images of the Fe_3O_4/Au colloid, showing the interplanar distances in gold (Au) and magnetite (Mag) particles.

3.2. Raman Spectra

After centrifugation of a portion of the bimetallic colloid, the precipitate was examined using a micro-Raman spectrometer and showed the typical Raman band of magnetite (Fe_3O_4) at 665 cm^{-1} (see Figure 3), in agreement with the literature [41,42], confirming the magnetic properties of the nanosystem.

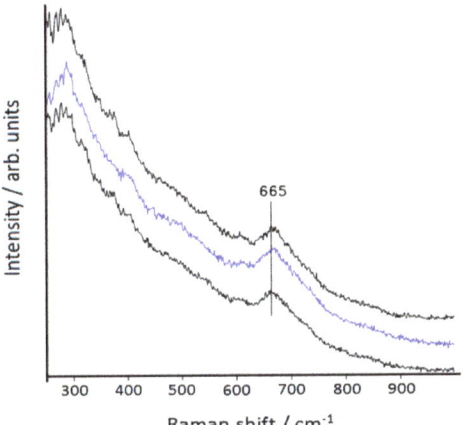

Figure 3. Micro-Raman spectra of the Fe_3O_4/Au nanoparticles deposited on Al plate at different points of the dry film. Excitation: 785 nm.

3.3. XPS Measurements

The XPS spectrum relative to the $f_{7/2}$–$f_{5/2}$ gold spectral region (see Figure 4) can be fitted by two components: the main one is located at 84.3 eV ($f_{7/2}$), while the subordinate is located at a higher energy value (85.4 eV). These components can be attributed to Au(0) and Au(I), respectively, as well as occurring in the case of gold laser-ablated in deionized water [43].

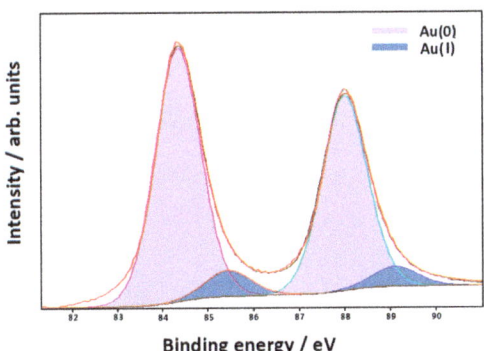

Figure 4. XPS spectrum of the bimetallic nanoparticles in the gold $f_{7/2}$–$f_{5/2}$ spectral region.

3.4. UV–Visible Extinction Spectra

Figure 5 shows the UV–visible absorption spectra of the colloidal samples obtained by laser ablation of iron (Spectrum A) and then laser ablation of gold (Spectrum B). The band observed in Spectrum B around 525 nm is attributable to the surface plasmon resonance of non-aggregated gold nanoparticles. By adding 2,2'-bipyridine (bpy), the plasmon band is shifted to 535 nm.

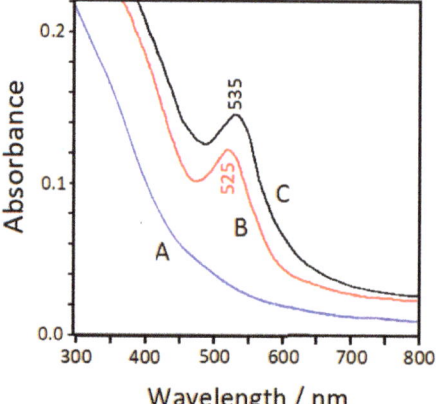

Figure 5. UV–visible extinction spectra of the Fe_3O_4 (A) and Fe_3O_4/Au (B) colloids. Spectrum C refers to the bimetallic colloid in the presence of 2,2′-bipyridine.

3.5. Surface-Enhanced Raman Scattering

The evidence of the surface plasmon band of nanosized gold particles (see Figure 5) suggests the possibility of SERS activity of this bimetallic system. However, the molecular ligand needs to be effectively adsorbed to provide a reliable Raman enhancement. Hence, in order to find confirmation of our hypothesis, we checked the SERS response of the bimetallic colloid in the presence of 10^{-4} M 2,2′-bipyridine (bpy). By activation with NaCl, we observed a satisfactory SERS spectrum of bpy in the bimetallic colloid (Figure 6), with results quite similar to those reported in the literature for the adsorption of bpy on pure gold colloidal nanoparticles [44]; those frequencies are reported in Table 1 for comparison. This similarity indicates that our SERS spectrum is attributable to 2,2′-bipyridine bound to the gold nanoparticles present in the Fe_3O_4 colloidal matrix, which does not impair the ligand adsorption on gold. In Table 1 the IR and Raman frequencies of solid bpy [45] are also reported.

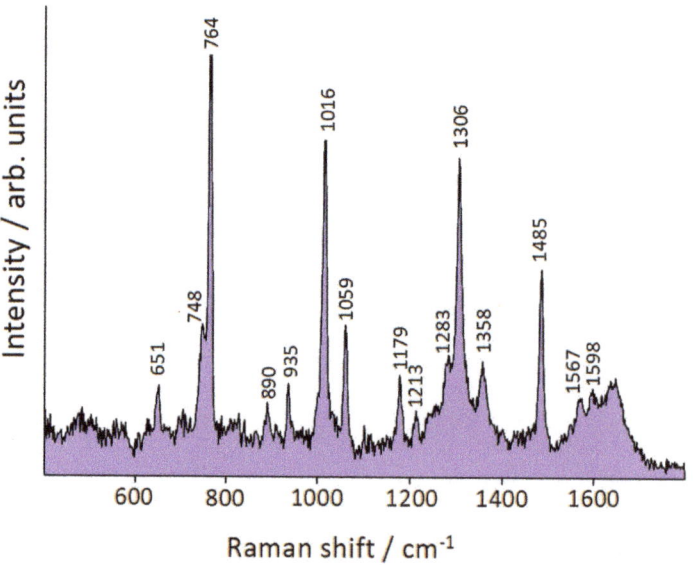

Figure 6. SERS spectrum of 2,2′-bipyridine in the bimetallic colloid. Excitation: 647.1 nm.

Table 1. Observed and calculated frequencies (cm^{-1}).

Symmetry Species [45]	bpy IR/Raman [45]	bpy/Au SERS	bpy/Au+ Calc.	bpy/Au° Calc.	bpy/Au SERS [44]
B$_u$	1575		1603	1592	
A$_g$	1589	1598	1598	1594	1586
B$_u$	1550		1590	1581	
A$_g$	1572	1567	1575	1570	1562
A$_g$	1482	1485	1491	1483	1479
A$_g$	1446		1469	1460	
B$_u$	1450		1445	1433	
B$_u$	1410		1429	1419	
B$_u$	1265	1358	1324	1323	
B$_u$	1250		1331	1298	
A$_g$	1309	1306	1306	1313	1301
A$_g$	1301	1283	1289	1286	
A$_g$	1236		1294	1274	
B$_u$	1140	1213	1206	1184	
A$_g$	1146	1179	1190	1176	1173
B$_u$	1085		1128	1113	
A$_g$	1094		1114	1096	
B$_u$	1065		1078	1077	
A$_g$	1044	1059	1062	1049	1057
B$_g$	—		1039	1032	
B$_u$	1040		1038	1023	
A$_u$	—		1033	979	
A$_g$	994	1016	1007	1022	1010
B$_u$	995		993	985	
A$_u$	975		993	983	
B$_g$	—		991	976	
B$_g$	909	935	923	913	
B$_u$	890		825	827	
B$_g$	815	890	908	913	
A$_u$	755		786	774	
A$_g$	764	764	760	767	761
B$_g$	742	748	746	754	
A$_u$	740		747	756	
B$_u$	655		658	656	
A$_g$	614	651	653	636	646
B$_u$	620		632	617	
B$_g$	550		555	561	
A$_u$	—		447	479	
A$_g$	440		441	415	
B$_g$	409		422	415	403
A$_u$	—		405	380	
A$_g$	332	356	353	327	353
B$_g$	224		226	241	

The addition of NaCl was necessary to obtain a satisfactory SERS spectrum of bpy. The presence of chloride anions, which strongly adsorb on the surface of the gold nanoparticles, has double validity because it can promote both the nanoparticle aggregation necessary for an efficient SERS response and the formation of active sites capable of strongly binding ligand molecules, similar to what occurs with silver nanoparticles activated by chloride anions [46–48]. In practice, in our bimetallic suspension it was not possible to obtain a valid SERS spectrum of 2,2'-bipyridine, even at 10^{-4} M concentration, unless we added NaCl. To induce particle aggregation or concentration, magnetic attraction could be employed, instead of adding chloride anions, in order to improve the SERS signal of the adsorbed ligands. In the future, this method will be tested by also evaluating the occurrence of possible problems in colloidal stability. In the present work, we used chloride activation to obtain an effective SERS response in a stable aqueous suspension in order to evaluate the possible use of these nanoparticles in biomedical applications.

However, one last problem remains to be solved: what kind of active site on the surface of the gold nanoparticles is involved in the interaction with the molecule, given that the XPS spectrum also showed the presence of ionized gold such as Au(I)? DFT calculations on the molecule linked to a neutral or a positively charged gold adatom can help in this purpose.

3.6. DFT Calculations

In Table 1 the experimental SERS frequencies of bpy are compared with those calculated for bpy/gold model complexes, along with the IR and Raman frequencies of solid bpy [45], whose molecules present a trans-planar structure. We observe that the prominent SERS bands (at 353, 651, 764, 1016, 1059, 1179, 1306, and 1485 cm^{-1}) correspond to the bpy Raman bands of A$_g$ symmetry species. For the simulation of the SERS spectra of the adsorbed bpy, we used the functional B3LYP, along with the Lanl2DZ basis set.

The choice to use this basis set was justified by the following considerations.

(a). This basis set has been widely employed in many literature articles to successfully reproduce both the structural and vibrational properties of different molecules. Here we report only a few very recent examples [49–55].
(b). Core electrons can be treated in an approximate way via effective core potentials (ECPs). This treatment includes scalar relativistic effects, which are important for the proper description of the geometric, electronic, and spectroscopic properties of heavy atoms. The LanL2DZ basis set is the best known basis set for molecular systems containing these atoms and for the efficient simulation of the Raman spectra of complexes with transition metals and the SERS spectra of molecules adsorbed on silver or gold nanoparticles, as demonstrated by many recent papers (for example, [38,50,52–55]).

We also tested the reliability of this basis set by examining the free 2,2′-bipyridine molecule in its typical *trans* conformation and comparing our DFT results with those reported in the literature [48] for the same molecule, with the same functional but with a different basis, 6-31+G*. The Lanl2DZ basis set used by us provided results generally comparable with those reported in the literature, as shown in the Supplementary Materials, regarding both structural parameters (Table S1) and vibrational frequencies (Table S2).

DFT calculations were performed for two gold complexes, where the bpy molecule in *cis* conformation is linked by means of the nitrogen atoms to a neutral Au atom or to a gold cation, Au$^+$. The complex bpy/Au$^+$ better reproduces the observed SERS frequencies than the complex bpy/Au$°$. In the first case, the average error between the calculated and observed frequencies is 7.75 cm^{-1}; in the second one, the average error is significantly larger at 13.27 cm^{-1}. In addition, the interaction of the molecule with a neutral atom is quite weak, in comparison with the interaction with Au$^+$, as shown by the bpy→gold electronic charge transfers and the N–gold bond distances reported in Table 2, with |e| being the unsigned electron charge. The Mulliken partial charges are reported in Table S1. Hence, it is possible to conclude that the ligand molecules, when they adsorb on gold, strongly interact with positively charged active sites of the nanoparticle surface. Figure S4 shows the calculated normal modes of the bpy/Au$^+$ complex relative to the prominent SERS bands. All these correspond to in-plane vibrations of the bpy molecule, in particular, the bands observed at 356, 651, 764, and 1016 cm^{-1} correspond to ring deformations, and those at 1306 and 1485 cm^{-1} to H bending modes.

To better quantify the charge transfer, we also employed a descriptor (called D$_{CT}$, charge transfer distance) [56] that was mainly proposed to describe electron–hole displacement in optical excitations (S$_n$→S$_0$, n = 1, 2 ... , with S being singlet electronic states). The D$_{CT}$ version adopted here is based on a partial charge (namely Mulliken's) approach, using the spreadsheet reported in the Supplementary Materials of reference [56] and already employed with success for electronic transitions [37,57]. With the D$_{CT}$ scheme, the difference between the electronic density of the ground state (S$_0$) and the excited state of interest (S$_n$) gives rise to a charge separation that can be modeled in a dipolar fashion due to a barycenter of reduced electronic charge (Q+ here) and a barycenter of increased electronic charge (Q− here). The vector connecting the two points gives a straightforward depiction of the direction and magnitude of the overall charge movement and allows for calculating the amount of charge transferred. While this powerful yet easy approach was mainly developed to model different electronic states of the same system, it can, in principle, also be adopted for ground states of systems with different components (as long as the geometry of common moieties of the relaxed systems does not change significantly); this is discussed in more detail in the Supplementary Materials.

Table 2. Calculated charge transfers and bond distances.

Model Complex	Bpy→Gold Charge Transfer	N–Gold Bond Distance
bpy/Au°	−0.232 $|e|$	2.62 Å
bpy/Au$^+$	−0.502 $|e|$	2.23 Å

To the best of our knowledge, this is the first time the D_{CT} index has been adopted to describe charge rearrangements due to surface effects and not to light excitations, and it is reported in Figure 7.

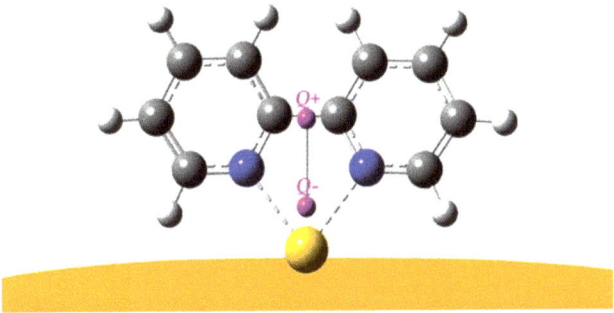

Figure 7. Adsorption model of bpy on Au$^+$ adatom. Points Q+ and Q− are the barycenters of the depletion and the increment of electron density, respectively, with respect to an isolated bpy molecule (*cis* conformation) and an isolated Au$^+$ cation.

With this approach, the computed charge transfer distance is about ~1.95 Å and the amount of charge moving is ~1.1 $|e|$, higher than that estimated from just the increase of electron charge on the Au atom; this is due to the fact that the D_{CT} takes into account the charge displacement over the whole system.

Finally, it is appropriate to define the limits of the DFT calculation model used by us, based on the chemical interaction between a bpy molecule and a single (positively charged) metal adatom. This complex correctly reproduces the positions of the SERS bands, because it is able to predict how the structure and, therefore, the force constants of the molecule change due to interaction with the metal. However, our model fails to satisfactorily reproduce the observed SERS intensities, as shown in the simulated SERS spectrum reported in Figure S5. Actually, in the case of 2,2'-bipyridine, which is linked to gold in a bidentate way by means of the lone pairs of the nitrogen atoms, our model cannot simulate the effect that the gold nanoparticles have on the polarizability of the adsorbed molecule and, therefore, on the intensities of the observed SERS spectrum.

4. Conclusions

Stable nanoparticles constituted of gold and iron oxide were obtained in an aqueous environment by means of laser ablation of Fe and Au targets in two successive steps, avoiding the presence of surfactants, stabilizers, residual reductants, and by-products which could interfere in both the adsorption and the detection of ligands. By using this technique, a mere mixture of two different metal colloids is not obtained, because gold nanoparticles are found to be embedded in the colloidal matrix of iron oxide. The latter was identified as magnetite by both microscopic and spectroscopic analyses. The plasmonic properties of the obtained colloidal nanosystem, as well as its capability of ligand adsorption, were tested by SERS spectroscopy using 2,2′-bipyridine (bpy) as a probe molecule. Thanks to the DFT calculations performed on model systems of gold/ligand complexes, it is possible to argue that positively charged active sites of the gold nanoparticles are responsible for the adsorption of ligand molecules when these approach the metal surface. In this way, strong interaction takes place between molecule and metal, with consequent efficient SERS enhancement, involving the charge transfer of one electron from the molecule to the metal.

Unlike the mixed Ag/Fe_3O_4 and Ag/NiO colloids previously prepared by two-step laser ablation [23,24], the present magneto-plasmonic nanoparticles are more biocompatible and are therefore potentially useful not only in the field of sensors, but also for biomedical applications. Our bimetallic colloidal suspensions are expected to have very low toxicity. Gold nanoparticles are known to be biocompatible and chemically stable, making them ideally suitable for biological applications [58]. Also, magnetite nanoparticles can exhibit low toxicity [59–61], which is closely dependent on the preparation method. In this respect, laser ablation in pure water represents the procedure of choice for the best biocompatibility properties.

Finally, a possible interpretation of the connection between Fe_3O_4 and Au nanoparticles can be proposed. In our sample, colloidal gold is intimately linked to the ferromagnetic material constituted of a mucilaginous matrix of small magnetite (Fe_3O_4) nanoparticles. Hence, all the bimetallic material can be completely separated by magnetic attraction from the aqueous environment wherein it is dispersed. In the literature [26], ultrafine Fe_3O_4 nanoparticles were employed to remove heavy metal ions from contaminated waters, thanks to their excellent adsorption performance. In a similar way, the magnetite nanoparticles obtained by laser ablation could act as adsorbents for the laser-ablated gold nanoparticles, forming a mixed bimetallic colloidal suspension. In fact, we verified by XPS measurements and DFT calculations that our gold particles have a positively charged surface. This could make them suitable to be captured by the magnetite nanoparticles, similarly to what happens with heavy metal ions.

Supplementary Materials: The Supplementary Materials are available online at http://www.mdpi.com/2079-4991/10/1/132/s1. Figure S1: Zeta potential data; Figure S2: Fe_3O_4/Au bimetallic colloidal sample, before and after application of magnetic field; Figure S3: EDX analysis; Figure S4: Calculated normal modes of the bpy/Au+ complex relative to the prominent SERS bands. The hydrogen atoms are omitted; Figure S5: Simulated SERS spectrum for the bpy/Au+ complex; Table S1: Mulliken partial charges in the bpy/gold complexes; Table S2: Observed and calculated vibrational frequencies (cm^{-1}) of bpy; Table S3: Mulliken partial charges.

Author Contributions: M.M.-M. and E.G. produced the bimetallic colloids; M.M.-M. recorded Raman and absorption spectra; E.G. performed the microscopic analysis; F.M.-M. carried out quantum chemical calculations and designed the D_{CT} analysis; M.M.-M. coordinated and designed the research; M.M.-M. and F.M.-M. improved the manuscript accordingly to the Reviewers' comments. All authors have read and agree to the published version of the manuscript.

Funding: This research received no external funding.

Acknowledgments: The authors acknowledge MDPI editorial's invitation to contribute to the *Nanomaterials* Special Issue dedicated to Laser Synthesis of Nanomaterials. They also wish to thank Stefano Caporali (UniFI) for his support in the XPS measurements.

Conflicts of Interest: The authors declare no conflict of interest.

References

1. Schlücker, S. *Surface Enhanced Raman Spectroscopy: Analytical, Biophysical and Life Science Applications*; Wiley-VCH: Weinheim, Germany, 2011.
2. Procházka, M. *Surface-Enhanced Raman Spectroscopy, Bioanalytical, Biomolecular and Medical Applications*; Springer: Basel, Switzerland, 2016.
3. Wang, Y.-X.J. Super paramagnetic iron oxide based MRI contrast agents: Current status of clinical applications. *Quant. Imaging Med. Surg.* **2001**, *1*, 35–40.

4. Cervadoro, A.; Giverso, C.; Pande, R.; Sarangi, S.; Preziosi, L.; Wosik, J.; Brazdeikis, A.; Decuzzi, P. Design maps for the hyperthermic treatment of tumors with superparamagnetic nanoparticles. *PLoS ONE* **2013**, *8*, e57332. [CrossRef] [PubMed]
5. Zhang, Y.; Qian, J.; Wang, D.; Wang, Y.; He, S. Multifunctional gold nanorods with ultrahigh, stability and tunability for in vivo fluorescence imaging, SERS detection, and photodynamic therapy. *Angew. Chem. Int. Ed.* **2013**, *52*, 1148–1151. [CrossRef] [PubMed]
6. Park, Y.I.; Kim, H.M.; Kim, J.H.; Moon, K.C.; Yoo, B.; Lee, K.T.; Lee, N.; Choi, Y.; Park, W.; Ling, D.; et al. Theranostic probe based on lanthanide-doped nanoparticles for simultaneous in vivo dual-modal imaging and photodynamic therapy. *Adv. Mater.* **2012**, *24*, 5755–5761. [CrossRef]
7. Yathindranath, V.; Rebbouh, L.; Moore, D.F.; Miller, D.W.; van Lierop, J.; Hegmann, T. A versatile method for the reductive, one-pot synthesis of bare, hydrophilic and hydrophobic magnetite nanoparticles. *Adv. Funct. Mater.* **2011**, *21*, 1457–1464. [CrossRef]
8. Li, Z.; Wei, L.; Gao, M.Y.; Lei, H. One-pot reaction to synthesize magnetite biocompatible magnetite nanoparticles. *Adv. Mater.* **2005**, *17*, 1001–1005. [CrossRef]
9. Zhang, Y.; Kohler, N.; Zhang, M. Surface modification of superparamagnetic magnetite nanoparticles and their intracellular uptake. *Biomaterials* **2002**, *23*, 1553–1561. [CrossRef]
10. Qu, H.; Lai, Y.; Niu, D.; Sun, S. Surface-enhanced Raman scattering from magneto-metal nanoparticle assemblies. *Anal. Chim. Acta* **2013**, *763*, 38–42. [CrossRef]
11. Sun, L.; He, J.; An, S.; Zhang, J.; Ren, D. Facile one-step synthesis of Ag@Fe_3O_4 core–shell nanospheres for reproducible SERS substrates. *J. Mol. Struct.* **2013**, *1046*, 74–81. [CrossRef]
12. Bao, Z.Y.; Dai, J.; Lei, D.Y.; Wu, Y. Maximizing surface-enhanced Raman scattering sensitivity of surfactant-free Ag-Fe_3O_4 nanocomposites through optimization of silver nanoparticle density and magnetic self-assembly. *J. Appl. Phys.* **2013**, *114*, 124305. [CrossRef]
13. Sharma, G.; Jeevanandam, P. A facile synthesis of multifunctional iron Oxide@Ag core-shell nanoparticles and their catalytic application. *Eur. J. Inorg. Chem.* **2013**, *36*, 6126–6136. [CrossRef]
14. Prucek, R.; Tuček, J.; Kilianová, M.; Panáček, A.; Kvítek, L.; Filip, J.; Kolář, M.; Tománková, K.; Zbořil, R. The targeted antibacterial and antifungal properties of magnetic nanocomposite of iron oxide and silver nanoparticles. *Biomaterials* **2011**, *32*, 4704–4713. [CrossRef] [PubMed]
15. Wang, L.; Wang, L.; Luo, J.; Fan, Q.; Suzuki, M.; Suzuki, I.S.; Engelhard, M.H.; Lin, Y.; Kim, N.; Wang, J.Q.; et al. Monodispersed Core-Shell Fe_3O_4@Au Nanoparticles. *J. Phys. Chem. B* **2005**, *109*, 21593–21601. [CrossRef] [PubMed]
16. Reguera, J.; Jimenez de Aberasturi, D.; Henriksen-Lacey, M.; Langer, J.; Espinosa, A.; Szczupak, B.; Wilhelm, C.; Liz-Marzan, L.M. Janus plasmonic-magnetic gold-iron oxide nanoparticles as contrast agents for multimodal imaging. *Nanoscale* **2017**, *9*, 9467–9480. [CrossRef] [PubMed]
17. Ovejero, J.G.; Morales, I.; de la Presa, P.; Mille, N.; Carrey, J.; Garcia, M.A.; Hernando, A.; Herrasti, P. Hybrid nanoparticles for magnetic and plasmonic hyperthermia. *Phys. Chem. Chem. Phys.* **2018**, *20*, 24065–24073. [CrossRef]
18. Tymoczko, A.; Kamp, M.; Rehbock, C.; Kienle, L.; Cattaruzza, E.; Barcikowski, S.; Amendola, V. One-step synthesis of Fe-Au core-shell magnetic-plasmonic nanoparticles driven by interface energy minimization. *Nanoscale Horiz.* **2019**, *4*, 1326–1332. [CrossRef]
19. Bertorelle, F.; Pinto, M.; Zappon, R.; Pilot, R.; Litti, L.; Fiameni, S.; Conti, G.; Gobbo, M.; Toffoli, G.; Colombatti, M.; et al. Safe core-satellite magneto-plasmonic nanostructures for efficient targeting and photothermal treatment of tumor cells. *Nanoscale* **2018**, *10*, 976–984. [CrossRef]
20. Pan, S.; Liu, Z.; Lu, W. Synthesis of naked plasmonic/magetic Au/Fe_3O_4 nanostructures by plasmon-driven anti-replacement reaction. *Nanotechnology* **2019**, *30*, 65605. [CrossRef]
21. Muniz-Miranda, M.; Caporali, S.; Marsili, P.; Giorgetti, E. Fabrication and characterization of Ag/Pd colloidal nanoparticles as stable platforms for SERS and catalytic applications. *Mater. Chem. Phys.* **2015**, *167*, 188–193. [CrossRef]
22. Giorgetti, E.; Marsili, P.; Canton, P.; Muniz-Miranda, M.; Caporali, S.; Giammanco, F. Cu/Ag-based bifunctional nanoparticles obtained by one-pot laser-assisted galvanic replacement. *J. Nanopart. Res.* **2013**, *15*, 1360. [CrossRef]
23. Muniz-Miranda, M.; Gellini, C.; Giorgetti, E.; Margheri, G. Bifunctional Fe_3O_4/Ag nanoparticles obtained by two-step laser ablation in pure water. *J. Colloid Interface Sci.* **2017**, *489*, 100–105. [CrossRef] [PubMed]

24. Gellini, C.; Deepak, F.L.; Muniz-Miranda, M.; Caporali, S.; Muniz-Miranda, F.; Pedone, A.; Innocenti, C.; Sangregorio, C. Magneto-plasmonic colloidal nanoparticles obtained by laser ablation of nickel and silver targets in water. *J. Phys. Chem. C* **2017**, *121*, 3597–3606. [CrossRef]
25. Gao, F. An Overview of Surface-Functionalized Magnetic Nanoparticles: Preparation and Application for Wastewater Treatment. *ChemistrySelect* **2019**, *4*, 6805–6811. [CrossRef]
26. Fato, F.P.; Li, D.-W.; Zhao, L.-J.; Qiu, K.; Long, Y.-T. Simultaneous Removal of Multiple Heavy Metal Ions from River Water Using Ultrafine Mesoporous Magnetite Nanoparticles. *ACS Omega* **2019**, *4*, 7543–7549. [CrossRef]
27. Frisch, M.J.; Trucks, G.W.; Schlegel, H.B.; Scuseria, G.E.; Robb, M.A.; Cheeseman, J.R.; Scalmani, G.; Barone, V.; Petersson, G.A.; Nakatsuji, H.; et al. *Gaussian 09*; Revision D.01; Gaussian, Inc.: Wallingford, CT, USA, 2009.
28. Lee, C.; Yang, W.; Parr, R.G. Development of the Colle-Salvetti correlation-energy formula into a functional of the electron density. *Phys. Rev. B* **1988**, *37*, 785–789. [CrossRef]
29. Becke, A.D. Density-functional thermochemistry. III. The role of exact exchange. *J. Chem. Phys.* **1993**, *98*, 5648–5652. [CrossRef]
30. Hay, P.J.; Wadt, W.R. Ab initio effective core potentials for molecular calculations. *Potentials for the transition metal atoms Sc to Hg. J. Chem. Phys.* **1985**, *82*, 270–283.
31. Wadt, W.R.; Hay, P.J. Ab initio effective core potentials for molecular calculations. Potentials for main group elements Na to Bi. *J. Chem. Phys.* **1985**, *82*, 284–298. [CrossRef]
32. Hay, P.J.; Wadt, W.R. Ab initio effective core potentials for molecular calculations. Potentials for K to Au including the outermost core orbitals. *J. Chem. Phys.* **1985**, *82*, 299–310. [CrossRef]
33. Grimme, S.; Ehrlich, S.; Goerigk, L. Effect of the damping function in dispersion corrected density functional theory. *J. Comp. Chem.* **2011**, *32*, 1456–1465. [CrossRef]
34. Muniz-Miranda, M.; Pagliai, M.; Muniz-Miranda, F.; Schettino, V. Raman and computational study of solvation and chemisorption of thiazole in silver hydrosol. *Chem. Commun.* **2011**, *47*, 3138–3140. [CrossRef] [PubMed]
35. Pagliai, M.; Muniz-Miranda, F.; Schettino, V.; Muniz-Miranda, M. Competitive Solvation and Chemisorption in Silver Colloidal Suspensions. *Prog. Colloid Polym. Sci.* **2012**, *139*, 39–44.
36. Muniz-Miranda, M.; Muniz-Miranda, F.; Caporali, S. SERS and DFT study of copper surfaces coated with corrosion inhibitor. *Beilstein J. Nanotechnol.* **2014**, *5*, 2489–2497. [CrossRef] [PubMed]
37. Muniz-Miranda, F.; Pedone, A.; Muniz-Miranda, M. Spectroscopic and DFT investigation on the photo-chemical properties of a push-pull chromophore: 4-Dimethylamino-4′-nitrostilbene. *Spectrochim. Acta A* **2018**, *190*, 33–39. [CrossRef] [PubMed]
38. Gellini, C.; Muniz-Miranda, F.; Pedone, A.; Muniz-Miranda, M. SERS active Ag-SiO$_2$ nanoparticles obtained by laser ablation of silver in colloidal silica. *Beilstein J. Nanotechnol.* **2018**, *9*, 2396–2404. [CrossRef] [PubMed]
39. Gracia Pinilla, M.Á.; Villanueva, M.; Ramos-Delgado, N.A.; Melendrez, M.F.; Menchaca, J.L. Au and Cu nanoparticles and clusters synthesized by pulsed laser ablation: Effects of polyethylenimine (PEI) coating. *Dig. J. Nanomater. Biostruct.* **2014**, *9*, 1389–1397.
40. Nedkov, I.; Merodiiska, T.; Kolev, S.; Krezhov, K.; Niarchos, D.; Moraitakis, E.; Kusano, Y.; Takada, J. Microstructure and Magnetic Behaviour of Nanosized Fe$_3$O$_4$ Powders and Poly crystalline Films. *Monatsh. Chem.* **2002**, *133*, 823–828. [CrossRef]
41. Shebanova, O.N.; Lazor, P. Raman spectroscopic study of magnetite (FeFe$_2$O$_4$): A new assignment for the vibrational spectrum. *J. Solid State Chem.* **2003**, *174*, 424–430. [CrossRef]
42. Li, H.; Qin, L.; Feng, Y.; Hu, L.; Zhou, C. Preparation and characterization of highly water-soluble magnetic Fe$_3$O$_4$ nanoparticles via surface double-layered self-assembly method of sodium alpha-olefin sulfonate. *J. Magn. Magn. Mater.* **2015**, *384*, 213–218. [CrossRef]
43. Sylvestre, J.-P.; Poulin, S.; Kabashin, A.V.; Sacher, E.; Meunier, M.; Luong, J.H.T. Surface Chemistry of Gold Nanoparticles Produced by Laser Ablation in Aqueous Media. *J. Phys. Chem. B* **2004**, *108*, 16864–16869. [CrossRef]
44. Joo, S.-W. Adsorption of Bipyridine Compounds on Gold Nanoparticle Surfaces Investigated by UV-Vis Absorbance Spectroscopy and Surface Enhanced Raman Scattering. *Spectrosc. Lett.* **2006**, *39*, 85–96. [CrossRef]
45. Ould-Moussa, L.; Castella-Ventura, M.; Kassab, E.; Poizat, O.; Strommen, D.P.; Kincaid, J.R. Ab initio and density functional study of the geometrical, electronic and vibrational properties of 2,2′-bipyridine. *J. Raman Spectrosc.* **2000**, *31*, 377–390. [CrossRef]

46. Sanchez-Cortes, S.; Garcia-Ramos, J.V.; Morcillo, G.; Tinti, A. Morphological Study of Silver Colloids Employed in Surface Enhanced Raman Spectroscopy: Activation when Exciting in Visible and Near-Infrared Regions. *J. Colloid Interface Sci.* **1995**, *175*, 358–368. [CrossRef]
47. Giorgetti, E.; Marsili, P.; Giammanco, F.; Trigari, S.; Gellini, C.; Muniz-Miranda, M. Ag nanoparticles obtained by pulsed laser ablation in water: Surface properties and SERS activity. *J. Raman Spectrosc.* **2015**, *46*, 462–469. [CrossRef]
48. Lopez-Tocón, I.; Valdivia, S.; Soto, J.; Otero, J.C.; Muniz-Miranda, F.; Menziani, M.C.; Muniz-Miranda, M. A DFT Approach to the Surface-Enhanced Raman Scattering of 4-Cyanopyridine Adsorbed on Silver Nanoparticles. *Nanomaterials* **2019**, *9*, 1211. [CrossRef]
49. Basha, M.T.; Alghanmi, R.M.; Shehata, M.R.; Abdel-Rahman, L.H. Synthesis, structural characterization, DFT calculations, biological investigation, molecular docking and DNA binding of Co(II), Ni(II) and Cu(II) nanosized Schiff base complexes bearing pyrimidine moiety. *J. Mol. Struct.* **2019**, *1183*, 298–312. [CrossRef]
50. Fiori-Duarte, A.T.; Bergamini, F.R.G.; de Paiva, R.E.F.; Manzano, C.M.; Lustri, W.R.; Corbi, P.P. A new palladium(II) complex with ibuprofen: Spectroscopic characterization, DFT studies, antibacterial activities and interaction with biomolecules. *J. Mol. Struct.* **2019**, *1186*, 144–154. [CrossRef]
51. Khodashenas, B.; Ardjmand, M.; Baei, M.S.; Rad, A.S.; Khiyavi, A.A. Gelatin–Gold Nanoparticles as an Ideal Candidate for Curcumin Drug Delivery: Experimental and DFT Studies. *J. Inorg. Organomet. Polym.* **2019**, *29*, 2186–2196. [CrossRef]
52. Sahan, F.; Kose, M.; Hepokur, C.; Karakas, D.; Kurtoglu, M. New azo-azomethine-based transition metal complexes: Synthesis, spectroscopy, solid-state structure, density functional theory calculations and anticancer studies. *Appl. Organomet. Chem.* **2019**, *33*, e4954. [CrossRef]
53. Hmida, W.B.; Jellali, A.; Abid, H.; Hamdi, B.; Naili, H.; Zouari, R. Synthesis, crystal structure, vibrational studies, optical properties and DFT calculation of a new luminescent material based Cu (II). *J. Mol. Struct.* **2019**, *1184*, 604–614. [CrossRef]
54. Maiti, N.; Malkar, V.V.; Mukherjee, T.; Kapoor, S. Investigating the interaction of aminopolycarboxylic acid (APCA) ligands with silver nanoparticles: A Raman, surface-enhanced Raman and density functional theoretical study. *J. Mol. Struct.* **2018**, *1156*, 592–601. [CrossRef]
55. Ricci, M.; Lofrumento, C.; Becucci, M.; Castellucci, E.M. The Raman and SERS spectra of indigo and indigo-Ag_2 complex: DFT calculation and comparison with experiment. *Spectrochim. Acta A* **2018**, *188*, 141–148. [CrossRef] [PubMed]
56. Jacquemin, D.; Le Bahers, T.; Adamo, C.; Ciofini, I. What is the "best" atomic charge model to describe through-space charge-transfer excitations? *Phys. Chem. Chem. Phys.* **2012**, *14*, 5383–5388. [CrossRef] [PubMed]
57. Ciofini, I.; Le Bahers, T.; Adamo, C.; Odobel, F.; Jacquemin, D. Through-Space Charge Transfer in Rod-Like Molecules: Lessons from Theory. *J. Phys. Chem. C* **2012**, *116*, 11946–11955. [CrossRef]
58. Shukla, R.; Bansal, V.; Chaudhary, M.; Basu, A.; Bhonde, R.R.; Sastry, M. Biocompatibility of gold nanoparticles and their endocytotic fate inside the cellular compartment: A microscopic overview. *Langmuir* **2005**, *21*, 10644–10654. [CrossRef]
59. Souza, D.M.; Andrade, A.L.; Fabris, J.D.; Valério, P.; Góes, A.M.; Leite, M.F.; Domingues, R.Z. Synthesis and in vitro evaluation of toxicity of silica-coated magnetite nanoparticles. *J. Non-Cryst. Solids* **2008**, *354*, 4894–4897. [CrossRef]
60. Auffan, M.; Rose, J.; Wiesner, M.R.; Bottero, J.-Y. Chemical stability of metallic nanoparticles: A parameter controlling their potential cellular toxicity in vitro. *Environ. Pollut.* **2009**, *157*, 1127–1133. [CrossRef]
61. Mahmoudi, M.; Simchi, A.; Milani, A.S.; Stroeve, P. Cell toxicity of superparamagnetic iron oxide nanoparticles. *J. Colloid Interface Sci.* **2009**, *336*, 510–518. [CrossRef]

© 2020 by the authors. Licensee MDPI, Basel, Switzerland. This article is an open access article distributed under the terms and conditions of the Creative Commons Attribution (CC BY) license (http://creativecommons.org/licenses/by/4.0/).

MDPI
St. Alban-Anlage 66
4052 Basel
Switzerland
Tel. +41 61 683 77 34
Fax +41 61 302 89 18
www.mdpi.com

Nanomaterials Editorial Office
E-mail: nanomaterials@mdpi.com
www.mdpi.com/journal/nanomaterials

www.ingramcontent.com/pod-product-compliance
Lightning Source LLC
LaVergne TN
LVHW070741100526
838202LV00013B/1282